Urban and Landscape Perspectives

Volume 9

AF182019

Series Editor

Giovanni Maciocco

For further volumes:
http://www.springer.com/series/7906

Translated by:

Samantha Cipollina, MCIL - Networld srl
Valerie Cleverton
Margaret Kersten

Aims and Scope

Urban and Landscape Perspectives is a series which aims at nurturing theoretic reflection on the city and the territory and working out and applying methods and techniques for improving our physical and social landscapes.

The main issue in the series is developed around the projectual dimension, with the objective of visualising both the city and the territory from a particular viewpoint, which singles out the territorial dimension as the city's space of communication and negotiation.

The series will face emerging problems that characterise the dynamics of city development, like the new, fresh relations between urban societies and physical space, the right to the city, urban equity, the project for the physical city as a means to reveal *civitas*, signs of new social cohesiveness, the sense of contemporary public space and the sustainability of urban development.

Concerned with advancing theories on the city, the series resolves to welcome articles that feature a pluralism of disciplinary contributions studying formal and informal practices on the project for the city and seeking conceptual and operative categories capable of understanding and facing the problems inherent in the profound transformations of contemporary urban landscapes.

Making Strategies in Spatial Planning

Knowledge and Values

Maria Cerreta · Grazia Concilio · Valeria Monno
Editors

Springer

Editors
Maria Cerreta
Department of Conservation of
Architectural and Environmental Heritage
University of Naples Federico II
Via Roma 402
80132 Naples
Italy
cerreta@unina.it

Grazia Concilio
Department of Architecture and Planning
Polytechnic of Milan
Via Golgi 39
20133 Milan
Italy
grazia.concilio@polimi.it

Valeria Monno
Department of Architecture and Town
Planning
Polytechnic of Bari
Via Orabona 4
70125 Bari
Italy
valeria.monno@libero.it

ISBN 978-94-007-3343-5 ISBN 978-90-481-3106-8 (eBook)
DOI 10.1007/978-90-481-3106-8
Springer Dordrecht Heidelberg London New York

Cover illustration: Aristotle University of Thessaloniki, Greece. "Urban Strategies: The city reclaims its park" (instructors: Titie Papadopoulou, Anastasios Tellios; students: Alkis Avramidis, Savina Derdelakou, Evita Fanou) in Spot on Schools exhibition, Beyond Media 2009 Visions.

Printed on acid-free paper

Springer is part of Springer Science+Business Media (www.springer.com)

Foreword

This provocative collection of essays challenges traditional ideas of strategic spatial planning and opens up new avenues of analysis and research. The diversity of contributions here suggests that we need to rethink spatial planning in several far-reaching ways. Let me suggest several avenues of such rethinking that can have both theoretical and practical consequences.

First, we need to overcome simplistic bifurcations or dichotomies of assessing outcomes and processes separately from one another. To lapse into the nostalgia of imagining that outcome analysis can exhaust strategic planners' work might appeal to academics content to study 'what should be', but it will doom itself to further irrelevance, ignorance of politics, and rationalistic, technocratic fantasies. But to lapse into an optimism that 'good process' is all that strategic planning requires, similarly, rests upon a fiction that no credible planning analyst believes: that enough talk will miraculously transcend conflict and produce agreement. Neither single-minded approach can work, for both avoid dealing with conflict and power, and both too easily avoid dealing with the messiness and the practicalities of negotiating out conflicting interests and values – and doing so in ethically and politically critical ways, far from resting content with mere 'compromise'.

Second, we must rethink the sanctity of expertise. By considering analyses of planning outcomes as inseparable from planning processes, these accounts help us to see expertise and substantive analysis as being 'on tap', ready to put into use, rather than being particularly and technocratically 'on top'. When we understand outcomes as often contingent not simply upon planning processes but upon shifting relations of authority and power, we make spatial planning more complex, but potentially more accountably democratic as well. Expertise becomes not the unassailable province of the academic but now politically accountable, subject to debate, an integral and contestable part of strategic planning processes rather than a privileged framing or decision-making element detached from mechanisms of local or regional voice and accountability.

Third, we might now see strategic planning not so much to provide answers in advance to development questions but rather to provide what we might call 'spaces of deliberative opportunities'. In such spaces, diverse local actors in diverse processes can bring forward creative, if opposing, ideas and suggestions and proposals in efforts to try to shape urban and regional futures. Here we see strategic planning, as several of the following essays suggest, as not simply a series of instrumental performances but as providing the occasions on and through which spatialised participants articulate their identities and traditions and interests and values and do more than that too: they transform their own and one another's imaginations of what's possible, and not least of all they may actually work to forge the coalitions and pressure and creative organisations to implement their strategic visions.

Fourth, this means we must give up the whipping boy or the scapegoat or the facile complaint that appealing to deliberation in planning must mean some idealistic or romantic appeal to argument and persuasion alone. For 30 years now, planners who have left argumentation largely aside have taught us that public deliberation means facing conflict and so engaging in analytically critical conflict resolution strategies as much as, or indeed more than, it means engaging in any stereotypically rational process of debate or persuasion. Indeed, deliberation itself easily encompasses three quite distinct processes that students of planning have failed to appreciate and that planners have confused practically as well: processes of dialogue, debate and negotiation. The first seeks understanding and mutual recognition through conversation, and it may be facilitated and enhanced to reach those ends and avoid the dangers of talk, talk, endless talk (to say little of disrespect). The second seeks justification and vindication of claims through arguments about what is right or wrong or true or false, and it may be moderated to reach those ends and avoid the dangers of escalation or damaged relationships (to say little of systematic bias). The third, and perhaps only the third, negotiation, seeks agreements on action, commitments to act, through refining and reframing proposals to meet parties' interests, and these processes can be mediated – not merely facilitated or moderated – to produce creative and mutually interest-serving ends and outcomes and avoid, then, the otherwise possible lose–lose agreements we rightly abhor and call 'lousy compromises'.

Fifth, then, this means we must understand strategic spatial planning not only to involve stakeholders 'interactions' and 'networks' but also to call for our careful and critical analysis of their practical engagements and actual negotiations too. As importantly, because processes of negotiation in turn contingently threaten pragmatic agreements that can be mutually inferior to other, quite possible, more mutually satisfying, 'mutual gains' agreements – we come to the essential and inescapable, conflict-addressing, critical role of mediation-like interventions. So we need to introduce some version, a culturally and spatially appropriate version, of mediation skills to be taught in all settings where strategic spatial planning forms part of the agenda at hand.

Strategic spatial planning will call for dialogues to assess traditions and identities, interests and values. If dialogic elements are ignored, recognition of identities and values will be flawed and planning processes will be deeply problematic: simply solving the wrong problems, for example. If elements of debate are ignored, expertise will be squandered and planning will suffer needlessly. If negotiations are ignored, planning will become just pious talk without connections to practical action.

Sixth, then, let us try to find a way amidst the complexities of strategic spatial planning not just to wring our hands, not to equate the presence of conflict with the impossibility of acting and planning well. In the political circumstances in which planning inevitably takes its place, planners must have practical capacities to work in the face of conflict. Conflicts present difficulties, not necessarily impossibilities. So strategic spatial planners must learn to distinguish and then to re-integrate not only distinct institutional moments and processes of dialogue, debate and negotiations but also the outcome-oriented practices of facilitating, moderating and mediating: to produce recognition and understanding, scientifically established bases for action, and actual commitments that serve substantive, spatially defined and rooted interests and values as well.

Seventh, then, let us not forget, in our scepticisms about planners playing de facto mediating roles (among others!), that in politically complex spatial planning processes, mediators no more make multiparty agreements than do mid-wives make babies. Let me repeat this, because it reframes an all-too-common misunderstanding of conflict resolution practice: mid-wives don't make babies; parents do. Mediators do not make agreements; stakeholders do. That's all the more reason that we should explore and refine mediators' roles: to pay more careful attention to the complex, practical opportunities of the diverse deliberative processes that arise systematically in contemporary strategic spatial planning efforts (Forester, 2009).

Ithaca, New York John Forester
January 2010

Reference

1. Forester, J. (2009). *Dealing with differences: Dramas of mediating public disputes*. New York: Oxford University Press.

Preface

The idea of this book originated in the international School in Evaluation for Planning 'Small-medium-sized Cities: perspectives of Strategic Planning', held in Campi Salentina (Southern Italy) in October 2006 and organised by the book's editors. The school was conceived as an occasion to explore synergies and complementarities between evaluation and planning within the framework of strategic planning.

At the beginning of this century, making strategic spatial plans (Healey, Khakee, Motte, & Needham, 1997) was foremost on planning agendas. It was seen as a proactive approach to 'govern' cities and regions facing new developments and challenges determined by globalisation processes under a tough neoliberism and the rise of a new and diffused awareness of environmental issues. The growing complexity and fragmentation of cities and regions determined by radical changes in production processes, the diffusion of new technologies, the crises of representative democracy, the increasing immigration flows, the raise of environmental concerns and the accruing of uneven developments required a fresh planning theory and practice inspired by renewed long-term thinking associated with a more realistic and effective approach (Albrechts, 2009). Its main goal was to produce new cities/regions based on the ideal of coexistence among humans and between humans and non-humans. These ideal cities/regions were seen as a collective actor and *demos* (Le Gales, 2002; Kazepov, 2005) able to creatively manage through complex governance processes their urban development by balancing the goal of economic growth with those of sustainability, inclusivity and enlarged democracy.

Although different kinds of strategic planning have been practiced in different contexts, they have flourished in the perspective of the so-called relational approach (Albrechts, 2009). By enabling a profound transformation in terms of both spatial imaginations and institutional innovations, this appeared as the most appropriate 'technology' to translate the need of linking long-term thinking with a more realistic and effective approach into practice. In fact, it recognises the need for governments to adopt a more entrepreneurial style of planning in order to enhance cities' competitiveness, abandoning bureaucratic approaches and involving skills and resources which are external to the traditional administrative apparatus. From being a comprehensive design, strategic spatial planning is reconceptualised as a social process developing in deliberative arenas through which a range of people in diverse institutional relations and positions come together to design plan-making processes and develop contents and strategies for the management of spatial changes (Healey, 2005).

This process is not thought to generate formal outputs in terms of policy and project proposals; it is conceived as a transformative practice producing a decision framework influencing relevant parties in their future investment and regulation activities. Spatial planning becomes the provider of strategic frames of reference (Albrechts, 2009; Healey, 2006). It is considered an emergent social product in complex governance contexts, with the power to 'frame' discourses and shape action through the persuasive power of their core concepts. Strategic frames which accumulate sufficient power to enrol others, which travel across significant institutional sites of urban governance and which endure through time can shape the future. Crucial to this articulation of planning theory is a conception of the relational complexity of physical and social space. It constitutes the basis of a theory which acts as a balance between what can be fixed and what is left to emerge while imagining better futures. This kind of spatial planning does not refer to the dimension of strategy just in terms of instrumental rationality in order to reduce and treat complex situations; it is able to explore the possible advantages of dealing with (anticipating and, most of all, playing with) the multiple and interacting actors' (and agencies') behaviours (see Chapter 3, this book).

However, by the mid-2000s strategic spatial planning was experiencing difficulties. A new way of looking at, listening to and feeling the relational complexity was suggested as being able to break the impasse (Healey, 2006). Already after a short period of experimentation in strategic spatial planning, some doubts on its efficacy started emerging. The emphatic atmosphere of the beginning of the new millennium showed some feeble but clear symptoms of a crisis about to come. Under a tough neoliberism, governance processes had been transformed into a smokescreen for powerful actors (Amin, 2006; Purcell, 2006; Swyngedouw, 2005). A technocratic-physicalist conception of planning kept on dominating planning practice, while cultural and institutional barriers slowed down the pace of institutional change which had been hoped and expected to be reached by means of strategic planning processes.

It was clear that the ideal of strategic planning could be easily used to favour the most aggressive neoliberal models of urban and regional development. The hoped-for institutional innovations and economic social and environmental improvements appeared really hard to reach in practice, even in the most innovative institutional contexts. Its being based on governance processes increasingly used by influential stakeholders compromised the translation of theory into practice. Growing attention was paid to issues concerning discourses in order to understand contextual factors preventing planning from being successful. In peripheral urban and regional areas, strategic planning did not seem able to produce any movements towards the new economy.

The Campi Salentina school on Evaluation in Planning was organised when the hope to change inspired by this new more effective way of planning started to be challenged by disillusion with strategic planning practice. The school was conceived as an intense dialogue among students and academicians with different perspectives on strategic planning. As usually happens, at the end of the school we realised that it had been much more than a period of training. It became an opportunity in which crucial issues had emerged through debates focusing on the interplay between practice and theory. Specifically, we realised that much of the debate which followed the key lectures challenged a unified vision of strategic planning. The reflections and discussions activated during the week showed that theories and experiences in local contexts were divergent, conflicting and not aligned to a codified concept of strategy and strategic planning. Furthermore, the debate had raised two questions that had only rarely been dealt with as clear way of interpreting the variegated landscape of strategic planning practices.

All the debates which developed during the week seemed to call for renewed attention to issues concerning knowledge and values underlying spatial strategies and strategic planning. In fact the relational strategic planning approach was criticised because of its creative posture which risked glossing over the epistemological and cognitive dimension underlying practices of knowledge production. The school had disclosed knowledge and values by considering them as the origin of diversification and fragmentation of experiences and visions in strategic planning.

Knowledge and values were seen to be effective components of an interpretative framework of such a theoretical and practical diversification and were debated as entities both inspiring and nurturing planning theory and flowing within the myriad planning practices. Analogously, their intriguing interplay made up of convergence, friction, resistance or irremediable distance was also under profound scrutiny. Knowledge and values were considered power engines of relational and communication mechanisms in and for human settlements, and the school started to break into these mechanisms while analysing spatial strategies and strategic planning.

Knowledge and values were also discussed as keys to reintroduce a vision of planning as a field of struggle. As Foucault maintains, changing knowledge implies changing power relationships (Crampton & Elden, 2007). Contrary to the win–win scenario depicted by many theoretical approaches to strategic planning in which the achievement of a shared vision is equalled to the right decision or a transformative change, a focus on knowledge and values allows planning to be reconceptualised as a contested field and to trace the direction in terms of costs and injustices implied by that change. Whose knowledge is it? Whose places are they? What are the values which a planning process is striving for?

We think that these questions are also crucial to reconnect planning to the dialectics of space, a dimension missing from many strategic planning accounts. Usually relegated in closed arenas, the dynamics producing places re-emerge as a field of forces always at work. Friction, resistance and residue are concepts that stand not only for individual subjects but also for collective ones (societies, communities, organisations of any type) and also for places that cannot let themselves be crossed without an imprint or trace being left (Maciocco & Tagliagambe, 2009a, p. 61).

The exploration of knowledge and values mobilised by strategic planning processes also had the power to re-emphasise their relevance in the link between evaluation and strategic planning. Despite the continuous appeal to integrate evaluation and strategy-making processes, a more mechanicistic approach is often adopted in practice, sometime favoured by the diffusion of environmental strategic assessment. The school confirmed that, as Khakee (1998) maintains, evaluation and planning are inseparable concepts. Yet, while in the field of strategic planning evaluation tends to be treated suspiciously because of its technocratic biases, nevertheless, it seems to be more open to hybridisation by the planning field and able to adapt its methods and tools to a more humanised changing world.

This book retains the strategic planning perspective which emerged during the school days because we believe that knowledge and values enable dialogues between different visions of strategy and strategic planning which could offer a new ground for a critical reflection on issues and challenges raised by both the diversity of theoretical interpretations and the theoretical incongruence arising from planning practices. It sees planning as an unstable landscape of theories and practices, constantly challenging the planning itself and being continually adjusted and invented: planning is seen as being 'on the move'. If planning is anything, it is an evolving field which has to change in order to respond to both external changes and changes produced by planning itself. This is a necessity rather than an interior need of planning.

This book, like many others, focuses on the problems and challenges which strategic planning has been facing in recent years; yet, it deals with these issues from a different perspective. It does not take an idea of strategic planning as a more or less good procedure to be followed but as an evolving and challenging critical dialogue between theory and practice. It explores this dialogue in terms of knowledge and value as resources necessary to re-think the concept of strategy in spatial planning critically by considering knowledge and values as key lenses for analysing theoretical positions, processes, practices and outputs of strategising mechanisms. The book is a journey constituted by macro and micro-explorations each of them interconnecting the micro and macro in a way in which the interplay between theory and practice can be evaluated. Consequently, it does not offer a new alternative perspective on strategic planning. It gathers traces and clues on how strategic planning could be reframed.

However, the book represents an evolution compared with the analysis carried out during the school week. It collects keynote lectures which were debated during the school as well as invited papers which improve the dialogue on the complexity and multiplicity of strategies and strategic planning visions and interpretations to be reconnected within the knowledge/values interplay. The discussion is carried out from both theoretical and practical points of view in order to re-conceptualise strategic planning practices as processes contextually 'architecturing' the evolution of value and knowledge structures.

The book is organised in four sections. The first two sections debate the strategic planning approach from a macro-perspective. The former is a disciplinary dialogue looking for normative directions and methods enabling a long-term and dynamic planning approach in relation to urban and regional structural changes. The second section critically observes strategic planning as the materialisation of a theory mediated by both planning tools and contexts to reflect on the working of its assumptions in practice.

The third and fourth sections turn to a micro-perspective to examine the architecturing of knowledge and values within spatial strategy-making mechanisms. The micro-dimension is explored as a complex environment where macro-phenomena can be generated and kept working both intentionally and not.

The first section of this book dissects the multiple meanings of 'strategy' coexisting in planning and their implication in terms of mobilisation and creation of knowledge and values. These range from the enthusiastic movement towards a relational approach up to more cautious views trying to rescue the dimension of land and conflict or move towards new and experimental ideas. At the same time, we think that such differences also reflect the not always easy interplay between ideal models and the local traditions of planning. Whereas northern European planning promotes a more consensual and procedural strategic shift as the right way of making places in the twenty-first century, the southern context appears more cautious and in some ways opposing the usual colonisation of ideas. These clashes of cultures also reveal the risks implicit in the ideal of the possible imitation and translation of ideas. Imitating can suddenly produce a success of the imitator, but in the long-term it can create idea and action deserts. Luis Albrechts and Klaus Kunzmann emphasise the creative dimension of strategic planning while Alessandro Balducci focuses on the possibility of producing a change through such a perspective.

Luis Albrecths in his chapter exposes the fundamental theoretical pillars at the basis of the relational strategic planning approach. He maintains that in the face of challenges posed by the rapid structural changes affecting urban and regional development a proactive planning is the only appropriate response. This calls for transformative planning practices focused on the structural problems in society, the construction of images or visions of a preferred outcome through scenario building processes and the identification of processes necessary to implement it. Without new ideas about how to tackle the developments and challenges, planning efforts seem futile. Transformative practices require an unconditioned openness to the multiple and different creativity intended as a result of a mixing of the critical analysis of history of places and the exploration of alternative futures and a clear political stand as knowledge and values mobilised in planning are not neutral. Specifically it traces the kind of governance which has the capacity to strengthen creativity, diversity and sustainability. Finally, it sketches challenges to be faced by planners in terms of attitudes and skills.

Klaus Kunzmann highlights how the creativity of planning has to be grounded on the territorial capital as a base for local and regional action. He maintains that, under the pressure of tough competition, a strategic planning approach can help small- and medium-sized cities find a profile enabling them to maintain their economic, social and cultural functions and contrast dangerous processes of peripherisation. In fact, strategic planning through the construction of partnerships of local and regional institutions can generate future-oriented initiatives to be developed and implemented. Such a perspective stresses the importance of the territorial capital as a base for local and regional action. People living in these towns, their competence and tacit knowledge, their community commitment and their international networks constitute the territorial the capital for creative governance, where local and regional institutions in a socio-political environment of mutual trust have to cooperate and complement each other.

Though reflecting on the specific case of the Milan strategic plan, Alessandro Balducci inquiries into the achievements of strategic planning in terms of collective intelligence improvements. However, he argues that if we want an answer to the direct question of what changes strategic planning has been able to introduce, we will only be able to indicate initial, provisional and probably fragile results. For him, it is too early to try to evaluate the outcome of such complex processes. He also wonders whether in order to appreciate the changes introduced by strategic planning it would be better to regard specific practices of strategic planning as an adaptive approach to it in situations of growing complexity and rapid change of dynamic urban regions rather than as deviations from a mainstream conception of strategic planning.

The chapters by Francesco Indovina and Luigi Mazza and Jean Hillier spell out alternative conceptions of strategy to the philosophy of 'sharing', underpinning the relational strategic planning approach from different points of view. Francesco Indovina reminds us that a strategy can only be justified by a situation of conflict. Should we refer to a situation of conflict or to collaboration? This choice demands the interpretation of relationships within society in general and the local community in particular. This re-locates the public administration at the centre of strategic planning. If urban and territorial changes are interpreted as constituting a locus of conflicts where 'contenders', by using their own power and attempting to neutralise their opponents, try to impose their own objectives (interests), then the main purpose of a strategic plan is not to identify general objectives but rather methods (strategic methods) to achieve interventions, actions and policies able to shape and realise pre-defined objectives through a process combining political intentionality, knowledge of the situation and participation. Whether this strategy will lead to some form of 'strategic planning' is less important than the imperative goal the public administration must set for itself: the governance of change.

Luigi Mazza debates strategic planning from a land use and mobility planning perspective. From this specific field of planning a plan represents the solution to land use conflicts. In order to solve them, strategic and land use plans have to be seen as reciprocally enforcing. The whole development process is to be conceived as the result of two parallel circular processes: the strategic one in which a vision is defined, a coalition built and some projects selected; and the (land use) planning one in which the selected projects interact with the land use plan. If a true strategic process is developed, the land use plan becomes the tool used by the locality to register and adjust the outcome of the strategic process. The relationship between the projects selected by the strategic process and the land use plan is not one way but interactive.

Jean Hillier abandons the actor-centred perspective shaping the previous positions to propose a multiplanar theory of planning. She draws on the work of Gilles Deleuze and Félix Guattari and looks at spatial planning as strategic navigation concerned with discovering the options people have as to how to live rather than as a process dealing with judgements and solutions. Planning should not be concerned with understanding the world in terms of practical effectiveness of classificatory representation, but the pragmatic Deleuzean *how*: not so much 'what does it mean?' as 'how does it work?' Governance, planners and other agents of governance become *experiments* or *speculations* entangled in a series of modulating networked relationships in both rigid and flexible circumstances, where outcomes are volatile, and problems are not 'solved' once and for all, but are constantly recast and reformulated in new perspectives.

The second section of the book is a critical analysis of the key concepts underlying strategic planning. It considers their working in practice in relation to goals such as democracy, inclusion, empowerment, equality and ecological development. It challenges fundamental assumptions and beliefs of strategic planning while hinting at the need to invent alternatives to the current planning conceptions and practices which can be not only more accountable and legitimate but also emancipatory. The society in this section suddenly assumes importance, not as "an idealised system, by an alleged compactness and idyllic cohesion, but as a fragmented multitude, divided, dispersed, ramified and broken into pieces, yet capable of finding and hunting out unusual modalities of comparison, convergence and mutual recognition, often based on the awareness of exclusion, rather than the illusion of inclusion. A crucial role is played by 'voiceless subjects' and border experiences" (Maciocco & Tagliagambe, 2009b, p. 225).

The chapter written by Rob Atkinson asks to what extent the turn to spatial planning has been able to produce a more integrated and coordinated approach to urban and regional policy. The UK experience seems to confirm the fact that economic development is still the primary strategic driver of regional development compared to environmental sustainability, equality, social inclusion and local empowerment. Furthermore, within 'integrationist' approaches lines of accountability and responsibility for policy are often unclear. Instead of being a tool which is magically able to integrate policies and actions in particular spaces and places, and somehow merge them into a nested interlocking hierarchy of policies, spatial planning, at best, can help to expose them and suggest alternative ways of addressing problems.

The transition from government to governance is examined by Panagotis Getimis and Thilo Lang. Getimis examines the shift as a re-orientation away from 'hierarchies' towards 'heterarchies'. He focuses on the opportunities and risks that may come from governance arrangements and on prerequisites for avoidance of governance failure. Governance and contemporary planning processes have not replaced government and conventional planning. It is important for policy-makers to be aware of the co-existence and complementarities of governance modes, avoiding risks and enhancing opportunities for participatory governance, thus ensuring both effectiveness and legitimacy. Thus, new complementarities between the old and the new should be sought, in order not to replace old problems with new ones in the pursuit of greater participation, effectiveness and legitimacy.

The understanding of governance proposed by Thilo Lang is built upon empirical findings in a wide set of urban regeneration cases and interviews carried out with public and private actors involved in urban development. This chapter compares the forms of governance adopted in medium-sized cities examined through different theoretical modes of urban governance and their specificities. None of the analysed initiatives in the mentioned study can be regarded as the output of such forms of strategic governance. However, the different forms of local governance must be considered as potentially helpful for the successful implementation and operation of local initiatives. Receiving support from individuals or organisations linked to local governance arrangements may also be helpful.

The central issue in Swyngedouw and Monno's chapters is that the consensual logic characterising the post-political city prevents emancipatory futures from emerging.

Valeria Monno argues that strategic planning risks functioning as a governing paralysing meta-cultural frames rather than enabling new urban imagination and institutional change. Within the theoretical framework of the relational strategic planning approach, the conceptualisation of the imagination as construction of executable possibilities offers spatial planning a comfort zone within which socio-economic and environmental crises can be anesthetised and treated as a set of more or less-known problems and solutions. The exclusion of the 'impossible' prevents the differences/tensions between what is considered possible and impossible from acting as a legitimate source and driver of an emancipatory change.

Erik Swyngedouw argues that alternative non-dystopian environmental futures as political achievements can only emerge from a new non-dualistic nature-society conception. This radically turns the question of sustainability into a question of democracy and the recuperation of the horizon of democracy as the terrain for the cultivation of conflict and the naming of different socio-environmental urban futures. To begin to unpack 'sustainability' we need to recapture the political as the decisive material and symbolic space, as the space from which different socio-environmental futures can be imagined, fought over and constructed. A radical socio-environmental political programme, therefore, has to crystallise around imagining new ways to organise processes of socio-metabolic transformation.

The third section of the book examines the practice of production of knowledge. The mobilisation, organisation and management of distributed knowledge is investigated as something problematic and in need of being managed effectively in plural contexts in order to transform multiple knowledge into a common good and whether this knowledge is used as a resource to shape a vision of the future, or as source for action strongly related to the 'doing' dimension of strategy-making.

The first two chapters of this section explore two different, very different, methodological frameworks considered crucial for capturing and mobilising cognitive resources in complex decision-making environments characterised by uncertainty and dynamicity of components and their related relational frames. Abdul Khakee maintains that use of future studies can make decision-makers aware of the great variety of possibilities lying ahead. The aim of his chapter is to examine some important aspects of the relationship between future studies and planning and to present some models where future studies have been developed as an integral part of urban planning. When used as an integral part of planning processes, future studies can throw light on hidden dynamics of change which risk being overwhelmed by macroscopic processes. In this respect, future studies are able to unveil many nuances in the polarised space of stereotyped future images, and thus display many unforeseen future possibilities through a recombination of identities and desires within non-hierarchical spaces of co-existence.

The contribution by Nikos Karacapilidis discusses a different methodological frame and explores technological perspectives of knowledge management in multi-actor deliberation processes. He discusses whether and how argumentative collaboration for policy and decision-making can be effectively supported by an appropriately developed information system and considers the relevance of making some portion of the mobilised knowledge explicitly represented and available to actors within and outside the deliberation process, thus enabling the re-telling of the deliberation story and possibly activating learning mechanisms.

Dino Borri introduces the problem of lack of robust scientific attention on knowledge and knowledge-in-action coordination in multi-agent environments. He argues that this limitation is particularly invalidating, as the current generation of spatial plans aims at democratising its traditional expert and top-down approach and enhancing its knowledge contents and multi-logic potentials. By reflecting on knowledge engineering experiments carried out in multi-agent environments, he discusses this topic in relation to two aspects of strategic interactive planning specifically concerning the change of frames and the appropriateness of planning rationales in dealing with multi-agent environments.

Milan Zeleny describes the experience of strategic planning in Zlin (Czech Republic) with the framework of a peculiar interpretation of strategy-making also related to the history of this urban context and strongly connected to the recognition of knowledge as a strategic resource for this specific context and for the planning process in general. According to Zeleny knowledge in complex environments can be looked at as ‘what is done and can be done’, and the example of Zlin is used to shape such a vision concretely at the urban strategy-making level.

Very close to the vision of knowledge and knowledge management in strategy-making given by Zeleny, Grazia Concilio describes strategy as a work in progress being modelled by a knowledge and practice ‘bricolaging’. Knowledge and practices are considered as reciprocally shaping and can be composed into a bricolage throughout an empirical exploration of knowledge and practices themselves; spaces for this kind of exploration are called ‘strategic episodes’. This vision of strategy and strategy-making is also investigated referring to the strategy-making process in a Natural Reserve in southern Italy. She proposes looking at strategies as macro-phenomena of strategy micro-foundations to be recognised as emergent and/or intentionally activated at the micro-level of complex spatial realities.

The fourth section of the book is concerned with exploring the role and the dynamics of values, taking into account differences in their relationship to knowledge. This section reflects on the role of values and evaluation processes and demonstrates how a ‘value-based approach’ (complex, multi-dimensional, tangible and intangible values) can affect strategic thinking and dialogue with diverse forms of knowledge involved in planning. Here evaluation seems to be a field open to hybridisation by the planning field and in search of adapting its methods and tools to a more humanised changing world.

Luigi Fusco Girard opens the section starting with the concept of complex social values and focuses on contextual material and immaterial relations as crucial elements for human sustainable development. Tangible and intangible values are the components of cultural resilience and creativity, and evaluation becomes a critical process supporting actions and producing new values. Fusco Girard assumes an 'interpretative' approach to evaluation which is able to transform experiences carried out all over the world into indispensable resources for collective learning and human sustainable development in local contexts.

Values and evaluations are also the focus of Giuseppe Munda's chapter which focuses on the opportunities offered by evaluation to deal with uncertainty and complexity of socio-spatial systems. Evaluation is discussed considering its implications in planning and referring especially to the limits of traditional evaluation models when dealing with the reflexive nature of the real world. He discusses how multi-dimensional approaches to evaluation can better respond to the need for learning and co-evolution of current social systems. The proposed Social Multi-Criteria Evaluation is part of this perspective.

Federico Sittaro and Begum Ozkaynak propose two different applications of the Social Multi-Criteria Evaluation underlying the need to develop bottom-up decision-making processes. Federico Sittaro, starting by describing the case of funds allocation in a complex multi-level and multi-organisation environment in an environmental sensitive area, discusses evaluation problems like the multi-scale issue and the problem of accountability, thus posing crucial questions: how can trust be operationally created and activated as a resource for development? To what extent can local actors or institutions be given power to make decisions in contexts where decision-making problems reach an international level?

The chapter by Begun Ozkaynak, looking at a Turkish example, directs attention to the fact that defining objectives and setting priorities of urbanisation is strongly challenged by the increasing influence of goals like global economic integration and competitiveness in the global marketplace. She considers the struggle for local strategies to be of any effectiveness with regard to local identity and culture but also to be formed and reformed according to the logic of macro-level factors which are not always compatible with the local ones.

Finally, starting from the conviction that a decision-making situation is an 'opportunity' and not a problem, Maria Cerreta focuses on the inseparability of evaluation from planning. Evaluation and planning are seen as interdependent and mutually shaping. Together they give rise to strategy-making processes rich in feedback and interaction where decisions can be nothing else than micro-decisions. Evaluation is a way to activate learning throughout planning and is conceptualised as 'thinking through complex values'. In this perspective, the evaluation/planning interplay can seize the 'opportunity' to make knowledge diversity and multiplicity activate a multiplicity of multi-dimensional values able to generate strategic objectives and actions. Strongly dependent on both the context and the decision situation, evaluation cannot be approached within the framework, however complex, of any methodological structure: a combination of techniques is envisaged to create an adequate and 'situated' evaluation environment.

At the time of writing this introduction cities and regions are in the middle of a tough economic crisis which makes future structural changes radically uncertain while reshaping forms of urbanisation. Many of the premises and hopes which were the foundation of the relational strategic planning approach are showing their limits. Among these: the idea that a more direct inclusion of the 'market' could solve most urban problems; the belief that governance processes would solve social justice issues concerning social justice of urban and regional development; and the conviction that ecological modernisation of the cities and regions would significantly reduce environmental impacts of urbanisation. Nevertheless, confidence that a relational strategic planning approach represents the most appropriate way to imagine and manage urbanisation processes, though with a more contextual sensibility, survives. This book questions this approach in the attempt to disclose its potentialities by analysing its most relevant problems and failures and also by looking for strategic approaches which are more sensitive to the complexity of places.

It discusses some unsolved issues that strategic spatial planning has to face. They concern: the role governments play in shaping spatial futures; how to expand the horizons of democracy; the need to return to the city, and in general to human settlements, by strengthening the link between strategic planning and critical spatial studies; a critical examination of some unquestioned planning goals such as sustainability; an action-oriented approach which ignores the differences and tensions among the imminent and immanent; the lack of paying robust scientific attention to knowledge and knowledge-in-action production and coordination in plural environments without which planning has difficulty in promoting learning and changes of frames; finally, the issue of and the need for rethinking evaluation less in terms of a monitoring and control system and more as fresh engagement with the issues concerned, thus enabling evaluation and planning to be dealt with as activities reciprocally shaping each other.

Napoli, Italy Maria Cerreta
Milano, Italy Grazia Concilio
Bari, Italy Valeria Monno

References

1. Albrechts, L. (2009). From strategic spatial plans to spatial strategies. *Planning Theory and Practice*, 10 (1), 133–149.
2. Amin, A. (2006) The good city. *Urban Studies*, 43(5–6), 1009–1023.
3. Crampton, J., & Elden, S. (Eds.). (2007). *Space knowledge and power: Foucault and geography*. Aldershot: Ashgate.
4. Healey, P. (2005). Editorial. *Planning Theory and Practice*, 6(1), 5–8.
5. Healey, P. (2006). Relational complexity and the imaginative power of strategic spatial planning. *European Planning Studies*, 14 (4), 525–546.

6. Healey, P., Khakee, A., Motte, A., & Needham, B. (Eds.). (1997). *Making strategic spatial plans. Innovation in Europe*. London: UCL Press.
7. Kazepov, Y. (Ed.). (2005) *Cities of Europe: Changing contexts, local arrangements, and the challenge to urban cohesion.* Oxford: Blackwell Publishing.
8. Khakee, A. (1998) Evaluation and planning: Inseparable concepts. Town Planning Review, 69(4), 359–374.
9. Le Gales, P. (2002) *European cities: Social conflicts and governance.* Oxford: Oxford University Press.
10. Maciocco, G., & Tagliagambe, S. (2009a) Civitas alone can save the urbs. In: G. Maciocco & S. Tagliagambe (Eds.), *People and space: New forms of interaction in the city project* (pp. 217–226). New York: Springer.
11. Maciocco, G., & Tagliagambe, S. (2009b) The dialectic of recognition: Places and friction. In: G. Maciocco & S. Tagliagambe (Eds.), *People and space: New forms of interaction in the city project* (pp. 61–75). New York: Springer.
12. Purcell, M. (2006) Urban democracy and the local trap. Urban Studies, 43(11), 1921–1941.
13. Swyngedouw, E. (2005) Governance innovation and the citizen: The janus face of governance-beyond-the-state. Urban Studies, 42(11), 1991–2006.

Acknowledgements

As already told in the preface, this book is the conclusive result of an effort started in 2006 in Campi Salentina (Southern Italy) where the School in Evaluation for Planning organised the 1-week course 'Small Medium Sized Cities: Perspectives of Strategic Planning'. This effort has been supported by many people and institutions we want to thank.

Many thanks are due to the municipality of Campi Salentina that funded the course and supported the whole logistics with personnel and accommodation-resources. Particularly our acknowledgements are due to Massimo Como and Riccardo Mattei respectively major and urban planning delegate of the city in 2006.

We also want to thank all those who accepted our invitation and joined the school: Louis Albrechts, Rob Atkinson, Dino Borri, Andreas Faludi, Panagiotis Getimis, Luigi Fusco Girard, Nikos Karacapilidis, Abdul Khakee, Thilo Lang, Luigi Mazza, Frank Moulaert, Giuseppe Munda, Begüm Özkaynak, Federico Sittaro and Mike Raco; and also those who assisted the students works: Adele Celino, Gianfranco Ciola and Carmelo Torre together with Thilo Lang and Giuseppe Munda; finally all the students who were able to make the school an intense reflexive and learning experience for all of us.

Our acknowledgements are also due to: Dipartimento di Architettura e Urbanistica and Dipartimento di Ingegneria Ambientale per lo Sviluppo Sostenibile of the Politecnico di Bari, SIT, MPS, Planetek Italia srl, that co-funded the whole experience; to Fondazione Giovanni Astengo, Dipartimento di Conservazione dei Beni Architettonici e Ambientali of the Università Federico II di Napoli, the Regional Administration of Puglia, URBING, AiSRE, Custodia consulenza e formazione, and to Unione Comuni Nord Salento for the sponsorship of the school initiative.

We also want to thank Fondazione Caripuglia and Dipartimento di Architettura e Urbanistica (Politecnico di Bari) who supported the book publication by paying for chapters translations and reviews (where needed) and for the whole editing work.

Our thanks are due to Margaret Kersten and Valerie Cleverton for their carefulness and expertise in reviewing many of the book chapters, and to Samantha Cipollina for the translations.

Our special thanks to Franco Lancio for creating the notable graphic design image of the school: an extraordinary added value to our work.

Last but not least, we are grateful to the Editor Giovanni Maciocco, who decided to guest this book in the book series 'Urban and Landscape Perspectives', to Silvia Serreli, Monica Johansson and Lisa Meloni for their useful suggestions and patience and to John Forester who wrote the foreword.

Finally, we want to thank many of the book's who were patient with us and tolerated the long time we needed to get the book published.

Napoli, Italy Maria Cerreta
Milano, Italy Grazia Concilio
Bari, Italy Valeria Monno

Contents

Contributors

Louis Albrechts Department of Architecture, Urbanism and Planning, Catholic University of Leuven, Kasteelpark Arenberg 51, 3001 Heverlee, Belgium, louis.albrechts@asro.kuleuven.be

Rob Atkinson Cities Research Centre, University of the West of England, BS16 1QY, Bristol, UK, rob.atkinson@uwe.ac.uk

Alessandro Balducci Department of Architecture and Planning, Polytechnic of Milan, 20133 Milan, Italy, sandro.balducci@polimi.it

Dino Borri Department of Architecture and Town Planning, Polytechnic of Bari, 70125 Bari, Italy, d.borri@libero.it

Maria Cerreta Department of Conservation of Architectural and Environmental Heritage, University of Naples Federico II, 80132 Naples, Italy, cerreta@unina.it

Grazia Concilio Department of Architecture and Planning, Polytechnic of Milan, 20133 Milan, Italy, grazia.concilio@polimi.it

John Forester Department of City and Regional Planning, Cornell University, Ithaca, 14853 NY, USA, jff1@cornell.edu

Luigi Fusco Girard Department of Conservation of Architectural and Environmental Heritage, University of Naples Federico II, 80132 Naples, Italy, girard@unina.it

Panagiotis Getimis Panteion University of Political and Social Sciences, 17671 Kalithea, Athens, Greece, getimisp@gmail.com

Jean Hillier School of Architecture, Planning and Landscape, Newcastle University, NE1 7RU Newcastle upon Tyne, UK, j.s.hillier@ncl.ac.uk

Francesco Indovina Faculty of Urban and Regional Planning, IUAV University of Venice, 30135 Venice, Italy, indovina@iuav.it

Nikos Karacapilidis Industrial Management and Information Systems Lab, University of Patras, 26 504 Rio Patras, Greece, nikos@mech.upatras.gr

Abdul Khakee Department of Urban Planning and Environment, Royal Institute of Technology, 100 44 Stockholm, Sweden, ablkhe@infra.kth.se

Klaus R. Kunzmann Department of Spatial Planning in Europe, Technical University of Dortmund, 44227 Dortmund, Germany, klaus.kunzmann@udo.edu

Thilo Lang Leibniz Institute for Regional Geography, 04329 Leipzig, Germany, mail@thilolang.de

Luigi Mazza Department of Architecture and Planning, Polytechnic of Milan, 20133 Milan, Italy, luigi.mazza@polimi.it

Valeria Monno Department of Architecture and Town Planning, Polytechnic of Bari, 70125 Bari, Italy, valeria.monno@libero.it

Giuseppe Munda Department of Economics and Economic History, University of Barcelona, 08193 Bellaterra (Barcelona), Spain, giuseppe.munda@uab.cat

Begüm Özkaynak Department of Economics, Boğaziçi University, 34342 Istanbul, Turkey, begum.ozkaynak@boun.edu.tr

Federico Sittaro Médecins Sans Frontières, 1090 Brussels, Belgium, federico.sittaro@gmail.com

Erik Swyngedouw School of Environment and Development, The University of Manchester, Manchester, M60 1QD, UK, erik.swyngedouw@manchester.ac.uk

Milan Zeleny Graduate School of Business Administration, Lincoln Centre, Fordham University, NY 100123 New York, USA, mzeleny@fordham.edu

Part I
Debating Spatial Planning in a Strategic Perspective

In recent years, city governments and other entities concerned with urban futures have been exhorted to produce spatial strategies, indicating how their areas might develop in the future. But many of the resultant strategies do little 'strategic work' in the sense of shaping future development trajectories.

Healey, P. (2009). In search of the "strategic" in spatial strategy making. *Planning Theory and Practice, 10*(4), 439–457.

Chapter 1
How to Enhance Creativity, Diversity and Sustainability in Spatial Planning: Strategic Planning Revisited

Louis Albrechts

1.1 Setting the Context

Most societies face major developments and challenges: the growing complexity (rise of new technologies, changes in production processes, crisis of representative democracy, diversity, globalisation of culture and the economy), increasing concern about the rapid and, apparently, random course of uneven development, the problems of fragmentation, the ageing of the population, a growing awareness of environmental issues (at all scales, from local to global), the long-standing quest for better coordination (both horizontal and vertical), the re-emphasis on the need for long-term thinking and the aim to return to a more realistic and effective method (Albrechts, 2004; 2006; Martufi, 2005; Breheny, 1991; Cars, Healey, & Mandanipour, 2002; Freestone & Hamnett, 2000; Friedmann, 2004; Gibelli, 2003; Harvey, 1989; Healey, Cameron, Davoudi, Graham, & Mandanipour, 1995; Landry, 2000; Le Galès, 2002; Newman & Thornley, 1996; Swyngedouw, Moulaert, & Rodriguez, 2002).

There appears to be a recognition of the need for governments to adopt a more entrepreneurial style of planning in order to enhance cities' competitiveness, as well as a growing awareness that a number of planning concepts (compact cities, liveable cities, creative cities, multi-cultural cities, fair cities) cannot be achieved solely through hard physical planning. Moreover, in addition to the traditional land use regulations, urban maintenance, production and management of services, governments are being called upon to respond to new demands, which imply the abandonment of bureaucratic approaches and the involvement of skills and resources that are external to the traditional administrative apparatus. All these expand the agenda.

We may consider four different types of reaction to these developments and challenges: *reactive* (the rear-view mirror), *inactive* (going with the flow), *preactive* (preparing for the future) and *proactive* (designing the future and making it happen)

L. Albrechts (✉)
Department of Architecture, Urbanism and Planning, Catholic University of Leuven, 3001 Heverlee, Belgium
e-mail: louis.albrechts@asro.kuleuven.be

M. Cerreta et al. (eds.), *Making Strategies in Spatial Planning*,
Urban and Landscape Perspectives 9, DOI 10.1007/978-90-481-3106-8_1,

(Ackoff, 1981). My thesis is that only the proactive reaction is appropriate, as it calls for the transformative practices that are needed to cope with the continuing and unabated pace of change driven by the (structural) developments and challenges. Transformative practices focus on the structural problems in society; they construct images or visions of a preferred outcome and indicate how to implement them (Friedmann, 1987). Transformative practices also require a clear political stand as they are not neutral; they deal with values: who do we involve? Which issues do we tackle first? Transformative practices without creativity, without new ideas about how to tackle the developments and challenges seem futile.

This chapter deals with four main questions. First, what kind of planning do we need to cope with the developments and challenges ahead? Second, how can we enhance creativity for diversity and sustainability? Third, what type of governance has the capacity to strengthen creativity, diversity and sustainability? Fourth, what does this mean for planners in terms of attitudes and skills?

1.2 What Kind of Planning Approach Is Suitable?

Traditional spatial planning is basically concerned with the location, intensity, form, amount and harmonisation of land development required for the various space-using functions. In the 1960s and 1970s, in a number of countries, spatial planning evolved towards a system of comprehensive planning – the integration of nearly everything – at different administrative levels. In the 1980s, when the neo-liberal paradigm replaced the Keynesian–Fordist paradigm and when public intervention retrenched in all domains, many countries witnessed a retreat from planning fuelled not only by the neo-conservative disdain for planning but also by post-modernist scepticism, both of which tend to view progress as something which, if it happens, cannot be planned. Accordingly, the focus of urban and regional planning practices shifted to projects (Secchi, 1986), especially those involving the revival of rundown sections of cities and regions, and to the development of land use regulations.

A positivist view of planning assumes that the best future follows automatically, if the analytical and forecasting techniques are rigorously applied. The same reasoning made modernist planners believe that the future could be predicted and controlled (Ogilvy, 2002).

Places are faced by problems and challenges that cannot be tackled and managed adequately with the old intellectual apparatus and mindset. Consequently, we have to reflect creatively and innovatively on the approaches (both in terms of process and substance), the concepts and the techniques that we use and the logics we apply in tackling these problems and challenges. We have to think afresh and, as it were, reinvent our places in order to secure a better future and to improve the quality of life for all citizens. Therefore, planning must involve a creative effort to imagine structurally different futures, and to bring this creative imagination to bear on political decisions and the implementation of these decisions. The challenge is to find a systematic approach that provides a critical interpretation of the existing reality and incorporates (or involves) creative thinking about possible futures and how to get there.

1.2.1 'New' Strategic Spatial Planning

The motivations for constructing a 'new' type of strategic spatial planning vary, but the objectives have typically been to construct a challenging, coherent and coordinated vision, to frame an integrated long-term spatial logic (for land use regulations, for resource protection, for sustainable development, for spatial quality, for equity, etc.), to enhance action-orientation and to create a more open, multi-level type of governance.

My definition of 'new' strategic planning contains three components: a *what*, a *how* and a *why*.

What? 'New' strategic spatial planning is a transformative and integrative, preferably public sector led, socio-spatial process through which a vision, coherent actions and means for implementation are produced that shape and frame what a place is and what it might become (Albrechts, 2004, 2006). The term 'spatial' brings into focus the 'where of things', whether static or dynamic, the creation and management of special 'places' and sites, the inter-relations between and among different activities and networks in an area and significant intersections and nodes in an area which are physically co-located (Healey, 2004a, p. 46).

Amin (2004, p. 43) argues that cities and regions possess a distinctive spatiality as agglomerations of heterogeneity locked into a multitude of relational networks of varying geographical reach. Strategic spatial planning processes with an appreciation of 'relational complexity' demand a capacity to 'hear', 'see', 'feel' and 'read' the multiple dynamics of a place in a way that can identify just those key issues that require collective attention through a focus on place qualities (Healey, 2005, 2006).

The focus on the spatial relations of territories allows for a more effective way of integrating different agendas (economic, environmental, cultural, social and policy) as these agendas affect places. It also carries a potential for a 'rescaling' of issue agendas down from the national or state level, and up from the municipal and neighbourhood level. The search for new scales of policy articulation and new policy concepts is also linked to attempts to widen the range of actors involved in policy processes by means of creating new alliances, partnerships and consultative processes (Albrechts, Healey, & Kunzmann, 2003). Moreover, a territorial focus seems to provide a promising basis for encouraging different levels of government to work together (multi-level governance) and in partnership with actors in diverse positions in the economy and civil society.

How? 'New' strategic spatial planning focuses on a limited number of strategic key issues. It takes a critical view of the environment in terms of determining strengths and weaknesses in the context of opportunities and threats. Strategic spatial planning focuses on place-specific qualities and assets (social, cultural, intellectual, qualities of the urban tissue, both physical and social) in a global context. It studies the external trends, forces and resources available. Strategic spatial planning identifies and gathers major actors (public and private). It allows for a broad (multi-level governance) and diverse (public, economic and civil society) involvement during the planning process. It creates solid, workable long-term visions (a geography of the unknown) and strategies at different levels, taking into account the power structures (political, economic, gender and cultural), uncertainties and competing values. Strategic spatial planning designs plan-making structures and develops content, images and decision frameworks for influencing and managing spatial change. It is about building new ideas and processes that can carry them forward, thus generating ways of understanding, ways of building agreements and ways of organising and mobilising for the purpose of exerting influence in different arenas. Finally, strategic spatial planning, both in the short and the long-term, focuses on framing decisions, actions, projects, results and implementation. It also incorporates monitoring, feedback, adjustment and revision.

Why? The why question deals with values and meanings of 'what ought to be'. Without the normative dimension, we risk adopting a pernicious relativism where anything goes (Ogilvy, 2002). In a conscious, purposive, contextual, creative and continuous process, new strategic planning aims to enable a transformative shift, where necessary, to develop openness to new ideas, and to understand and accept the need and opportunity for change. Transformative practices oppose a blind operation of the market forces and involve constructing 'desired' answers to the structural problems of our society. Normativity indicates the relations with place-specific values, desires, wishes or needs for the future that transcend mere feasibility and that result from judgements and choices formed, in the first place, with reference to the idea of 'desirability' and 'betterment', and to the practice of the good society (Friedmann, 1982). To will particular future states is an act of choice involving valuation, judgement and the making of decisions that relate to human-determined ends and to the selection of the most appropriate means for coping with such ends. This is contrary to futures as extensions of the here and now. 'Futures' must symbolise some goods, some qualities and some virtues that the present lacks (diversity, sustainability, equity, spatial quality, inclusiveness and accountability). Considering quality, virtues and values is a way of describing the sort of place we want to live in, or think we should live in.

1.2.2 Four-Track Approach

The 'new' strategic spatial planning approach is operationalised in a four-track approach (Fig. 1.1). The four tracks (Albrechts, van den Broeck, Verachtert, Leroy, & van Tatenhove, 1999; van den Broeck, 1987, 2001) can be seen as working tracks: the first for the vision, the second for the short-term and long-term actions, the third for the involvement of the key actors and, finally, the fourth for a more permanent process (mainly at the local level) involving the broader public in major decisions. The proposed tracks may not be viewed in a purely linear way. The context not only forms the setting of the planning process but also takes form from and undergoes changes in the process (Dyrberg, 1997).

The four-track approach (Fig. 1.2) is based on the inter-relating four types of rationality: *value rationality* (the design of alternative futures), *communicative rationality* (involving a growing number of actors in the process, both private and public), *instrumental rationality* (looking for the best way to solve the problems and

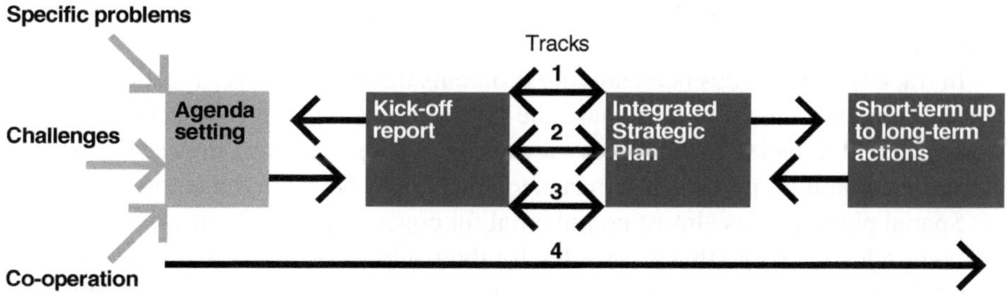

Fig. 1.1 Possible macro-structure for the overall strategic planning process

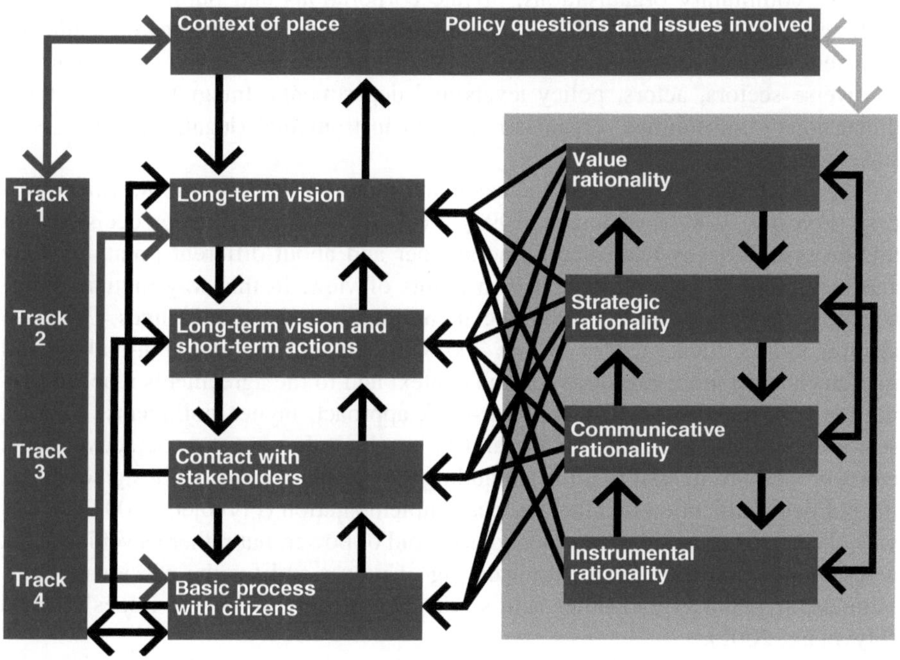

Fig. 1.2 Tentative integration of different concepts of rationality

achieve the desired future) and *strategic rationality* (a clear and explicit strategy for dealing with power relationships) (Albrechts, 2003a).

In the first track, the emphasis is on the long-term vision. In this sense, the long-term constitutes the time span one needs to construct/realise the vision. The envisioning process translates complex inter-relations between place qualities and multiple space–time relational dynamics into multiplex, relational spatial imaginations (Healey, 2006).

The vision (the product of envisioning) is constructed in relation to the social values to which a particular environment is historically committed (Ozbekhan, 1969).

By introducing envisioning, 'new' strategic planning transcends mere contingency planning.

In track two, the focus is on solving problems through short-term actions. It concerns acting in such a way as to make the future conform to the vision constructed in track one and to tackle problems in view of this vision. Tackling concrete problems during the planning process is a means to create trust between the actors.

Spatial planning has almost no potential for concretising strategies, so track three involves relevant actors that are needed for their substantive contributions, their procedural competences and the role they might play in acceptance, in getting basic support and in providing legitimacy. This stresses the need to find effective connections between political authorities and implementation actors (officers, individual citizens, community organisations, private corporations and public departments) (Albrechts, 2003b; Hillier, 2002). Both the technical skills and the power to allocate sufficient means to implement proposed actions are usually spread over a number of diverse sectors, actors, policy levels and departments. Integration in its three dimensions – substantive, organisational and instrumental (legal, budget) – is at stake here.

The fourth track is about an inclusive and more permanent empowerment process (Forester, 1989; Friedmann, 1992) involving citizens in major decisions. In this process, citizens learn about one another and about different points of view and they come to reflect on their own points of view. In this way mutual understanding can be built up, a sort of 'social and intellectual capital' (Innes, 1996; see also the more critical view of Mayer, 2003). To make formal decision-making and implementation more responsive to the context and to the agreements reached during the plan-making process, the four-track approach invites politicians, citizens, sector experts and all actors of the arenas in which they meet to be active in the entire process, from start to finish, including the agenda setting, the design of plans, the political ratification and the practical implementation (Flyvbjerg, 2002). In this way, the arenas are used not as locations devoid of power, but rather as vehicles that acknowledge and account for the working of power and for the passionate commitment of planners and other actors who care deeply about the issues at hand (Flyvbjerg, 2002).

The proposed four-track approach cannot change the power relations, but we are confident (Forester, 1989; Healey, 1997a; Innes, Grüber, Thompson, & Neuman, 1994; Sager, 1994) that empowerment, as developed in track four, supports wider, collective efforts to change such relations.

The end product consists of an analysis of the main processes shaping our environment, which amounts to a dynamic, integrated and indicative long-term vision (frame), a plan for short-term and long-term actions, a budget and a strategy for implementation. It constitutes a consensus or (partial) (dis)agreement between the key actors. For the implementation, credible commitments to action engagement (commitment package) and a clear and explicit link to the budget are needed, where the citizens, the private sector, different levels of governance and planners enter fair, administrative and financial agreements to realise these actions (collective spatial agreement).

1.2.3 Differences with Traditional Planning

This 'new' strategic spatial planning is presented not as a new ideology preaching a new world order but as a method for creating and steering a range of better futures for a place based on shared values (Ogilvy, 2002).

Its normative viewpoint produces quite a different picture than that resulting from traditional planning in terms of plans (strategic plans versus master plans or land use plans), type of planning (providing a framework versus technical/legal regulation), type of governance (government-led versus government-led but negotiated form of governance) and content (a vision and concrete actions that accept the full diversity of a place, while focusing on local assets and networks in a global context, social-spatial quality and a fair distribution of the joys and burdens).

The normative point of view may seem to some people (Mintzberg, 1994) too broad a view of strategic spatial planning. However, the many experiences documented in the planning literature (Albrechts, 2006; Albrechts, Alden, & Da Rosa Pires, 2001, 2003; Hamnett & Freestone, 2000; Healey, Khakee, Motte, & Needham, 1997; Martinelli, 2005; Pascual & Esteve, 1997; Pugliese & Spaziante, 2003) support, at least partially, this broader view.

This view also implies that strategic spatial planning is not a single concept, procedure or tool. In fact, it is a set of concepts, procedures and tools that must be tailored carefully to whatever situation is at hand, if desirable outcomes are to be achieved. Strategic spatial planning is as much about the process, the institutional design and mobilisation as it is about the development of substantive theories. Content relates to the strategic issues selected in the process. The capacity of a strategic spatial planning system to deliver the desired outcomes is dependent not only on the system itself but also on the conditions underlying it. These conditions – including political, cultural and professional attitudes towards spatial planning (in terms of planning the content and the process) and the political will on the part of the institutions involved in setting the process in motion – affect the ability of planning systems to implement the chosen strategies.

Strategic spatial planning is selective and oriented to issues that really matter. It is not just a contingent response to wider forces, but it is also an active force in enabling change, however, its approaches and practices cannot be considered neutral with respect to class, gender, age, race and ethnicity (Albrechts, 2002; Sandercock, 1998).

Strategic spatial planning does not flow smoothly from one phase to the next. It is a dynamic and creative process. New points of view and facts that come to light today might very well alter certain decisions made yesterday.

1.3 How to Enhance Creativity for Diversity and Sustainability?

In 'new' strategic spatial planning, we envision spaces we want to live in. Envisioning is the process by which individuals – or preferably groups – develop visions of future states for themselves, their organisation, their city or their region that are sufficiently clear and powerful to arouse and sustain the actions necessary for (parts of) this vision to become a reality (Goodstein, Nolan, & Pfeiffer, 1993).

Envisioning does not claim to eliminate uncertainty of predictions; instead, it seeks to manage uncertainty as well as possible, and to enable people to make decisions in view of the desired futures. According to Godet (2001, p. 8), envisioning is above all a state of mind (imagination and anticipation) that leads to behaviour (hope and will).

Envisioning possible futures involves a conscious, purposive, contextual, creative and continuous process of representing values and meanings for the futures. In strategic planning, envisioning enables a transformative shift, where necessary, to develop openness to new ideas (diversity, sustainability) and to understand and accept the need and opportunity for change. Since the envisioning of discontinuous futures involves change, all the usual forms of resistance to change (and definitely to structural change) are present.

Envisioning assumes that the city, or the region, has experienced or is experiencing a need for transformation: that is, the city or region now understands – at least at some level – that its future must be discontinuous with its past and present (Goodstein et al., 1993).

With envisioning, we focus on 'what ought to be'. In the final analysis, we must come back to what 'is', if we want to present ideas and concepts that are solid, workable and of testable value. To get to these ideas, we need both the solidity of the analysis and the creativity of the design of alternative futures. To avoid naïve thinking, all of this must be rooted in an understanding of the basic processes that shape places. This must be done recognising the conditions of power, inequality and diversity. 'Whose vision will be created?' is a basic question that needs to be asked.

As futures are not just 'out there', waiting to be discovered, we have to construct them. This is not a linear, but rather a dialectic (backcasting and forecasting) process as visions should not be seen as static descriptions of futures. They have to encompass and portray the dynamic nature of development, changing challenges and contexts. We cannot confront complex dynamic realities with a language designed for simple static problems (Senge, 1990). Hence the need for ways of thinking and for tools that help planners to cope with change in a dynamic environment (Winch, 1998).

The future results from judgements and choices formed with reference to the idea of desirability and betterment. One central concept of our age –sustainability – provides a new lens which can help sharpen our vision of the substance, that is, desirability and betterment. It is a rich concept that needs to be extended beyond environmentalism to reconfigure conceptions of the economy, the social, the cultural, the political and the spatial dimensions. Our concept of sustainability cannot be imagined without acknowledging the politics of difference and spatial quality. This implies that a clear statement must be made against any notion of a purely quantitative approach to growth (Hamilton, 2003) and in favour of the need for a 'just' use of resources and social cohabitation. If we look at plans today, most, if not all, of them embrace some unspecified notion of sustainability, though almost none of them question growth as such.

1.3.1 Backcasting and Transformative Practices

Strategic spatial planning processes achieve transformative power by developing new concepts and new ways of thinking that change the way resources are used, (re)distributed and allocated, and the way the regulatory powers are exercised.

They mobilise all necessary resources, they develop the power to 'travel' and 'translate' into an array of practice arenas and they transform these arenas, rather than merely being absorbed within them. Those concepts and ways of thinking that accumulate sufficient power to become routinised may then 'sediment' down into the cultural ground, which sustains ongoing processes and feeds into new strategic spatial processes (Albrechts & Lievois, 20 04; Hajer, 1995; Healey, 2005, pp. 147–148; Healey, 2006, p. 532). Transformative change rarely occurs in instant revolutions. Changes evolve in many small ways, building a ground of understanding and experiences which, over time, eventually come together in what history may then describe as 'a transformative moment' (Healey, 2005, p. 158; Healey, 2006, p. 541). Transformative practices simply refuse to accept that the current way of doing things is necessarily the best way; they break free from concepts, structures and ideas that persist only because of the process of continuity. It is precisely the discontinuity that forces us outside the usual boundaries of 'reasonableness' (de Bono, 1992). Discontinuity is at odds with a concept of the future as an extension of the here and now. Discontinuity builds on a contrary assumption based on a psychological notion of time (a normative approach; Berger, 1964; de Jouvenel, 1964; Jantsch, 1970; Ozbekhan, 1969). Normativity indicates the relations with specific values, desires, wishes or needs for the future. So the normative approach creates (Ozbekhan, 1969), invents (Gabor, 1969) or constructs (Massé, 1965) more desirable futures. This intellectual tradition yields the willed futures (Berger, 1964; Dubos, 1969; Jantsch, 1970; Ozbekhan, 1969 and the French intellectual tradition of 'les futuribles', de Jouvenel, 1964), that is, a conception of 'futures' that transcends mere feasibility and that results from judgements and choices formed, in the first place, with reference to the idea of 'desirability', to the idea of 'betterment' and to the practice of the good society (Friedmann, 1982). To will particular future states is an act of choice involving valuation, judgement and the making of decisions that relates to human-determined ends and to the selection of the most appropriate means for coping with such ends. This is contrary to futures as extensions of the here and now. Such 'futures' must be imagined as differing radically and structurally from the present reality. They must represent situations which are not merely temporal extensions of the here and now. 'Futures' must symbolise some good, some qualities and some virtues that the present lacks. Speaking of quality, virtues and values is a way of describing the sort of place we want to live in, or think we should live in.

We are not bound by our past to live out a future that is predetermined and therefore predictable. A willed future is a clear reaction against the future as a mere extension of the here and now. On the other hand, the future cannot be so open that anything is possible, as if we could achieve anything we wanted to achieve (Berger, 1964; Ogilvy, 2002; Ozbekhan, 1969). Conditions and constraints on 'what is' and 'what is not' possible are placed by the past and the present. These conditions and constraints have to be questioned and challenged in the process, given the specific context of place and time. So, in order to imagine the conditions and constraints for the future differently, we need to both deal with history and overcome history. Therefore, we also need an exploratory approach. The inter-relation between the normative and the exploratory approach (so-called prospective thinking, see Ozbekhan, 1969) defines the boundaries of a fairly large space between openness and fixity. Thus transformative planning becomes the activity whereby (within certain boundaries) that which can be willed is 'imposed' on that which is, and it is 'imposed'[1] for the purpose of changing what is into what is willed. It differs from the established or traditional way of thinking, in which there is no choice and we are not even aware of other possibilities. The normative approach invents, or creates, futures – in relation to the context, the social and cultural values to which a particular place/society is historically committed – as something new rather than as a solution arrived at as a result of existing trends. It is only by working backwards ('reverse thinking', 'backcasting') that we are able to open up and use other directions. If a place/society develops a vision of where it wants to be and then looks backward, the barriers appear.

Envisioning is linked to values, to choices, and therefore it is far from neutral. This puts the finger on the ideological role that envisioning may play and the danger of manipulation that is inherent in the process. The danger of manipulation will be lessened once envisioning is accessible to everyone, that is, to all those concerned, and not just to the planners and the leadership who can participate (Godet, 2001).

Transformative practices involve constructing 'desired' answers to the structural problems of our society. Our imagination possesses the ability to escape mentally from established patterns of thinking, and it makes us keep exploring and connecting our thoughts.

1.3.2 Envisioning as a Collective Process

Since envisioning is so central to the strategic planning process and so all invasive, it cannot be confined to a single actor or institution in the process. We consider envisioning to be a collective process that concerns futures for which citizens are themselves responsible. Their vision, then, is more than a wish list: it involves commitment to the realisation of the vision through practice (Friedmann, 1987).

A vision provides citizens with views of the future that can be shared, a clear sense of direction, a mobilisation of energy and a sense of being engaged in something important (Goodstein et al., 1993). A vision is 'communicatively rational' to the degree that it is reached consensually through deliberations involving all relevant stakeholders, where all are equally empowered and fully informed, and where the conditions of ideal speech are met (Innes, 1996). The images provided in a vision involve dynamic interaction among the actors relevant in the process rather than a unidirectional flow. The reiterative process occurs at the moment of creating the vision as well as throughout the process of its implementation. The values and images of what a society wants to achieve must be discussed in the envisioning process. Values and images are not generated in isolation but are socially constructed, given meaning and validated by the traditions of belief and practice; they are reviewed, reconstructed and invented through collective experience (Foucault, 1980; Ozbekhan, 1969, p. 11; Elchardus, Hooghe, & Smits, 2000; Hillier, 1999, p. 24).

We must be aware of the impact on the social and psychological milieu of the consumer society, which teaches citizens how to think about themselves and their goals. Citizens' tastes, priorities and value systems are, to a large degree, manipulated by the very markets that are supposed to serve them (Hamilton, 2003, p. 66). Within (and constrained by) this established framework of the market society, places and communities face the challenge of constructing (or rejecting) and implementing the discourses of cultural diversity, sustainability and place quality and, hence, of creatively transforming their own functioning and practice.

Envisioning reveals how things can be different, how things could be truly better, how people can be innovative, how we can unlock the natural creativity of the citizens to improve our cities and regions and how we can legitimise these natural tendencies that are typically inhibited or suppressed by the daily demands of our governance systems. The construction of different futures, which lies at the very heart of the transformative practices, requires creativity and original synthesis (Ozbekhan, 1969, p. 87).

It is for this reason that we need to look for tools and traditions that stimulate creativity.

Envisioning is an attempt to imagine that which by its very nature cannot be defined in advance, given that the effort to envision has to take place within the (unsatisfactory) conditions of value systems beyond which it is attempting to move (Pinder, 2002). Creative people (see among others, Craft, 2000; Plsek, 1997) are curious. They purposefully move out of their comfort zone, they seek to open their minds and those of others to the new, divergent thinking. They think for themselves and show a sustained openness to integrating thinking with experience. The ability to practically shape and develop an idea is just as important as the ability to imagine the idea in the first place. Imagination and analysis are equal partners in creativity (Plsek, 1997, p. 30).

1.3.3 Creative Thinking Tools

Plsek (1997), Landry (2000) and Michalko (2001) teach us that to create original ideas and creative solutions we must use appropriate techniques. Landry (2000) argues that all authors dealing with creativity, imagination and visioning make three points: (1) patterns of thought can change, (2) ideas can be liberated through tools and (3) new solutions can be found.

They do so by utilising techniques to increase the number of ideas, to generate new ideas and to reframe old ideas. Plsek (1997), in line with de Bono (1992), argues that the many tools for creative thinking are variations on the themes of mental attention, escape and movement. Creativity requires that we first focus our attention on something. This is in line with the characteristic of strategic planning that focuses on key issues. For Michalko (2001), these techniques vary from seeing what no one else is seeing to thinking what no one else is thinking.

The first category involves knowing how to see and making thoughts visible. The second category involves thinking fluently, thinking flexibly, making novel combinations, connecting the unconnected, looking at the other side, looking into other worlds, finding what you are not looking for and awakening the collaborative spirit. The first category brings us to look at problems and challenges from different perspectives: the perspectives of the elderly, the young, the poor, migrants, women, shopkeepers, business people and so on. Turn a problem (a cost) into an asset. Offering the actors the possibility to express themselves in more than one language and communicative form (written, oral, drawing, maps, music) could help to remove barriers to creativity when the actors are taking part in debates and decisions about places. The second category is about generating new and more ideas, for example, brainstorming and jamming (Kao, 1997) as well as combining and recombining ideas, knowledge, images and thoughts into different combinations, and about focusing on the collective intelligence of the group as being greater than the intelligence of the individual. All these techniques match very well with the nature of scenarios.

1.3.4 Focus on Design

The agenda setting (Albrechts, 2004, 2006; Albrechts & van den Broeck, 2004; Bryson & Crosby, 1992) and the creation of images must be looked upon as contextual, conscious and purposive actions to represent values and meanings for the future. Hence the need to shift from analysis, which seeks to discover a place that might exist, towards design (in its broadest sense), which creates a place that would not otherwise exist. This is similar to Habermas's knowing (understanding the challenges and the options available) and steering (the capacity to take action to deal with the challenges; Habermas, 1996). The steps required to deliver and to implement the wished for spatial outcome vary according to the underlying structure. Imaginaries are not neutral. They are based on context, values, current drivers and trends. Scenario building turns out to be an excellent tool for designing possible futures and for determining how to get from here to there, what has to be changed first and what next.

Scenarios augment understanding by helping us to see what possible futures might look like. They help us to think about how places/institutions will operate under a variety of future possibilities and they enable decision-makers/civil society to detect and explore all or as many alternative futures as possible in order to clarify present actions and subsequent consequences. For Schwartz (1991, p. 192), this is 'rehearsing the future'. Moreover, scenarios are a way of understanding the dynamics that are at work shaping the future; they are an attempt to identify the primary 'driving forces' (social, economic, technological, cultural and political) that are at work in the present. Scenarios identify contingent decisions by exploring what places/institutions might do, if certain circumstances were to arise; they enable us to reflect on a series of 'what if' stories.

Some of the driving forces are fixed in the sense that they are completely outside our control and will play out in any narrative about the future. Therefore, the 'possible futures' must be placed within a specific context (economic, social, cultural, political and power), place, time and scale regarding specific issues that are of interest and within a particular combination of actors. The context provides the setting for the process, though it also takes form and undergoes change in the course of the process.

1.4 Governance

In the field of governance, there is a pervasive struggle among a variety of pluralistic democratic tendencies, each of which seeks to acknowledge a wide range of actors in policy-making and techno-corporate tendencies. This is a struggle to maintain control over the management of a place using the tools of technical analysis and management, and following the standardised rulebooks or recipes of conventional collaboration between government, major business organisations and trade unions (Albrechts, 1999; Healey, 1997a).

I argue that a feasible and efficient planning process should be centred on the elaboration of a mutually beneficial dialectic between top-down structural policies and bottom-up local uniqueness. Both a bottom-up approach rooted in the conditions and potentialities of diversity (interpreted in their broadest sense) and a complementary multi-level top-down approach aimed at introducing fundamental and structural changes are indispensable. Indeed, a mere top-down and centrally organised approach runs the danger of overshooting the local, historically evolved and accumulated knowledge and qualifications, while a one-dimensional emphasis of a bottom-up approach tends to deny – or at least to underestimate – the importance of linking local differences to structural macro-tendencies (Albrechts & Swyngedouw, 1989). This dialectic constitutes the bare essence of multi-level governance.

1.4.1 Pluralist and Inter-culturalist Places

Some politicians are reluctant to involve the public in decision-making, because it involves giving up some control, and people who hold power are usually not inclined to give it up or to share it. In other places, there is a tendency to involve major actors in the process. As spatial planning has almost no potential for concretising strategies, the relevant actors get involved.

Potentially, planning has an impact on and links to a very wide range of issues (from citizens with an interest in a place to nature). These interests can be diverse and conflicting. Citizens must claim a role in the political system (Mathews, 1994). Some citizens have the knowledge, the skills, the power and the networks through which they are able to influence or even steer planning proposals and policy decisions. Others lack the means and the cultural codes to participate in the system. Their voice has hardly any impact on decisions. As class, gender, race and religion matter in terms of whether citizens are included in the process (Young, 1990), the future is/will be created under conditions of inequality and diversity. Any change must deal with issues of power and resistance, and the irreconcilability of certain forms of interests. This requires a democratic policy that can encompass the realities of difference, inequality and so on. (Huxley, 2000). The core is a democratic struggle for inclusiveness in democratic procedures, for transparency in government transactions, for accountability of the state and planners to the citizens for whom they work, for the right of citizens to be heard and to have a creative input in matters affecting their interests and concerns at different scale levels and for reducing or eliminating unequal power structures between social groups and classes (Friedman & Douglas, 1998). Pluralist democratic tendencies develop in the wake of a crisis of representative democracy and a demand to transform the state in ways that will serve all of its citizens, especially the least powerful. Out of this shift towards a more hybrid democracy, in some places, a type of governance has emerged that expands practical democratic deliberations rather than restricts them, that encourages diverse citizens' voices rather than stifles them and that directs resources to basic needs rather than to narrow private gain. This type of approach uses public involvement to present real political opportunities. Actors learn from action not only what works but also what matters. Involving citizens (and especially weak groups) in socially and politically relevant actions is intended to give them some degree of empowerment and a sense of ownership that results in acceptance (Friedmann, 1992).

Increased personal mobility has made places more mixed. This can be seen either as a threat or as an opportunity. On the one hand, it can destabilise a place as 'others' bring in habits, attitudes and skills alien to the original society; on the other hand, it can enrich and stimulate the potential of a place by creating hybrids, crossovers and boundary blurring (Landry, 2000, p. 264). Places must be creative, with mutual understanding between cultures based on the ideal of equity; (this is nothing less than a claim to full citizenship – see Sandercock, 2003, p. 98). Inter-culturalism (Landry, 2000) builds bridges, helps foster cohesion and conciliation and produces something new out of the multi-cultural patchwork of places (Landry, 2000) so that the views of a place held by minorities or otherwise socially excluded groups are taken into account and their ideas are brought to bear on the planning, political decision-making and implementation processes.

1.4.2 Learning Processes

Society as a whole (both citizens and politicians) feels uncomfortable when challenged to think beyond the short-term and to reflect on multiple futures; consequently, it takes an unconsciously deterministic view of events. How can citizens, politicians and planners be convinced that they can have meaningful choices and will not have to be a complete prisoner of circumstances? How can different groups in a place be made aware that they are inter-dependent, that they share the same physical space and may therefore face similar problems and that there are some problems that they cannot solve on their own? How can they be made aware that they may loose, if they do not cooperate? How can they be persuaded to consider the alternative to what they feel in their heart? Yet, when the sustainability, quality and equity of places are at stake, then this is exactly what we may need to do: to imagine alternative futures in order to master change.

The building of scenarios can become a learning process, if it looks in an open way to the future, if it integrates the knowledge of what might happen with an understanding of the driving forces and a sense of what it means to a place and its citizens. Active participation in a collective action of scenario building may generate trust as participants in the process are likely to find that (and to understand why) some scenarios present a future that they would like, while others would be highly undesirable.

The process helps the participants to think more broadly about the future and its driving forces and to realise that their own actions may move a place towards a particular kind of future. The process allows participants to step away from entrenched positions and identify positive futures that they can work at creating. It allows for a high degree of ownership of the final product and illustrates that citizens do have a responsibility for the(ir) future. So the real test is not whether one has fully achieved the 'conceived' future, but rather whether anyone has changed his or her behaviour because he or she saw the future differently (Schwartz, 1991).

1.4.3 Institutionalisation

Government systems of development, control and regulation have often been fixed for a long time, yet are seldom reviewed and adapted to changing circumstances. The life of an institution often seems to be more important than what it does. Hence the need to view governance institutions not as a set of formal organisations and procedures established in law and 'followed through', but rather as referring to the norms, standards and mores of a society or social group that shape both formal and informal ways of thinking and acting (Healey, 2004b, p. 92). Our notions of nature are also inextricably entangled in different forms of social life (Macnaghten & Urry, 1998). In some places the process of 'discourse structuration' and its subsequent 'institutionalisation' becomes perhaps more important than the plan as such (Albrechts, 1999, 2003a, 2003b; Albrechts & van den Broeck, 2004; Hajer, 1995). In this way new discourses may become institutionalised, that is, embedded in the norms, attitudes and practices, thus providing a basis for structural change. From this point, a shared stock of values, knowledge, information, sensitivities and mutual understanding may spread and travel through an array of regional, provincial and local governmental arenas, sector departments and consultants. New approaches and new concepts can be sustainably embedded via institutionalisation (Gualini, 2001; Healey, 1997b). However, this takes time and dedication. A Government may call upon this intellectual capital (Innes et al., 1994) when using its control function to reframe people's ways of thinking.

1.4.4 Multi-level Governance

A multi-level governance approach would offer the potential to tease out causal linkages between global, national, regional, metropolitan and local change, while also taking account of the highly diverse outcomes of such interactions. The dialectic between shifts in institutional sovereignty towards supranational regulatory systems and the principle of subsidiarity, which entails the rooting of policy action in local initiatives and abilities, illustrates the embeddedness of place policy-making in the multiple institutional domains and interaction arenas that blur the meaning of hierarchical settings in the development of policies (Gualini, 2001).

Tensions occur between the well-known scale (and related government structure) of a nested hierarchy (from large to small or from top to bottom) and the scale (in terms of the reach) of relationships in time and space (Albrechts & Lievois, 2004; Healey, 2004b).

In a new governance culture the construction of arenas (who has to be involved, and what issues must be discussed), their timing (links to the strategic momentum), the definition of which arenas and issues seem fixed and the awareness that the meaning of 'fixed' may be relative in some contexts, all need careful reflection and full attention.

1.5 Impact for Planners?

All of these factors, of course, have an impact on the role, the position and the skills of strategic planners. On several occasions, strategic planners have acted as catalysts (Albrechts, 1999; Mintzberg, 1994), as counterweights and as initiators of change (Albrechts, 1999; Krumholz, 1982). They mobilise and build alliances. They present real political opportunities, learning from action not only what works but also what matters. They substantiate change and refuse to function smoothly as neutral means to given and presumably well-defined ends.

The developments and challenges society is facing are forcing planners to look for new thinking and new approaches, which are not predicated on successful formulas from the past. Planners need a robust culture of creativity to construct new mindsets, new tools and new attitudes. Hence, creativity must be an agenda item in planning education and in strategic planning practice.

1.5.1 Planning Versus Politics

Planners need to know their role as a planner. They must come to terms with the fact that planning is not an abstract analytical concept but a concrete socio-historical practice, which is indivisibly part of the social reality. As such, planning is in politics (it is about making choices) and it cannot escape politics (it must make values and ethics transparent), but it is not politics (it does not make the ultimate decisions). Since the planning actions themselves are a clear proof that such planning is not only instrumental, the implicit responsibility of planners can no longer simply be to 'be efficient' or to function smoothly as a neutral means of obtaining given, and presumably well-defined, ends. Planners must be more than navigators keeping their ship on course. They are necessarily involved with formulating that course (Forester, 1989).

To give power to the range of possible creative ideas/images in a planning process requires the capacity to listen, not just for an expression of material interest, but for what people care about, including the rage felt by many who have grown up in a world of prejudice and exclusion, being outside, being 'the other' (Forester, 1989; Healey, 1997a). Planners must use the power and the imagination available to them to anticipate and to counter the efforts of interests that threaten to make a mockery of a democratic planning process by misusing their power. It must be clear that planners can (and do) use their power and imagination also in the opposite way (Forester, 1989).

1.5.2 Need to Strengthen Creativity

Planning education and planning practice must activate imagination. The imagination is like a muscle: it strengthens through use. And nothing uses it better than looking beyond the domain of what we know (Kao, 1997). Unfortunately, planners for too long have been (and still are?) trained just to react to problems and difficulties. They are focused on reproducing answers on the basis of similar problems encountered in the past. They ask, 'What have I been taught in planning school or work that will solve this problem?' Then they analytically select the most promising approach based on experience, excluding all other approaches, and they work in the clearly defined direction towards the solution of the problem.

A change in this attitude is crucial for creativity. Planners must think productively (Michalko, 2001). Hence, the need to challenge their 'mental models' about places and lift the 'blinkers' that limit their creativity so their resourcefulness can be used as a building block for designing and formulating structurally new concepts and discourses (Schwartz, 1991). When confronted with a problem, planners have to ask themselves in how many different ways they can look at the problem, how they can rethink it and in how many different ways they can tackle it, instead of asking how they have been taught to solve it. Planners must be able to grasp the momentum and they must try to come up with many different responses, some of which may be unconventional, and possibly unique (Michalko, 2001, p. 2).

Hence planners need a mindset that is willing to force their thinking into the unknown, to trigger insights and unleash ideas, to explore new concepts and new ideas and to look for alternatives (to the settlement hierarchy, to the clear division between town and country, to the existing administrative boundaries and to the traditional Euclidean perspectives). Alternatives mean structurally different futures and not just variations on the same theme. This means that the planner must look for a transformative agenda (see Friedmann, 1987, p. 389, for transformative theory and

Sandercock, 2003, pp. 157–179, for transformative practices). Change is the sum of a great number of acts (individual, group and institutional) of re-perceptions and behaviour change at every level. This takes decision-makers, planners, institutions and citizens out of their comfort zones and compels them to confront the key beliefs, to challenge conventional wisdom and to look at the prospects of 'breaking-out-of-the-box'. Planners must help to create empathy for the difficulty of change. Not every one (individual planners, groups, institutions and citizens) wants to give up power associated with the status quo. The creative challenge should balance freedom and discipline, unite all stakeholders behind the creative effort and evince empathy for the difficulties of the creative process.

1.5.3 Preconditions for Creativity

In planning systems and governance structures, a climate and environments conducive to new ideas in which creative people (planners, politicians, civil servants and citizens) can flourish must be created. Planners, civil servants and governments need to think beyond customary job descriptions and traditional government structures, need to address problems in new ways and need to accept that the past is no blueprint for how to go forward.

Governments and planners need to trust the creativity of residents. They must acknowledge that there are multiple publics and that planning and governance in a new multi-cultural era require a new kind of multi-cultural literacy and a new kind of democratic politics, which is more participative, more deliberative and more agonistic. In order to build trust and confidence, an adequate and timely response is required to address serious problems being faced by the community. This community consists of social entities of citizens who are engaged with their place. Creativity in the long-term perspective is important and possible as long as it is combined with creativity in short-term actions. This combination of long-term perspective with short-term actions makes creativity tangible and enables it to react almost immediately to certain urgent problems with a clear perspective as to where to go and what the likely impacts of decisions are. It also promotes the building of trust, understanding and confidence in the process and among the actors. This means that we need visions that embody what is willed (this is the long-term strategy), we need concrete actions in response to the everyday problems, and we need longer-term actions for the realisation of possible futures.

1.6 Concluding Comments: What Difference Does 'New' Strategic Planning Make?

The first difference is related to time. It means that time flows from the 'invented' future, which challenges conventional wisdom, towards and into the experienced present. This means inventing a world that would otherwise not be. New strategic planning 'creates' a future environment, but all decisions are made in the present.

Second, this 'created future' has to be placed within a specific context (economic, social, political, cultural and power), place, time and scale regarding specific issues and a particular combination of actors. It not only provides the setting for the process but also takes form and undergoes changes in the process.

Third, new strategic planning is centred on the elaboration of a mutually beneficial dialectic between top-down structural developments and bottom-up local uniqueness.

Fourth, new strategic planning is selective and oriented to issues that really matter. As it is impossible to do everything that needs to be done, 'strategic' implies that some decisions and actions are considered more important than others and that much of the process lies in making the tough decisions about what is most important for the purpose of producing fair, structural responses to problems, challenges and aspirations (diversity, sustainability, equity, spatial quality, etc.).

Fifth, new strategic planning is about joint decision-making and integrated action. Space may provide an effective way of integrating agendas and actions at different levels of governance, and for integrating actors.

Sixth, new strategic planning relates to implementation. Things must get done! This is seen as the pattern of purpose, policy statements, plans, programs, actions (short-, medium- and long-term), decisions and resource allocation that defines what a policy is in practice, what it does, how it does it and why it does it – from the points of view of various affected publics. This stresses the need to find effective connections between political authorities and implementation actors (officers, individual citizens, community organisations, private corporations and public departments).

I now return to my four initial questions. First, I presented a planning approach that avoids two traps. Planning is usually confronted with the trap of linearity and the trap of being stuck in regulations. This planning approach combines the strategic force of reverse thinking with a critical analysis of the driving forces at work in the present. It constructs 'better' futures for overcoming the resistance of the established powers in the realisation of desired outcomes. Second, creativity for diversity and sustainability match seamlessly with our planning approach. Creativity opens up the minds of people and envisioning can serve as a learning device for rehearsing qualitative and sustainable futures and how to get there. Third, the proposed governance culture opts for a more hybrid mode of democracy open to diversity and structural change embedded in norms, attitudes and practices. This culture makes it possible for ideas, concepts and discourses to travel to other departments, consultants, agencies, political levels, citizens' associations and so on. Fourth, the plea for a transformative agenda challenges existing knowledge, conventional wisdom and practices, and the attitudes and skills of planners. Some of the ideas and suggestions I have made already exist (in cases and practices) (see Landry, 2000 for 'best' practices and Albrechts, 2004,2006 for references on strategic planning), which means that they are 'realistic' in certain specific contexts. They provide a fertile laboratory at the present time for experimenting with different planning and governance cultures.

Note

1. Although 'imposed' may refer to a top–down jargon, I use the term very deliberately. As soon as directions based on an emancipatory practice are agreed upon, they must be imposed for action.

References

Ackoff, R. (1981). *Creating the corporate future*. New York: John Wiley.

Albrechts, L. (1999). Planners as catalysts and initiators of change. The new structure plan for flanders. *European Planning Studies*, *7*(5), 587–603.

Albrechts, L. (2002). The planning community reflects on enhancing public involvement. Views from academics and reflective practitioners. *Planning Theory and Practice*, *3*(3), 331–347.

Albrechts, L. (2003a). Planning and power: Towards an emancipatory approach. *Environment and Planning C*, *21*(6), 905–924.

Albrechts, L. (2003b). Planning versus Politics. *Planning Theory*, *2*(3), 249–268.

Albrechts, L. (2004). Strategic (spatial) planning reexamined. *Environment and Planning B*, *31*(5), 743–758.

Albrechts, L. (2006). Shifts in strategic spatial planning: Some evidence from Europe and Australia. *Environment and Planning A*, *38*(6), 1149–1170.

Albrechts, L., Alden, J., & Da Rosa Pires, A. (2001). In search for new approaches for planning. In L. Albrechts, J. Alden, & A. Da Rosa Pires (Eds.), *The changing institutional landscape of planning* (pp. 1–7). Aldershot: Ashgate.

Albrechts, L., Healey, P., & Kunzmann, K. (2003). Strategic spatial planning and regional governance in Europe. *Journal of the American Planning Association*, *69*(2), 113–129.

Albrechts, L., & Lievois, G. (2004). The Flemish diamond: Urban network in the making. *European Planning Studies*, *12*(3), 351–370.

Albrechts, L., & Swyngedouw, E. (1989). The challenges for a regional policy under a flexible regime of accumulation. In L. Albrechts, F. Moulaert, P. Roberts, & E. Swyngedouw (Eds.), *Regional policy at the crossroads* (pp. 67–89). London: Jessica Kingsley.

Albrechts, L., & van den Broeck, J. (2004). From discourse to facts. The case of the ROM project in Ghent, Belgium. *Town Planning Review*, *75*(2), 127–150.

Albrechts, L., van den Broeck, J., Verachtert, K., Leroy, P., & van Tatenhove, J. (1999) *Geïntegreerd Gebiedsgericht Beleid: een Methodiek*. Research report by KU-Leuven and KUNijmegen en AMINAL, Ministry of the Flemish Community, Leuven.

Amin, A. (2004). Regions unbound: Towards a new politics of place. *Geografiska Annaler*, *86*(B), 33–44.

Berger, G. (1964). *Phénoménologie du Temps et Prospective*. Paris: PUF.

Breheny, M. (1991). The renaissance of strategic spatial planning? *Environment and Planning B: Planning and Design*, *18*(2), 233–249.

Bryson, J., & Crosby, B. (1992). *Leadership for the common good. Tackling public problems in a shared-power world*. San Francisco: Jossey-Bass.

Cars, G., Healey, P., & Mandanipour, A. (Eds.). (2002). *Urban governance, institutional capacity and social milieux*. Aldershot: Ashgate.

Craft, A. (2000). *Creativity across the primary curriculum*. London: Routledge.

de Bono, E. (1992). *Serious creativity. Using the power of lateral thinking to create new ideas*. New York: Harper Business.

de Jouvenel, B. (1964). *L'Art de la Conjecture*. Monaco: Du Rocher.

Dubos, R. (1969). Future-oriented science. In E. Jantsch (Ed.), *Perspectives of planning* (pp. 157–175). Paris: OCDE.

Dyrberg, T. B. (1997). *The circular structure of power*. London: Verso.

Elchardus, M., Hooghe, M., & Smits, W. (2000). De Vormen van Middenveld Participatie. In M. Elchardus, L. Huyse, & M. Hooghe (Eds.), *Het Maatschappelijk Middenveld in Vlaanderen* (pp. 15–46). Bruxelles: VUB Press.
Flyvbjerg, B. (2002). Bringing power to planning research. One researcher's praxis story. *Journal of Planning Education and Research, 21*(4), 357–366.
Forester, J. (1989). *Planning in the face of power*. Berkeley, CA: University of California Press.
Foucault, M. (1980). *The history of sexuality*. New York: Vintage.
Freestone, R., & Hamnett, S. (2000). Introduction. In S. Hamnett & R. Freestone (Eds.), *The Australian metropolis* (pp. 1–10). St Leonards: Allen and Unwin.
Friedmann, J. (1982). *The good society*. Cambridge: MIT Press.
Friedmann, J. (1987). *Planning in the public domain: From knowledge to action*. Princeton, NJ: Princeton University Press.
Friedmann, J. (1992). *Empowerment. The politics of alternative development*. Oxford: Blackwell.
Friedmann, J. (2004). Strategic spatial planning and the longer range. *Planning Theory and Practice, 5*(1), 49–62.
Friedmann, J., & Douglas, M. (1998). Editor's Introduction. In M. Douglas & J. Friedmann (Eds.), *Cities for citizens* (pp. 1–6). Chichester: John Wiley and Sons.
Gabor, D. (1969). Open-ended planning. In E. Jantsch (Ed.), *Perspectives of planning* (pp. 329–347). Paris: OCDE.
Gibelli, M. C. (2003). Flessibilità e Regole nella Pianificazione Strategica; Buone Pratiche alla Prova in Ambito Internazionale. In T. Pugliese & A. Spaziante (Eds.), *Pianificazione strategica* (pp. 53–78). Milano: Angeli.
Godet, M. (2001). *Creating futures*. London: Economica.
Goodstein, L., Nolan, T., & Pfeiffer, J. (1993). *Applied strategic planning*. New York: McGraw-Hill.
Gualini, E. (2001). *Planning and the intelligence of institutions*. Aldershot: Ashgate.
Habermas, J. (1996). Normative content of modernity. In W. Outhwaite (Ed.), *The Habermas reader* (pp. 341–365). Cambridge: Polity.
Hajer, M. (1995). *The politics of environmental discourse*. Oxford: Oxford University Press.
Hamilton, C. (2003). *Growth fetish*. London: Pluto Press.
Hamnett, S., & Freestone, R. (Eds.). (2000). *The Australian metropolis*. St Leonards: Allen and Unwin.
Harvey, D. (1989). From managerialism to entrepreneurialism: Formation of urban governance in late capitalism. *Geografiska Annaler B, 71*(1), 2–17.
Healey, P. (1997a). *Collaborative planning, shaping places in fragmented societies*. London: Macmillan.
Healey, P. (1997b). The revival of strategic spatial planning in Europe. In P. Healey, A. Khakee, A. Motte, & B. Needham (Eds.), *Making strategic spatial plans* (pp. 3–19). London: UCL Press.
Healey, P. (2004a). The treatment of space and place in the new strategic spatial planning in Europe. *International Journal of Urban and Regional Research, 28*(1), 45–67.
Healey, P. (2004b). Creativity and urban governance. *Policy Studies, 25*(2), 87–102.
Healey, P. (2005). Network complexity and the imaginative power of strategic spatial planning. In L. Albrechts & S. Mandelbaum (Eds.), *The network society: a new context for planning?* (pp. 146–160). New York: Routledge.
Healey, P. (2006). Relational complexity and the imaginative power of strategic spatial planning. *European Planning Studies, 14*(4), 525–546.
Healey, P., Cameron, S., Davoudi, S., Graham, S., & Mandanipour, A. (1995). *Managing cities: The new urban context*. Chichester: Wiley.
Healey, P., Khakee, A., Motte, A., & Needham, B. (Eds.). (1997). *Making strategic spatial plans. Innovation in Europe*. London: UCL Press.
Hillier, J. (1999). What values? Whose values? *Philosophy and Geography [now published as Ethics, Place and Environment], 2*(2), 179–199.

Hillier, J. (2002). *Shadows of Power*. London: Routledge.
Huxley, M. (2000). The limits of communicative planning. *Journal of Planning Education and Research, 19*(4), 369–377.
Innes, J. (1996). Planning through consensus-building: A new view of the comprehensive planning ideal. *Journal of the American Institute of Planners, 62*(4), 460–472.
Innes, J., Grüber, J., Thompson, R., & Neuman, M. (1994) *Coordinating growth management through consensus building: Incentives and the generation of social, intellectual and political capital*. California Policy Seminar, Institute of Urban and Regional Development, University of California, Berkeley, CA.
Jantsch, E. (1970). From forecasting and planning to policy sciences. *Policy Sciences, 1*(1), 79–88.
Kao, J. (1997). *Jamming. The art and discipline of business creativity*. New York: HarperCollins Publishers.
Krumholz, N. (1982). A retrospective view of equity planning, Cleveland, 1969–1979. *Journal of American Planners Association, 48*(4), 163–174.
Landry, C. (2000). *The creative city: a toolkit for urban innovators*. London: Earthscan.
Le Gales, P. (2002). *European cities: Social conflicts and governance*. Oxford: Oxford University Press.
Macnaghten, P., & Urry, J. (1998). *Contested natures*. London: Sage.
Martinelli, F. (2005). Introduzione. In F. Martinelli (Ed.), *La Pianificazione Strategica in Italia e in Europa: Metodologie ed Esiti a Confronto* (pp. 11–31). Milano: Angeli.
Martufi, F. (2005). Pesaro futuro con vista. In F. Martinelli (Ed.), *La Pianificazione Strategica in Italia e in Europa: Metodologie ed Esiti a Confronto* (pp. 92–122). Milano: FrancoAngeli.
Massé, P. (1965). *Le Plan ou l'Anti-hasard, Collection idées*. Paris: Gallimard.
Mathews, D. (1994). *Politics for people. Finding a responsible public voice*. Urbana, IL: University of Illinois Press.
Mayer, M. (2003). The onward sweep of social capital: Causes and consequences for understanding cities, communities and urban movements. *International Journal of Urban and Regional Research, 27*(1), 110–132.
Michalko, M. (2001). *Cracking creativity*. Berkeley, CA: Ten Speed Press.
Mintzberg, H. (1994). *The rise and fall of strategic spatial planning*. New York: The Free Press.
Newman, P., & Thornley, A. (1996). *Urban planning in Europe*. London: Routledge.
Ogilvy, J. (2002). *Creating better futures*. Oxford: Oxford University Press.
Ozbekhan, H. (1969). Towards a general theory of planning. In E. Jantsch (Ed.), *Perspective of planning* (pp. 45–155). Paris: OECD.
Pascual, I., & Esteve, J. (1997). *La Estrategia de las Ciudades. Planes Estratégicos como Instrumento: Métodos, Téchnicias y Buenas Prácticas*. Barcelona: Diputacion de Barcelona.
Pinder, D. (2002). In defence of utopian urbanism: Imagining cities after the end of Utopia. *Geografiska Annaler B, 84*(3–4), 229–241.
Plsek, P. (1997). *Creativity, innovation and quality*. Wisconsin: ASQ Quality Press.
Pugliese, T., & Spaziante, A. (Eds.). (2003). *Pianificazione Strategica per le Città: Riflessioni dalle Pratiche*. Milano: Angeli.
Sager, T. (1994). *Communicative planning theory*. Aldershot: Avebury.
Sandercock, L. (1998). *Towards cosmopolis. Planning for multicultural cities*. Chichester: John Wiley and Sons.
Sandercock, L. (2003). *Cosmopolis ll. Mongrel cities in the 21st century*. London: Continuum.
Schwartz, P. (1991). *The art of the long view*. New York: Doubleday Currency.
Secchi, B. (1986). Una Nuova Forma di Piano. *Urbanistica, 82*, 6–13.
Senge, P. (1990). *The fifth dimension*. New York: Century/Doubleday.
Swyngedouw, E., Moulaert, F., & Rodriguez, A. (2002). Neoliberal urbanization in Europe. Large-scale urban development projects and the new urban policy. *Antipode, 34*(6), 542–577.
van den Broeck, J. (1987). Structuurplanning in de Praktijk: Werken op drie Sporen, in Losbladige Uitgave. *Ruimtelijke Planning, A2C*, 53–119.

van den Broeck, J. (2001) *Informal arenas and policy agreements changing institutional capacity*. Paper of the first world planning school congress, Shanghai.

Winch, G. (1998). Dynamic visioning for dynamic environments. *The Journal of the Operational Research Society*, *50*(4), 354–361.

Young, I. (1990). *Justice and the politics of difference*. Princeton, NJ: Princeton University Press.

Chapter 2
Medium-Sized Towns, Strategic Planning and Creative Governance

Klaus R. Kunzmann

2.1 Introduction

Medium-sized towns located beyond metropolitan regions in Europe are among the victims of the current metropolitan fever. Despite all political rhetoric and European efforts to promote territorial cohesion, regions outside metropolitan regions are and will continue to be effected by globalising forces and strong regional competition. While future-oriented creative and knowledge industries flourish in a few metropolitan regions and in the core of Europe, regions and towns beyond such conurbations, and in the periphery of Europe, are increasingly struggling to maintain their economic, social and cultural functions. Medium-sized towns in such regions are hit by the increasingly competitive global economy. In order to secure employment and to maintain service functions for a stagnating regional population, these medium-sized towns are forced to find their own profile between international orientation and local embeddedness.

The chapter explores ways and means of stabilising the economic, social and cultural development functions of medium-sized towns. It stresses the importance of the territorial capital as a base for local and regional action. People living in these towns are seen as a relevant part of the territorial capital. Their competence and tacit knowledge, their community commitment and their international networks are the capital for creative governance, where local and regional institutions in a socio-political environment of mutual trust have to cooperate and complement each other. Only in such partnership of local and regional institutions future-oriented initiatives can be developed and implemented.

K.R. Kunzmann (✉)
Department of Spatial Planning in Europe, Technical University of Dortmund, 44227 Dortmund, Germany
e-mail: klaus.kunzmann@udo.edu

M. Cerreta et al. (eds.), *Making Strategies in Spatial Planning*,
Urban and Landscape Perspectives 9, DOI 10.1007/978-90-481-3106-8_2,

2.2 Strategic Planning for Medium-Sized Towns

Since the beginning of the twenty-first century, mega-cities and metropolitan regions have been getting attention of politicians, planners, city marketing managers and the international media. They draw on a plethora of academic literature that looks across Europe and beyond, on the role of world cities and metropolitan regions for development. The message is that only global cities or least large metropolitan regions can sustain Europe's competitiveness against Asian and American mega-towns. Governance in metropolitan regions has to be improved to make such regions more competitive.

During the late 1990s, the European Commission has supported a series of studies for the future development of large European cities such as Vienna, London, Berlin and Marseille to explore appropriate policies for metropolitan development. Since then, all over Europe metropolitan city regions have become a favourite area of academic and political interest. Consequently, most recommendations of the Lisbon Agenda, the highly praised policy paper of the European Commission, showed the corridors for future political and economic arenas towards strengthening competitiveness of Europe. The Agenda suggested, albeit only indirectly, to focus future policies on the promotion of innovation and knowledge industries in metropolitan regions. The Gothenburg declaration, which tries to cushion the economic focus of the Lisbon Agenda and European mainstream policies by raising the contrasting, or at least complementary issue of sustainability, does, regrettably, find much less political interest.

In 1997, the conference of German Ministers responsible for spatial planning followed such mainstream thinking and assigned nine German city regions the status 'European metropolitan region'. After protests from city regions, excluded from the champion's club, four more regions were granted the desired status. It enables city marketing managers to better market their cities internationally.

In this climate of 'metropolitan fever', areas in the shadow of metropolitan regions tend to be neglected. They seem to be the negligible victims of mainstream policies in times of globalisation and regional competition. While, as a rule, small- and medium-sized towns within metropolitan regions in Europe clearly benefit from the growing economy, those beyond the geographically disadvantaged hinterland of thriving metropolitan regions, seem to loose out. This is the case in Western, Central and Southeast Europe, though even more so in Eastern Europe, where most economic development is concentrated in a few capital city regions only. This is also true for the South Baltic Arc (Baltic, 1994).

However, voices of concern increasingly draw attention to the role of medium-sized towns, or secondary towns, as they are labelled in the Anglo-American world, for regional economic and social development. In July 2006, *Newsweek*, the American weekly, published a special report claiming that "The last century was the age of the mega-town. The next will belong to their smaller, humbler urban relations." And at a recent Expo Real in Munich, a workshop was dedicated to 'Cinderella towns', indicating that even the real estate industry has started to review its focus on big cities.

In 2006, the European Spatial Planning Observation Network (ESPON) published a study of a transnational team under the leadership of the Austrian Institute for Regional Studies and Spatial Planning entitled *The Role of Small- and Medium-Sized Towns in Europe*. This study explores the role of small- and medium-sized cities in regional development at the beginning of the twenty-first century and identifies a number of related research issues (ESPON, 2006).

The interest in the promotion of medium-sized towns is not new. It has a long tradition. When the World Bank in the late 1960s started to show an interest in urbanisation and urban development, development of secondary towns became a much acclaimed strategy for balanced regional development in developing countries. At that time a number of studies and books were published. They analysed the role of medium-sized urban centres for regional development and gave recommendations of how to promote the development of such towns (Hennings, Jenssen, & Kunzmann, 1981; Rondinelli, 1983). Between 1980 and 2000, German Technical Assistance favoured the project strategy and initiated a number of secondary towns' projects, among others in Bolivia, Yemen, Nepal and Malawi. The activities were supported by a document, which described the rationale, the principles and the elements of such a strategy (Drewski, Kunzmann, & Platz, 1989).

At the beginning of the twenty-first century, metropolitan concentration, spatial specialisation, spatial fragmentation and spatial polarisation are some of the consequences of globalisation and technological change in Europe (Kunzmann, 2007a). The fierce competition among city regions in Europe for investment, talent and creativity, nurtured by policy advisors, business consultants, researchers and ambitious city leaders, has nurtured 'metropolitan fever' (Leber & Kunzmann, 2006). This fever has resulted in the development of ambitious urban projects and mega-events to attract tourists and media. Such metropolitan fever tends to leave territories behind, territories which are geographically disadvantaged or do have less economic strength and political power. This chapter (i) defines and categorises medium-sized towns with respect to their function and geographical location; (ii) sketches their most important development challenges; describes the potential of medium-sized cities for regional development and stabilisation and (iv) gives first policy recommendations to promote medium-sized towns, in the context of strategic local and regional planning and creative governance.

2.3 Categorising Medium-Sized Towns

What is a medium-sized town? The definitions vary. The most common definition is that of a town with a population of 20,000 up to 200,000, depending on population density and the respective urban system in a country (European Foundation, 1994; Rivkin & Rivkin, 1982; Rondinelli, 1983). Such towns usually have a mix of supply, development and relief functions for the region in which they are geographically embedded. If located at an inner-European border or at the edge of the European Union, they may also have an additional function, that is, exchange or gateway. Medium-sized towns can be located:

- within larger metropolitan regions;
- on the edge of or in between metropolitan regions;
- in the geographical periphery of Europe.

Even within these three categories differences occur. They may stem from micro-locational advantages, local assets and cultural traditions or adjacent borders, or even politico-administrative factors, stemming from various historical events over centuries. The geographical location has strong influence on the respective function of a medium-sized town for the regional hinterland, though geography alone does not explain or determine functions of a medium-sized town. Such functions can be:

- a supply and stabilising function, that is, the task to sustain the role of a town as an economic, social and cultural centre in a region including the provision of goods and services for the households, local firms and enterprises;
- a development function, that is, the role of a medium-sized town as an engine for regional spatial development;
- a relief function, which means that a town is being chosen as a location for functions, decentralised for economic or political reasons from the metropolitan core;
- a border, exchange and gateway function, that is, the additional function of a town at an inner or outer European border as a gateway centre and a centre of cultural exchange.

Medium-sized towns in the Baltic Sea region usually encompass a mix of the above functions, albeit often a single function dominates. Gown towns, such as Greifswald, for example, function as well as central places for their rural hinterland. This is similarly true for ports, such as Szczecin in Poland. Medium-sized towns in the neighbourhood of metropolitan cores, which once used to be central places in a rural region, function as residential towns for the metropolitan population. They are as well attractive targets for services, which are farmed out from the core town and for institutions which search for affordable sites for their back offices, or are selected as pioneer locations for inward investment. The growing concentration of economic development in metropolitan regions affects each of the three categories of medium-sized towns quite differently (Fig. 2.1).

Mediums-sized towns within metropolitan regions are the most likely winners of ongoing territorial development trends (Fig. 2.2). They offer a combination of the advantages of living in the metropolitan core and in the countryside. Usually, such towns have a long history, a strong identity and a high degree of liveability, which is reflected by deeply rooted local traditions, good schools and public services, a high degree of security, accessibility to nature and leisure grounds, *Übersichtlichkeit* and *Langsamkeit*, and, last but not least, affordable real estate. Benefiting from their excellent connectivity by road and rail, they are favoured locations for households, who esteem more traditional life styles or are driven out from metropolitan core by the real estate market. In addition, easy access to the metropolitan airport makes 1-day business trips within Europe possible. As a rule such medium-sized towns

Function / Location	Supply and stabilisation	Development	Decentralisation and relief
In a metropolitan region	**Daily and weekly consumer goods and services** **Public services** (education, health, social services, affordable housing, justice, security, culture, leisure, etc.)	**SMEs** **Knowledge industries** **Creative industries**	**Back offices** **Qualified services for national customers** **Creative industries**
In between or on the edge of a metropolitan region	**Daily and weekly consumer goods and services** **Public services** (education, health, social services, affordablehousing, justice, security, culture, leisure, etc.)	**Single large industries in traditional location** **SMEs** **Special functions based on local profile and assets** (furniture, food, health) **Local potentials** (tourisms, recreation, leisure)	*Not relevant* *However* **Erosion of function in small towns, shifted to the medium-sized town**
In the periphery of metropolitan regions	**Daily and weekly consumer goods and services** **Public services** (education, health, social services, affordable housing, justice, security, culture, leisure, etc.)	**SMEs** **Knowledge industries** **Creative industries**	*Not relevant* *However* **Erosion of function in small towns, shifted to the medium-sized town**
At an inner or outer European border	**Logistic distribution, cultural exchange, intercultural communication**	**Logistics** **Customs services**	*Not relevant*

Fig. 2.1 Towards a typology of medium-sized towns

thrive. They grow in terms of population and economic development. Their budget is healthy and they can afford to maintain high standards of public infrastructure. Public management is efficient and public private partnerships can be organised at ease.

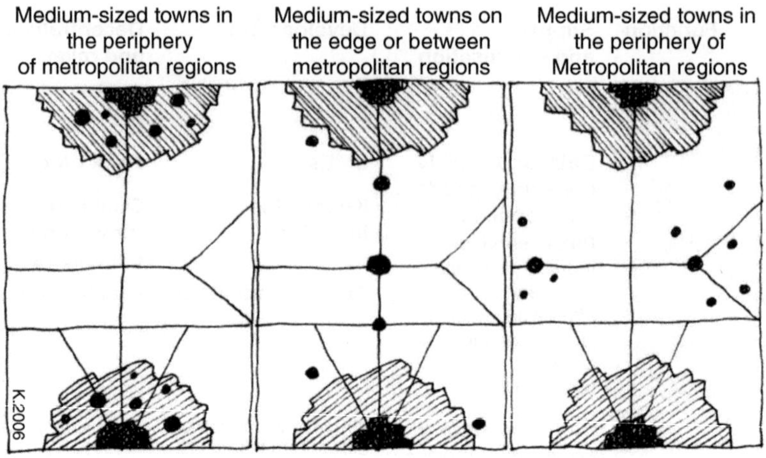

Fig. 2.2 A typology of medium-sized towns

Yet, another type of medium-sized towns within metropolitan regions are former rural villages on the edge of the core town, which have rapidly grown over the last decades, due to urban expansion and development pressure. Benefiting from their right to self-government, they have successfully opposed to become incorporated into the core town.

Medium-sized towns in between or on the edge of metropolitan regions are in a different position (Fig. 2.3). They can benefit from the development of the

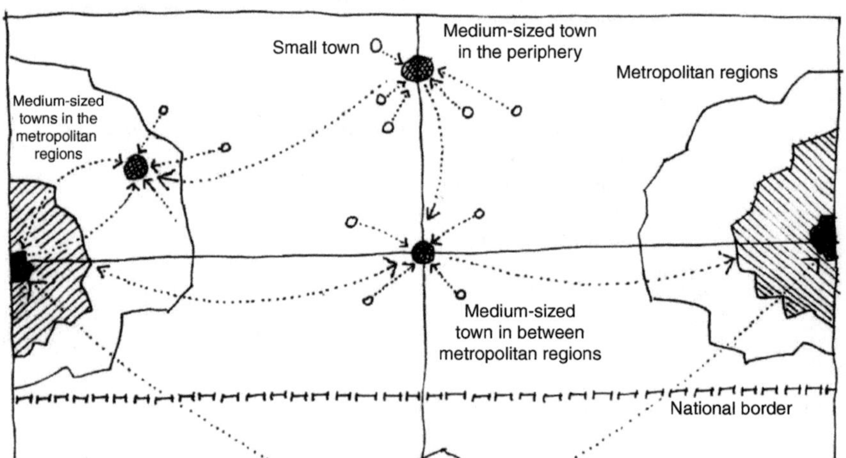

Fig. 2.3 Medium-sized towns in metropolitan regions

metropolitan region, if they are linked to the core by efficient and frequent metro-services or convenient and congestion-free connections to the dense metropolitan motorway system. Under such conditions, these towns function like an exterritorial island outside the metropolitan region in a rural environment, with all its natural amenities, attractions and environmental potentials. At the same time, urban attractions of the metropolitan core can be reached within 1 h commuting time. Real estate here is even cheaper than in the metropolitan region. Young families, attracted by the diversity of job opportunities in the metropolitan region and accepting the long commuting distances, may find here appropriate and affordable property. However, if such towns are poorly connected to the metropolitan core, they face similar difficulties as towns in the periphery. Though potentially located in the hinterland of the metropolitan region, they are less attractive for households and firms. The economically more active population tends to leave the remaining population ages and public infrastructure erodes. Lobbying for a better physical connectivity to the metropolitan core is one chance to reverse the negative trend in the long run; mobilising the endogenous territorial capital is another, probably more promising one.

Medium-sized towns in the periphery of Europe are the relative losers of globalisation. Their connectivity to the national and European transport network (air, rail and road) is poor. Hence the local economy suffers from the locational disadvantage. Consequently very little inward investment is made, and if it is made, then only due to enormous public subsidies and regulatory concessions given to attract such investment. Consequently, the job market looses its former attractivity and the training opportunities for school leavers diminish. Long-term unemployment increases as young, economically active households leave the town and migrate to metropolitan regions with their more diversified job markets.

People are aging. Primary and secondary schools are being closed, reducing variety and choice. The local tax base is eroding. Public services are being reduced mainly due to financial constraints. Gradually, local social and economic disparities are growing, followed by social tensions and security problems. Another consequence of the competitiveness mainstream in Europe is the gradual erosion of public and private services in small towns located in the immediate hinterland of the medium-sized town, which contributes to the further marginalisation of rural areas.

2.4 The Challenges of Medium-Sized Cities in Metropolitan Peripheries

Given the overall demographic, economic and environmental conditions of territorial development in Europe in a globalised world, the implications of China's economic growth for cities and regions in Europe, and the concentration of economic strength in a few metropolitan regions, medium-sized cities beyond metropolitan regions may have only modest prospects to strive economically. However, as central places in their respective regions, they undoubtedly have an essential role in stabilising the regional economy and in providing appropriate public services to the people, who wish to stay in or even to settle down in such places. The term stabilisation is explicitly used here to signal that traditional economic development, which implies economic growth, may not be the right starting point for local and regional action. It also signals that expectations for economic growth similar to that of the metropolitan areas may be unrealistic. The key task of strategic planning both at the local and the regional level of planning and decision-making in such regions is to support the necessary processes of territorial stabilisation.

A range of local challenges hampers the efforts to achieve such stabilisation processes in medium-sized towns. These challenges are widely known. They have been explored in many case studies. Though the challenges differ from town to town, a few generalisations can be made, independent of shortcomings related to geographic location, environmental circumstances and local economic history, or present and past politico-administrative conditions.

Such structural changes, affecting the provision of public services, economic development and employment in medium-sized cities, include the following:

- *demographic change and aging* – declining fertility and the aging of population have affected many nations and regions in Europe with considerable implications for social infrastructure;
- *concentration of economic power* – globalisation and technological change lead to growing concentration of economic development in metropolitan regions with considerable consequences for inter-regional logistic networks, knowledge and creative industries;
- *changing values and location preferences* – together with technological innovations, changing values, attitudes and preferences of individuals and households influence location preferences of firms and enterprises;
- *political complexity* – in a four- to five-tier system of planning and decision-making in Europe, it is increasingly difficult to insist on clear local or regional development positions; lobbying at higher tiers becomes increasingly difficult;
- *cosmopolitan communities* – increasing migration leads to growing social and economic polarisation in cities with considerable consequences for the provision of public services, local labour markets and security. Border and gateway cities are additionally burdened by their role as logistic exchange centres and national windows of cultural exchange.

Efforts to address these challenges at the local or regional level are hampered, among others, by:

- *fragile strategic consensus* – local governments tend to wait for strategic guidance from above; substantially defined programmes are linked to structural funds; however, regional governments prefer to rely on the strength of local self-government; such unfilled expectations tend to mutually block effective collaborative action;
- *intraregional conflicts* – sharing responsibilities between central medium-sized cities and surrounding suburban or rural local governments are hampered by

political manoeuvring and excessive justification of the right to self-government; such conflicts tend to hinder efficient use of urban–rural linkages;
- *lack of confidence and visionary power* – it is not easy for medium-sized cities to sharpen their local profile; insufficient knowledge of local assets and the territorial potential, fragile local coalitions, lethargy and negative attitudes towards the political system, the absence of a strong civil society or the courage to look beyond election periods, aggravate or even impede the development of strategic visions;
- *gridlocked decision-making processes* – politically motivated attitudes of local and regional administrations, meandering or parochial opinion leaders, mistrust among local leaders, vested interest of influential local stakeholders and not transparent decision-making processes make it difficult to find easy consensus on local development principles and projects.

Consequently, any local strategic planning has to set off from a careful analysis of the respective local shortcomings and the explorations of the local territorial capital, taking local socio-political milieus into account. It is quite obvious that strategies that focus on the improvement of physical conditions in urban districts or on employment initiatives (infrastructure) do not suffice to create local milieus for sustainable city development and the structural stabilisation of local conditions.

2.5 Territorial Capital of Medium-Sized Cities in Metropolitan Peripheries

The above typology shows that medium-sized towns, in addition to regional development and real and potential relief, have a supply and stabilisation role to play. In internal and external border regions of Europe they have the additional task to serve as points of cultural exchange and laboratories of inter-cultural cohesion in multi-cultural environments. Outside metropolitan areas, they are clearly the focal point of regional economies and engines of territorial development. This implies that they have an immensely important role for stabilising regions as life spaces for increasingly heterogeneous regional communities.

In this context, the concept of 'territorial capital', that was introduced into the discourse on European Spatial Development, is very useful. Following an Organization for Economic Co-operation and Development (OECD) definition (2001), the preparatory document to the Territorial Agenda defines a region's territorial capital as follows: "A region's territorial capital is distinct from other areas and is determined by many factors (which) (...) may include (...) geographical location, size, factor of production endowment, climate, traditions, natural resources, quality of life or the agglomeration economies provided by its cities (...) Other factors may be 'untraced inter-dependencies' such as understandings, customs and informal rules that enable economic actors to work together under conditions of uncertainty, or the solidarity, mutual assistance and co-opting of ideas that often develop in small- and medium-sized enterprises (SMEs) working in the same sector (social capital). Lastly there is an intangible factor, 'something in the air', called the 'environment' and which is the outcome of a combination of institutions, rules, practices, producers, researchers and policy-makers that make a certain creativity and innovation possible. This 'territorial capital' generates a higher return for certain kinds of investments than for others, since they are better suited to the area and use its assets and potential more effectively (...)" (CEMAT, 2007, p. 7).

The territorial capital is the asset or talent of a region, which shapes its regional economic potential, its socio-political culture, its environment of the arts and crafts, its visual appearance and its identity. And last but not least, it is the territorial capital that appeals to others, who live outside the region. The reference to territorial capital in the Territorial Agenda of European Conference of Ministers Responsible for Spatial/Regional Planning (CEMAT) opens the door for new efforts to promote endogenous territorial development and regional economic circuits in Europe. Regional development paradigms have been discriminated too long and shelved for being ineffective and naïve.

However, what is the specific territorial capital of medium-sized cities in regions in the South Baltic Arc, and beyond metropolitan areas, on which strategic stabilisation strategies could build upon?

- *Cultural traditions and local identity* – Most medium-sized towns have a long history. Centuries, sometimes even more than 1,000 years have shaped the local identity and the visual appearance of the town, and they formed local cultural traditions. Such traditions frame the annual calendar of public life and motivate local initiatives. They are an essential dimension of the local community spirit.
- *Tacit knowledge of the community* – Knowledge, competence, skills and special qualifications of people in medium-sized cities are an important asset of local economies. Handed down from family to family, from business to business and from entrepreneur to entrepreneur over centuries, and embedded in a regional environment, such knowledge is strongly rooted in the local economic history. Competences of the past are a good base for local strategies that aim at forming-up new fields of local competence. There is no old technology that does not have a modern, future-oriented equivalent.
- *Embeddedness of local businesses and firms* – A traditional strength of medium-sized towns are family enterprises, rooted in the town over generations and contributing to the local identity. Although such businesses are eroding as a consequence of globalising markets and aggressive franchising policies, they continue to have a key role in the local economy. They play an essential social role in the community and their supply chains are rather regional than international. Good examples for the vigour of such local economies are medium-sized Italian cities with thriving economies, where local production complexes are successfully serving international markets.
- *Easy informal networking* – Decision-making processes in medium-sized towns tend to be faster than in large cities. The smaller size of local administrations and

political committees allows easy communication. The agendas of local politics are less complex and more transparent. And the local community (everybody knows everybody) facilitates informal networking, which helps the decision-making processes. In addition, short distances in a medium-sized city make personal communication easy.

- *Übersichtlichkeit* – The size of a medium-sized town makes it easy to maintain civic traditions, to guarantee a certain local *Übersichtlichkeit.* As a rule, there is a good balance between urban anonymity and civic visibility. Social control is high, sometimes even disturbing, and security is less of a problem. The local *Übersichtlichkeit* facilitates individual orientation, it fosters civil courage, and it welcomes visitors to a place.
- *Entrepreneurship* – More than once, single creative and innovative entrepreneurs have succeeded to place their products or services in national or even global markets. Usually, the success emerged from a combination of advanced technologies, traditional endogenous knowledge and skills and the drive of an entrepreneur. What is also essential is a certain personal commitment to the locality. In partnership with successful entrepreneurs, local success stories could be used as a starting point for developing a local cluster of firms in branches along forward and backward linkages (in French: *filières*), offering opportunities for buy-outs and young start-ups from the town or a university in the metropolis.
- *Urban-rural relationship* – Traditionally medium-sized towns have good relations with the immediate rural hinterland. In the past, farmers sold their agricultural products at the towns market. The closest town has been the first target of young people wishing to leave rural life behind. In the twenty-first century, such relationships are economically and socially less important. Today, they are replaced by linkages which have more to do with experiencing nature or enjoying leisure or sports activities. However, with the challenge of resource conservation and the renaissance of bio-food and health considerations, such traditional linkages and food chains are being valued and revitalised.
- *International networks of memories* – A mostly untapped potential in larger as well as in smaller cities are the tacit international networks of citizens. Such networks are family connections, and networks stemming from inter-cultural marriages, linkages to former countries of residence, reminiscences of private longer and shorter stays in another country, business relations or memories of school exchange and studies abroad. Such individual networks are windows of opportunities for international communication and networking, from which the whole community can benefit (Kunzmann, 2000a).

There is, obviously, a backside to such local potentials, too. The more inward looking assets, as the ones sketched above, can easily turn against future-oriented urban development. They can close up a community against outside influence, they may foster parochial attitudes and hinder innovation processes and they may be exclusive in social terms, particularly, when it comes to integrating migrants. Consequently, strategies to promote local stabilisation and development will have to find the right balance between local traditions and global challenges. Therefore local educational institutions and newspapers have a key responsibility to break parochialism and open-up the local community. However, there is much evidence that neglecting such traditions and following mainstream trends do not lead to new sustainable economic development.

Overall, once employment is secured, most medium-sized cities in the shadow of metropolitan regions offer a significantly higher quality of life (in a healthy environment) at affordable cost. Compared to larger cities, it is this quality that can compensate for some of the deficits, of medium-sized towns; when it comes to educational choices, job opportunities and entertainment options.

2.6 Strategic Planning in and for Medium-Sized Towns

The many efforts to stabilise the development of medium-sized towns beyond metropolitan regions require efficient strategic planning. Strategic planning is what every large enterprise does to envision the future and to secure its position in a globalising world and a competitive market. Such strategic planning has to be done by medium-sized towns too. Strategic planning is a social process through which a range of people in diverse institutional relations and positions come together in plan-making process to develop contents and strategies for the management of spatial and structural change (Kunzmann, 2000b). This process generates not merely formal outputs in terms of policy and project proposals but also a decision-framework of principles (concerning, e.g. mobility, resource conservation or local partnerships) that may influence relevant parties in their future investment and regulatory activities (Healey, 1997).

Strategic spatial planning is predominantly a public sector led process, which aims to combine planning with implementation. Thus strategic planning has visionary and pragmatic dimensions. A strategic plan is not an ambitious spatial *leitbild* which has been developed by a planning department to guide spatial development processes, it is rather a framework for strategic decisions with a set of principles for guiding day-to-day development in a city. Such a framework requires collaboration (Healey, 1997) in order to create positive decision-making environments, which have been characterised as fertile *milieus for collective action* (Cars, 2002). This implies that strategic planning is more than land use planning, more than just assigning uses to spaces in a city, and waiting patiently for public and private investors to realise such assignments. Obviously, strategic planning is more than an exercise to set up a shopping list for public or public–private projects.

Strategic planning for medium-sized towns means bringing together intermediate stakeholders in a city, both public and private so that they may explore the endogenous territorial capital, and decide how priorities can be set, compromises made and forces joined. This is necessary to respond to local challenges, to secure jobs locally and to maintain a good quality of life for all citizens.

The stabilisation of medium-sized towns outside metropolitan regions can only be successful if all tiers of planning and decision-making are willing and committed to cooperate in such strategic development processes. Thereby, each tier of planning and decision-making has a particular role to play. It is essential, however, that strategic planning at the local level ('planning from below') is continuously concerted with strategic planning at the regional tier ('planning from above').

Local governments have to show initiative, be creative and proactive in using the local territorial capital for developing the local economy and the local community. They have to secure jobs at the local level for the people of the city and maintain services for households and local enterprises. In the context of their unanimous right to self-government, they are both free and responsible for responding to local challenges and developing an integrated strategic framework for local spatial, economic, cultural and social development. Much can be done and should be done at the local level, where initiatives have to be taken, visions developed, consensus among local citizens and stakeholders sought and implementation organised.

Below is a list of essentials of strategic planning that address local development problems and are aimed at overcoming shortcomings and preparing the local community for the future.

- *Base stabilisation strategies for the local territorial capital* – There is no other way to stabilise and develop a medium-sized town located beyond a metropolitan region than to rely on the endogenous territorial capital. The knowledge of the particular local capital is essential. It has to be carefully researched, evaluated, documented and locally communicated. Only with such knowledge mainstream fashions in economic development can be adequately assessed with respect to their relevance for the local economy and longer-term employment strategies.
- *Use and promotion of local knowledge and competence* – The more a town succeeds in using endogenous potential, that is, local merchants, crafts and firms or regional competence and tacit knowledge, the more it sharpens the local identity and its urban profile. This in turn attracts external interest and strengthens local commitment. The promotion of SMEs in areas of local knowledge and competence is a logical strategic consequence of action. A distinct local profile could attract knowledge industries, creative citizens and talent. Supporting local entrepreneurs to adapt to changing technologies and markets to form local production clusters and service networks must have priority over excessive promotion of inward investment.
- *Building on local quality of life* – The quality of the built and natural environment is a key factor in community building. Citizens who enjoy living in town are proud of the town and are more likely willing to stay. There are many ways to support the liveability of a town. The beauty of a town, the cultural heritage, local architectural traditions and attractive public parks are key elements of liveability, as well as individual security and leisure opportunities.
- *Engaging the local civil society* – Local communities are used to get supplies from the state, but the public sector is no longer able to meet all the needs. Hence it is indispensable to engage local communities in strategic planning and development, not as consumers of public services, but as actors in local efforts to improve liveability in the town. Involving migrants in such efforts could facilitate and speed-up

their integration. Though their visions may be different from the traditional local ones, they may be in the end more realistic and more future-oriented. And they may contribute their language skills and international networks to bring in new strands in local economic development.

- *Targeting the young generation* – Young people in a medium-sized town are likely to be attracted by metropolitan opportunities and promises. Their early active involvement in local projects could contribute much to reducing their motivation to leave the town after school. The more they feel that their concerns are taken seriously, the more they are willing to get involved in community projects. Costs for such involvement and for small projects are marginal compared with infrastructure costs or subsides for attracting volatile inward investment. In the end, they may change consumer attitudes into more pro-active collaboration.
- *International orientation* – Internationalisation is a key survival strategy for local communities in times of globalisation. Medium-sized towns can easily add an international component to local development strategies. Traditional marketing is one way to reach international attention; international benchmarking is another. Both are very much linked to the international image of a town as a tourist destination. Culturally justified twin city arrangements have a long tradition, though they are often seen as a burden rather than an opportunity. Export–import linkages of local businesses and enterprises are rarely seen as an asset for strategic urban development. And a town can easily benefit from the broad range of international networks of a community, once individuals are encouraged to participate in the exploration of related economic opportunities.
- *Involving local media in communicating values and vision* – As a rule local media are invited to report about local political, social or cultural events. And they do it from a more or less neutral position and with a journalistic ethos of opening-up and controlling political decision-making processes. It may be useful to invite key editors of local media to participate early in city profiling efforts, in developing city visions and in communicating visions to the local community. Being involved in development processes may help to mobilise community participation and local commitment and contribute to build up trust in local decision-making.
- *Forming sub-regional strategic alliances* – Conflicts between cities and surrounding local governments may lead to gridlocked situations, where decisions are unduly blocked. Carefully selected catalytic projects, which bring win-win situations for both, may be an appropriate means to build up strategic alliances, from which both, the central city and the neighbouring rural communities, can only benefit. It cannot be the task of regional governments to moderate sub-regional conflicts. Joint local brainstorming will certainly help to identify appropriate projects. Incentives from the regional government to promote sub-regional cooperation may help to accelerate consensus-building processes.
- *Promoting local economic circuits* – One regional response to globalisation are local and regional economic circuits. Such circuits support forward and backward linkages or supply chains in a region that rely on regional rather than on international production. In some areas, such as food, construction or cultural industries, it may be easy to promote, establish and sustain such circuits. They

have to find political support and local banks' support (Gaertner, 2007), attract financial incentives and remove institutional constraints. Economic circuits may also be the one outcome of successful regional cluster management. Regional economic circuits rely very much on long-term trust. This trust will have positive effect in a community, beyond economic rationales (Magnaghi, 2000).
- *Promoting learning processes* – Urban development processes are processes where participants articulate their objectives and interests, listen to the arguments of others, seek and find consensus and, if necessary, negotiate compromises (Stein, 2006). Processes where the participating stakeholders learn to understand others, communicate their interests and sharpen their arguments create excellent opportunities for learning. These innovative urban projects, considered catalytic projects, are initiated and implemented to test new development approaches; they are perfect grounds to promote local learning and to continuously qualify local planning and decision-making processes.
- *Forging inter-regional networks* – In recent years, INTERREG projects have been very successful in building inter-regional networks among cities and regions. Based on successes of such projects and related experiences, it could make much sense to forge sustainable inter-regional and inter-cultural networks across national borders. Such networks could involve students and teachers, sports and business clubs and choirs and youth orchestras. And they could be strengthened by un-bureaucratic temporary exchange of professional staff in public and semi-public institutions, as well as the exchange of trainees and apprentices. Thematic networks, such as the successful launch of the European route of brick architecture, should encourage the establishment of such networks (Pienkoß, 2007). Joint thematic fairs (food, arts, health) could be another option, as well as sport events, where teams of the participating cities compete for an annual trophy.

In the end, a responsive local government of a medium-sized city will have to explore their own appropriate strategy to stabilise local economic and social conditions to the benefits of all citizens. And, not to forget, people matter.

The regional level, such as the Länder in Germany, is of particular importance, when it comes to encouraging local planning processes. Regional governments, themselves locked into a complex politico-administrative system of supra-regional institutional bodies have to guide and support local governments in their efforts to carry out given and chosen tasks, and in the use of their local territorial capital. It is particularly this tier that has to enable local governments of medium-sized towns to be proactive, creative and efficient. Supportive actions 'from above' have to assist medium-sized towns to better use their territorial capital. Thereby three factors are very much in the hands of regional institutions:

1. *secure connectivity* – connectivity to regional national and international flows of goods and information is the key to economic success; connectivity, a combination of comfortable and financially competitive linkages to regional transport networks (rail, road, air), has become the essential location factor for both households and firms; consequently, lobbying for connectivity is the prime task

of policy-makers at all tiers of planning and decision-making. Inter-linking medium-sized cities, and connecting such networks to larger trans-European transport networks, must be the principle;
2. *promotion of knowledge industries and human capital* – local availability of skilled labour has always been an important location factor; today, even more so than in the past, rapid technological change requires a flexible and creative labour force, which is prepared and willing to continuously learn to adapt to new challenges, technologies and procedures; high-quality education and training facilities are indispensable to guarantee the availability of skilled labour locally; while the local government, in order to keep or to attract such a labour force in the town, can do much to promote local knowledge and skills, and to improve local liveability, only consistent policy frameworks at higher tiers prepare the ground for better training and education;
3. *promotion of gradual change of mainstream development paradigms* – there is much scientific evidence and political insight that medium-sized cities beyond metropolitan regions will have to rely to a large extent on the endogenous potential of the location; significant inward investment will only occur, if a specific profile of the region or single localities can attract outside interest, if technologically advanced local enterprises or specialised firms are a strong local asset or if investors with a personal link or interest in the location are willing to invest; consequently, local decision-making arenas and communities of practice in medium-sized towns will benefit much from a regional discourse environment, where endogenous local opportunities are given more attention.

Other essentials of regional governance to support strategic planning and wise management of medium-sized cities are as follows:

- *providing special funds for innovative action* – if, for whatever political or local reasons, local governments are neither proactive nor creative, the regional institution has to find appropriate ways and means to enhance a town's initiative strength; competitions among towns for special programmes (such as the *Regionale* in North Rhine-Westphalia, the competition for the Cultural City of Europe or the competition for casino development in the United Kingdom) have proved to be such an instrument. In this context it could also make sense to support applications of medium-sized cities by a small budget to overcome initial hesitation; it would, however, be important that local administrations, which have to manage and implement the programme later, write the application themselves and do not ask external consultants to do the job;
- *lowering expectations for unlimited financial support* – it has become the habit that local (as well as regional) governments expect continuous support for local initiatives and projects from European and central government programmes; this has led to a new dependency culture, and it has fostered an attitude where applications for funding follow the interest of the programme managers, rather than local needs and requirements;

- *encouraging the formation of urban networks* – medium-sized towns can benefit much from continuous engagements in intra-regional, inter-regional and international urban networks; the decision to participate actively in such networks will have to be made by the towns themselves; however, regional administrations can encourage towns with various forms of incentives to consider such involvement seriously; the formation of such networks can also be supported by policies that aim at a certain spatial division of labour among medium-sized cities in the larger region; medium-sized towns, located beyond capital and metropolitan regions, should be encouraged to sustain their inter-regional cooperation networks beyond INTERREG programmes, to learn from each other, to benefit from economic and cultural relations and to form inter-regional strategic alliances, as a means to withstand the metropolitan fever, which is spreading all over Europe;
- *sharpening the profile of the macro-region* – any medium-sized town will benefit from the image of the macro-region in which the town is located; any promotional effort to communicate the economic profile, the urban heritage, the cultural events, the quality of life and the beauty of the landscape will automatically draw the attention of business visitors, investors, tourists as well as knowledge workers and students to the towns of the region; medium-sized towns will benefit much from regional efforts made to enhance the image beyond regional boundaries and from efficient lobbying at higher tiers of planning and decision-making.

In the end, it is a balanced combination of bottom-up and top-down processes, which can support the development of medium-sized towns. Waiting for and relying on European funds for project development and implementation is certainly not the right policy.

2.7 Medium-Sized Towns and Creative Governance

Routine procedures of decision-making within and in between institutions in an established regulatory system have a tendency to filter out alternative ways of problem-solving. In a complex multi-tiers system of governance in Europe, it has become more and more difficult to change the regulatory system. In addition, with tightening local budgets and lean public management structures, the willingness to experiment with new strategies and processes is decreasing. Very often, only upon initiative and financial support of higher tier governments, new approaches to strategic development are being explored. The city networks initiative, the 2030 programme of the Federal German Government, the *Regionale* initiative of the State of North Rhine-Westphalia (Kunzmann, 2007b) or the INTERREG programmes of the European Union are good examples (Pinkoß, 2007).

As a rule, however, given the established political environment in medium-sized towns with their local rituals as well as personal networks and commitments, there is only limited space for new faces and fresh thoughts. And, with few exceptions, established institutions, local governments or regional public or semi-public institutions have a tendency to concentrate on routine management of day-to-day affairs.

Their willingness to change the routine path of institutional action, to experiment with new approaches to urban and regional development or to recruit new creative staff is limited. Their profound institutional knowledge of implementation is a key reason for seeing difficulties rather than opportunities. In-built institutional learning and innovation are scarce. Consequently, new tasks in a medium-sized town may occasionally even require the establishment of a new institution, as the existing one does not any longer have the innovative strength, the visionary power or the credibility of local stakeholders. Then only newly established institutions, with new persons in leading positions, for coping with new challenges. The establishment of such new and small agencies or moderation units staffed with handpicked professionals from within and outside the town or the region could be done for a limited time only.

With more creative governance, medium-sized towns can address and successfully cope with the challenges of globalisation they are facing (Hans Seidel Stiftung, 2007; Kunzmann, 2004). Creative governance is more than urban management based on routine procedures and responding to top-down commands and financial contributions. Implementing the above agenda for creative governance requires political goodwill and strong leadership supported by visionary politicians and professionals, who know how and when to start implementing one or the other project along the lines sketched above. In particular, it requires multiple creativity, creative institutions and creative actors, as well as creative holistic and thematic strategies and processes.

Acknowledgements The paper has originally been written for SEBco (http://www.sebco.eu), an INTERREG IIIB project [http://bsrinterreg.net] initiated and managed by the Ministry for Transport, Building and Regional Development/Mecklenburg-Vorpommern [http://www.vm.mv-regierung.de] and part-financed by the European Union under European Regional Development Fund (ERDF).

References

Baltic Institute. (1994). *Vision and strategies around the Baltic Sea 2010. Towards a framework for spatial development in the Baltic Sea Region*. Karlskrona: Baltic Institute.

Cars, G. (2002). Editorial introduction: Creating milieux for collective action. In G. Cars, P. Healey, A. Madanipour, & C. de Magalhaes (Eds.), *Urban governance, institutional capacity and social milieux*. Aldershot: Ashgate.

Drewski, L., Kunzmann, K. R., & Platz, H. (1989). *Promotion of secondary cities: A strategy for development cooperation*. Schriftenreihe der Deutschen Gesellschaft für Technische Zusammenarbeit (GTZ), 211, Eschborn.

ESPON (European Spatial Planning Observation Network). (2006). *ESPON Atlas: Mapping the structure of the European territory*. Bonn: Federal Office for Building and Regional Planning.

CEMAT (European Conference of Ministers Responsible for Spatial/Regional Planning). (2007). *The territorial state and perspectives of the European union. Towards a more competitive Europe of diverse regions*. European Conference of Ministers Responsible for Spatial/Regional Planning, Laxenbourg.

European Foundation. (Ed.). (1994). *Visions and actions for medium-size cities: Reports from the European workshops of Alicante, Volos and Oviedo*. European foundation for the improvement of living and working conditions, Luxemburg.

Gaertner, S. (2007). *Ausgewogene Strukturpolizit, Sparkassen aus Regional-Ökonomischer Perspektive*. Dissertation an der Fakultät Raumplanung der Universität Dortmund, Dortmund.

Hans Seidel Stiftung. (Ed.). (2007). *Zukunftsorientierte Kommunalpolitik*. Politische Studien. Themenheft, 1.
Healey, P. (1997). *Collaborative planning; shaping places in fragmented societies*. London: MacMillan Press.
Hennings, G., Jenssen, B., & Kunzmann, K. R. (1981). The promotion of relief poles. A strategy for the deconcentration of metropolitan region development in developing countries. *Applied Geography and Development, 18*, 7–29.
Kunzmann, K. R. (2000a). Network Europe: A Europe of city regions. In L. Bekemanns & E. Mira (Eds.), *Civitas Europa: Cities, urban systems and cultural regions between diversity and convergence* (pp. 119–131). Bruxelles: Peter Lang Verlag.
Kunzmann, K. R. (2000b). Strategic spatial development through information and communication. In W. Salet & A. Faludi (Eds.), *The revival of strategic spatial planning* (pp. 259–266). Amsterdam: Royal Netherlands Academy of Arts and Sciences.
Kunzmann, K. R. (2004). An agenda for creative governance in city regions. *DISP, 158*(3), 5–10.
Kunzmann, K. R. (2007a). The ESDP, the new territorial agenda and the peripheries in Europe. In N. Farrugia (Ed.), *The ESDP and spatial development of peripheral regions*. Valetta: Malta University Publishers Limited.
Kunzmann, K. R. (2007b). Von Außen Betrachtet: Standortimpulse durch Strategische Regionalentwicklung. In H. Beierlorzer & A. Kolkau (Eds.), *Dreiecksgeschichten: Eine Region in Bewegung* (pp. 74–80). Remscheid: Buchverlag.
Leber, N., & Kunzmann, K. R. (2006). Entwicklungsperspektiven Ländlicher Räume in Zeiten des Metropolenfiebers. *DISP, 166*(3), 58–70.
Magnaghi, A. (2000). *le Projet Local*. Mardaga: Sprimont/Belgique.
OECD (Organisation for Economic Co-operation and Development). (2001). *OECD territorial outlook*. Paris: OECD Publishing.
Pienkoß, C. (2007). *Sustainability of interreg projects: A Utopia? The meaning of intercultural project management for the success of strand B (Transnational Cooperation) of the Interreg III Community Initiative, with the Example of the Project 'European Route of Brick Gothic'*. Diplomarbeit am Istitut für Stadt und Regionalplanung der TU Berlin, Berlin.
Rivkin, G. W., & Rivkin, L. D. (1982). *Approaches to planning for secondary cities in developing countries*. Washington, DC: Rivkin Associates Inc.
Rondinelli, D. A. (1983). *Secondary cities in developing countries: Policies for diffusing urbanization*. Beverly Hills, NJ: Sage.
Stein, U. (2006). *Lernende Stadtregion: Verständigungsprozesse über ZwischenStadt*. Dissertation an der Fakultät Raumplanung der Universität Dortmund, Dortmund.

Chapter 3
Strategic Planning as a Field of Practices

Alessandro Balducci

3.1 Introduction

Beyond the appearance the dominant conception of strategic planning is still rooted in the rational comprehensive paradigm of planning. We have added sophistication, that is the consideration of the plurality of actors as a constitutive character of the process, the need to construct consensus among different subjects, the selectivity and the attention towards implementation. But the idea is still that of defining objectives and trying to design a set of actions which allow to pursue them.

We have been induced to choose a different approach in the experiment conducted in Milan. We did not have a strong power to support the plan. The Provincial Institution is quite weak, and within weak institutions the power of the politician in charge is not particularly relevant. The territory is not well defined: we have been aware since the beginning that the territory of the Province is just an administrative section of the Milan urban region which, by any definition, is larger than the province. Somehow we have been forced to adopt a much less linear approach. This approach is characterised by an indirect connection between a structure of argumentation which indicates a direction and a possible evolution of the current situation and a set of actions at different levels which are tentative, experimental and which try to push a very fragmented governance environment in the desired direction using various means.

This approach can define strategic planning as a field of practices rather than as a coherent sequence of coordinated actions.

My question is the following: is this way of conceptualising strategic planning just the result of a series of specific circumstances, or is this a promising approach which could be more effective in coping with situations where power is fragmented and strong leadership non-existent – an approach fostering innovation and change?

In order to respond to this question I have to first describe the context and the planning process and then link this to what I consider relevant literature.

A. Balducci (✉)
Department of Architecture and Planning, Polytechnic of Milan, 20133 Milan, Italy
e-mail: sandro.balducci@polimi.it

M. Cerreta et al. (eds.), *Making Strategies in Spatial Planning*,
Urban and Landscape Perspectives 9, DOI 10.1007/978-90-481-3106-8_3,

3.2 The Context

The starting point of this process was the request submitted by the Province of Milan to our Department of Architecture and Planning[1] at Polytechnic of Milan to develop a strategic plan.

In Italy, we have a three-tier system of local government based on Regions, Provinces and Municipalities. In the specific case in question there is the Region of Lombardia, with about 9 million people, and a Province of Milan, with about 4 million people distributed across 189 Municipalities. Among these is the Municipality of Milan, which accounts about 1.2 million inhabitants. All three levels have statutory land use or spatial planning powers, although the strongest powers remain with those of the municipalities, which are responsible for land use plans, and those of the regions, which are responsible for planning legislation. The Provinces, which are in charge of the Territorial Coordination of Provincial Plans, are a rather weak link in the chain of land use planning.

The Provincial Government elected in June 2004 put forward the idea of developing a strategic plan as an important point in its electoral programme. Accordingly, a special political head, 'Assessore al Piano Strategico' (sort of chancellor), was nominated by the President of Milan Province to take responsibility for the strategic plan.

The Provincial Government comprises of 15 *Assessori*, each with different functions and heading different departments with Territorial Planning, Mobility, Economic Development and Environmental Protection being the most relevant from our perspective. The establishment of specific and separate responsibility for the strategic plan is a sign of political commitment and will. The strategic plan was intended to be different from the statutory territorial plan.

It is important to reflect on the reasons for this choice. Strategic planning in Italy does not have formal recognition (Fedeli & Gastaldi, 2004). No planning law at the national or regional level defines or includes strategic plans among the planning tools. Nonetheless, in the last 10 years in particular a fair number of Italian cities have promoted strategic plans. Turin was the first, starting during the mid-1990s with a process which formed the basis for rethinking the potential of a former 'one-company town' that had been hit by the crisis in the automobile industry (Ave, 2005). This was very much inspired by the experience of Barcelona, which was one of the first and most successful in Europe. Then followed Florence, Rome and many other medium-sized cities which are now linked into a 'network of strategic cities'. Furthermore, in the North of Milan's urban region some Municipalities had got together at the beginning of the present century to develop a joint strategic plan as a means of coping with problems like infrastructure and transport and economic development and also as a response to the image of external Municipalities in the periphery of Milan.

These experiences spread the idea of strategic planning as an administrative innovation. The first reason for the decision by the Province of Milan was thus that strategic planning was considered as an innovative, proactive form of planning within the realm of political communication. It is planning designed for action and development, different than the idea of 'planning as control', linked to statutory territorial planning. And for the new centre-left coalition which had won a very uncertain electoral victory and wanted its activities to be seen as a fresh start for progressive policies, making a strategic plan was part of the picture.

I would add here that all this took place with no precise idea of what kind of strategic plan they wanted to promote. There were various interpretations and expectations within the Provincial Government: the *Assessore* for Economic Development had in mind a plan centred upon infrastructure and new development poles; the *Assessore* for Territorial Planning was looking for the strategic vision which was lacking in the Territorial Plan inherited from the previous government. The others had less clear ideas and thought of an instrument to coordinate different sectoral policies. The ambiguity of the idea was not necessarily a problem, in so far as the design of the process could cope with different intentions and interpretations.

Clearly, a symbolic dimension is assigned to the decision to initiate a process of strategic planning and, as Edelman considers, the symbolic value of a decision or a policy is deeply connected with its ambivalence (Edelman, 1985).

A second reason can be recognised in the fact that provinces, as stated earlier, have weak governments. Provinces situated between strong Regions and strong Municipalities, particularly in a situation like that of Milan with a big city at its core, have to fight for their political space. This does not result simply from the sum of formal powers, which are fragmented and articulated in many fields of competences. The institution of the Province is a very old one. Historically it precedes the Region, and is responsible for the provincial road system, for providing infrastructure for higher education, for the production of a provincial territorial plan, for leisure and culture and for some other residual functions. It is quite clear that its powers are many and dispersed, and also that in any specific field of public action they are not so crucial because there are other prevailing powers situated above or below the provincial level.

The government elected in 2004 had chosen to present its political programme under the slogan 'Province of Municipalities' (rather than a higher government body above the Municipalities). The slogan was intended to underline the intention of looking for the source of power not in the limited areas in which the Province could impose its decisions over other actors, but in an institution which is at the service of Municipalities and helps them to deal with the many problems that go beyond their individual capacity. The President then elected had many years of experience as mayor of one of the biggest Municipalities in the urban region, and many of the *Assessori* had experience of having served as mayors. In the context of the relationship with other actors and with the Municipalities, a particularly weighty problem was that of the relationship with the Municipality of Milan. Historically, though particularly in the last 15 years, one of the main obstacles to the Provincial Government's ability to act within its mandate had been the conflict with Milan. Not being able to cooperate with or to obtain cooperation from the Municipality of Milan, the Provincial Government has been the government of a territory with a 'big hole' in the middle. Since most of the problems have their cause or effect in the core city, this difficult relationship turned out to be a major weakness of the provincial power.

The strategic plan was therefore seen as a tool to engender new relationships with other levels of government, with Municipalities and particularly with Milan. It was seen as an experiment in governance, which could strengthen the Provincial Government in its capacity to cooperate, rather than impose decisions in residual areas.

We can conclude here that in this specific situation the decision to prepare a strategic plan was linked to many concurrent reasons, that is the ambiguity of its content, the image of innovation attached to its symbolic value, the interactive character of the process of its preparation and its open nature.

These contextual factors gave us a big responsibility to design a planning process that could be appropriate in the specific situation because, as Albrechts (2004) suggests strategic spatial planning is not a single concept, procedure or tool; it is a set of concepts, procedures and tools that must be tailored carefully to the specific situation. And this is what we have attempted to do.

Another important contextual factor in devising effective policies for the urban region of Milan was the Provincial border. What territory did we have to consider in order to handle things adequately? In recent years, many voices have raised the issue of the growing inter-dependency of an ever-wider territory in the central part of the Lombardy region (Balducci, 2004; Lanzani, 1991; Secchi, 2003). This area has been described as the 'Infinite City' (Bonomi & Abruzzese, 2004), a post-metropolitan region which creates space for the building of new territorial relationships. Comparing current images of this area with those of 30 years ago, it is all evident that a deep process of restructuring has taken place.

Firstly, the urban region of Milan, physically, now extends far beyond not just the Municipality but also the Province of Milan, and if we want to get a glimpse of the territorial complexity we probably have to consider as a minimum a region that includes ten provinces belonging to three different regions. Secondly, this expanding urban region is composed of conurbations which appear to have their own territorial form, not just as a result of a sprawl effect of Milan.

This territorial feature is confirmed by population trends: the ten provinces that are totally or partially included in the urban region have an overall population of almost 8 million people. The territory underwent moderate but continuous growth in the years 1981–2001. During that period of time the loss of population from the core city and the Province of Milan (–3.4%) was offset by significant growth in the surrounding provinces of the North – Como (+5.1%), Lecco (+8.7%), Varese (+3.1%), Bergamo (+11.3%) – and the nearer South, Lodi (+10.4%).

A simple observation of territorial phenomena in recent years shows (a) growth of many external areas pushed by the strength of Milan and also by a significant autonomous attraction capacity; (b) the relocation from the city of Milan of populations which belong to different social groups; (c) the localisation of new metropolitan functions in the field of commerce, production and leisure in this enlarged urban region. The above factors gave rise to a new and integrated geography of development.

Furthermore, it can be said that this area altogether also has strong relations with other more distant poles like Turin and Genoa in the west and Brescia and Verona in the east, with which it forms, what Peter Hall has called, a 'Mega-City Region' (Hall & Pain, 2006).

According to Hall and Pain a Mega-City Region is formed by "series of anything between 10 and 50 cities and towns, physically separate but functionally networked, clustered around one or more larger central cities, and drawing enormous economic strength from a new functional division of labour. These places exist both as separate entities, in which most residents work locally and most workers are local residents, and as parts of a wider functional urban region connected by flows of people and information carried along motorways, high-speed rail lines and telecommunications cables" (Hall & Pain, 2006; p. 3).

These tendencies have certain implications and they pointed us in two different directions. First, with the objective of designing a strategic plan for the Province, we had to be aware that the Province is just the core part of a large urban region which in turn is part of a Mega-City Region. This should be reflected in the design of the planning process. Second, within its boundaries we cannot consider the Province of Milan as a coherent territory which can be interpreted only along a linear relationship between the core city and its periphery; the Province itself is a polycentric region in which new territorial aggregations expand beyond institutional borders of Municipalities which are physically and socially visible.

In the last 10 years, there have been instances of coordination among Municipalities in this area, with the aim of coping with inter-communal problems and promoting new, more significant territorial identities. These processes are yet to be sustained by some form of institutional recognition. The strategic plan could be part of this process of recognition, bringing these bottom-up experiences into the realm of governance practices (Healey, 2004, 2007).

3.3 The Planning Process

Given that the Province is not a city, but rather the core of the Milan urban region, right from the start, we discarded the consolidated and successful model used for Barcelona, Lyon and Turin, where the strategic plan was based on the idea of the city as a unitary actor.

The Milan urban region does not have a single institution with the authority to take decisions over an area where there is a thick web of overlapping jurisdictions. Nor did we consider it useful to invest energy in trying to establish an authority for the city region as had been done many times in the past attempts to plan the metropolitan area (Balducci, 2003). This, in the light of what I have illustrated so far, would in any case be partial and insufficient. Wherever the boundary is traced, it would be crossed by territorial phenomena now or in the near future, given the strong integration already occurring at the level of the Mega-City Region. From this perspective the only viable alternative to the establishment of a jurisdiction is to foster cooperation among the existing actors in position of power with their powers, trying to influence choices rather than impose choices from above. It is what has been described as the evolution of urban leadership from 'power over' to 'power to' influence (Hambleton, 2007) in complex governance contexts.

If we shift our viewpoint in this direction we can see, on the one hand, that the flexibility of the boundaries is not a problem and should be turned from a possible weakness into a strength; and, on the other hand, that while for a statutory territorial plan it is almost impossible to conceive of this kind of flexibility, the open nature of the strategic plan is particularly appropriate to serve a more enabling, proactive and experimental process (Hillier, 2007). Of course this adds complexity to the already complex situation and makes it necessary to conceive of the strategic planning process as a 'field of practices' rather than as a set of rules or a precise sequence of actions.

We therefore designed a planning process in which the Provincial Government was to act as the promoter of a cooperative effort intended to prevent the tendencies towards fragmentation of the population and of its territory as well as offer support to the valorisation of its assets.

From the very beginning, we decided to call it a 'strategic project' rather than a 'strategic plan' in order to emphasise the difference between this and other strategic planning processes. This was a controversial choice, very much discussed in our group. The strategic project is promoted by the Province but belongs to various actors; it consists of many different actions that could eventually give rise to a strategic plan in a dynamic form, that is as a progress report rather than as a final document. The term 'project' gives the initiative the more modest but, at the same time, proactive character that we wanted to ensure.

Secondly, we immediately began working on the production of a new vision for the area. We wanted to bring in all the research work we had been doing on the urban region a new synthetic description of the area capable of: (1) making all the actors aware of the ongoing profound transformation processes; (2) offering new representations capturing the main trends, the internal articulation of the area and the development trajectories; and (3) being a constitutive communicative action in a situation in which all the traditional descriptions appeared to be outdated.

The commission given to the OECD (2006) by the *Assessore* for Economic Development to conduct a territorial review of metropolitan development helped us because it made clear the division of labour between their focus on the fundamentals of regional economy and the associated governance problems and our focus on the more general challenges imposed by the spatial change of the urban region. We saw the two strategic activities as complementary, as they effectively have been.

Our starting point was a highly challenging assumption: the welfare of the urban region can be achieved. The well-being of its inhabitants but also, indirectly, the competitiveness of its economy, is linked in Milan not to the expansion of infrastructure or to big projects, but rather to its capacity to achieve greater *liveability* by recovering compromised environment and overcoming difficulties that emerge in the daily life of individuals and businesses, consequent results of the strong economic development of the past. This is today the biggest limitation to further development. Therefore the strategic project must aim at promoting a city region that is more comfortable, is more friendly towards its inhabitants and businesses and is capable of rediscovering its environmental quality and preventing social exclusion.

We called this multi-dimensional notion of liveability *habitability*. We wanted to introduce a term which is not in common use and which might therefore raise public awareness of the general objective of the planning process.

We wanted to underline the fact that for the first time in the history of Milan's urban development, the problem of habitability is affecting citizens and businesses at the same time. We know in fact that new production does not have to take place in functionally and technically separate places, and above all that the development of the economy needs a city which, on the one hand, is attractive to high-quality workers and, on the other hand, is a place for accumulating creative capital, a complex system of interactions between companies, risk capital services, media, informal economies, private and public institutions, artists' communities, associations, social networks, the diffusion of know-how and cultures (Dematteis, 2006). The city, the urban region, is the habitable territory which is capable of hosting these rich interactions.

We defined the habitability theme in six different ways:

1. *residing* – finding a stable or temporary home, improving the common spaces and the connections with the public space and welcoming new populations;
2. *moving and breathing* – moving by different means, in different directions, finding comfortable waiting spaces for public transport and reducing congestion and pollution;
3. *space sharing* – connecting people in new public spaces of different types, widening ability to find silence to slow down the frantic pace of life, creating excitement in other places, allowing space for unplanned activities and bringing back nature where it has disappeared;
4. *making and using culture* – promoting culture in various places, stimulating institutions to engage in dialogue with informal producers of arts and creative culture and sustaining their networks;
5. *promoting new local welfare* – supporting voluntary actions and solidarity actions, boosting citizen participation and promoting social services for people facing difficulties;
6. *sustaining innovation* – attracting new talents, developing a policy for human capital and creating a new responsibility for business vis-à-vis the local community in which they operate.

This multi-dimensional definition of liveability attempts to describe the field of activities that we propose as the components of the strategy. To identify these practices it is necessary to look at the processes of de-territorialisation and re-territorialisation which affect the heart of the urban region: on the one hand, we can see the emergence of 'distance communities' (Amin & Thrift, 2001), communities of activities, populations relating to each other through new network connections without being rooted in a specific territory: students, immigrants, commuters, groups of young people with common interests in music, sport and so on;

all those groups who challenge the traditional relationship between community and place. On the other hand, we can see new territorial rooting processes which link inhabitants not only (or no longer) to Municipal boundaries but also to significantly wider areas, such as North Milan, Brianza, Alto Milandes and Adda Martesana – areas, strongly integrated by mobility development, where we have seen cooperation develop between Municipalities. Strengthening these relationships is the objective of proposing the image of the 'City of Cities' as an essential part of a description orientated towards the project.

It is an interpretive image which allows us to say that these conurbations which are found on the maps are not just concentrations of urban development, but can become rich histories of cooperation between communities, enabling them to face problems which go beyond their individual capacity, from environmental protection to land use, or the management of complementary services.

In this sense they are 'cities'. 'Milan City of Cities' is an image that can help public, private and third-sector parties to work towards creating better habitability.

Starting with this set of argumentations, we conceived different streams of action. The entire process was intended to be quite compact in terms of time and was marked by a series of products (strategic document, project atlas and the final version of the plan) and events (conferences, forums, exhibitions and workshops), which it was hoped would bring the plan out of the laboratory and into the city region.

The project was broken down into a series of steps that together were designed to activate a strategic planning process. The first step was a strategic document introducing *City of Cities*, the strategic project for the Milan urban region (Provincia di Milano & Politecnico di Milano-DiAP, 2006) presented as part of a public initiative in February 2006: a sort of White Paper on the themes of change in the urban region, rich in data and information. It launched the theme of habitability and presented the vision and the strategy. The second step was to initiate a call for projects and good practices (Provincia di Milano & Politecnico di Milano-DiAP, 2007a) which could contribute to the improvement of habitability in the Milan urban region. The idea of the competition was borrowed from a well-known European experience, that of IBA Emscher Park, which used, as a planning strategy, the innovative means of a project competition, through which a series of plans were selected and then guided to realisation. In our case, similarly, we received a huge response from Milanese society: foundations, universities, associations, individual or joint communes, non-profit organisations and private citizens all participated. At the end of a two-stage process of selection we had 259 proposals for good practices and project ideas, which covered all the facets of habitability indicated above and which portrayed a local community that was not only rich and lively but was also keen to enter into a relationship with institutions in order to contribute to the development of relevant public programmes.

The third step was the preparation of an atlas of policies and projects for habitability in the Province of Milan (Provincia di Milano & Politecnico di Milano-DiAP, 2007b), the result of a dialogue with the other 14 *Assessori*, delegated advisors and their managers. This was, on the one hand, an exercise in self-reflection and reciprocal internal information within the Provincial structure and across the sectors and, on the other hand, an exercise in external communication and information about what the Province was already doing in the field of habitability. There were already 52 projects and policies which could build another network of projects and policies. These in turn could interact with the network of projects and practices coming out of the competition.

The fourth step was the launch of a limited number of *pilot projects* which were designed to intervene in particularly relevant areas such as the realisation of a peri-urban woodland and the trying out of innovative policies for housing access, or a project for upgrading production spaces (Provincia di Milano & Politecnico di Milano-DiAP, 2007c).

The fifth step was an exhibition organised at the *Triennale di Milano*, a nationally and internationally recognised institution for the promotion of planning, architecture and design. The exhibition was held in the period May–July 2007 and provided information about the changes in the Milan urban region to a wider audience (10,000 people visited the exhibition) and translated the objectives of the project into a communicative language. It was jointly supported by the Province, the Municipality of Milan and the Chamber of Commerce. The lay-out had at its core the 'City of Cities Theatre', a meeting place where for 2 months an uninterrupted series of initiatives were held to construct, both literally and metaphorically, an arena in which people and decision-makers could meet and discuss the future of the urban region.

The final step of this first phase was the presentation in June 2007 of a final document presenting scenarios, visions and ideas for the habitable Milan urban region (Provincia di Milano & Politecnico di Milano-DiAP, 2007d), in which all the streams of action initiated in the planning process were presented at the conclusion of this first phase to illustrate what had been achieved at the different levels and what the project's aims were for the future.

3.4 Interpreting Strategic Planning

The conceptualisation of strategic planning has been quite influential since mid-1980s. Strategic planning tried to respond to the need of finding new non-hierarchical modes of planning (Bryson & Roering, 1987; Bryson, 1988), to deal with an uncertain future and to the need to provide an approach capable of 'planning under pressure' (Friend & Hickling, 1987).

This need to move from a traditional planning approach (based on a top-down and single-actor-centred activity of comprehensive planning, an un-contested use of technical knowledge and a linear concept of time and space) found promising materials to deal with the uncertainty and complexity of the contemporary world, in the tradition of private sector strategic planning, which is based on a predefined sequence of operations – (a) initial agreement, (b) stakeholders dialogue, (c) swot analysis, (d) definition of the vision, (e) strategy formulation and (f) listing of actions (Bryson & Roering, 1987).

As a matter of fact, European and American literature shows that a strategic approach could imply and allow different perspectives on planning. One which does not refer to the dimension of strategy just in terms of instrumental rationality in order to reduce and treat complex situations, but rather as one able to explore the possible advantages of dealing with (anticipating, but most of all playing with) the multiple and interacting actors' behaviours (and agencies).

In fact what seems to be at stake, and what leads a possible and necessary 'inquiry on' planning through the eyes of a strategic approach, is the wider crisis of the general framing of public action underlying planning processes. Some of the keywords of planning are in fact losing their consolidated meaning and are challenged by the changing landscape of contemporary society (Albrechts, 2001).

Several authors have tried to describe strategic planning as a field of practices able to elaborate new answers to emerging urban problems: the lessons of Lindblom (1975) in the 1970s are now not so far from those of scholars like Healey (2007), Albrechts (2004), Kunzmann (2004), Hillier (2007) or Dente (2007). These researchers offer relevant contributions to planners interested in bringing together an approach to the strategic dimension different from the conventional perspective.

3.4.1 Strategic Planning as Using the Intelligence of the Society

Charles Lindblom, in an insufficiently well-known essay about planning in which he compares conventional planning with what he calls 'strategic planning', holds that this "is a method that treats the competence to plan as a scarce resource that must be carefully allocated, not overcommitted. (...) It is planning that picks its assignments with discrimination, that employs a variety of devices to simplify its intellectual demands, that makes much of interaction and adapts analysis to interaction" (Lindblom, 1975, p. 41). And furthermore: "strategic planning is then systematically adapted in several specific strategic ways to interaction processes that take place of analytical settlements of problems of organisation and change. (...) Strategic planning plans the participation of the planners (or of the government for which they plan) in interaction processes, rather than replacing the processes. (...) Strategic planning tries to make systematic use of the intelligence with which individuals and groups in the society pursue their own preferences by moulding their pursuit, rather than substituting the planners' intelligence wholly for individual's or groups'. (...) Strategic planning attempts to develop and plan, in the light of a rationale for deciding which effects are to be achieved through decision and which only as epiphenomena" (Lindblom, 1975, pp. 44–45).

The discussion proposed in very abstract form by Lindblom is full of practical implications. We have to be aware as planners, and to convince our 'clients', of the limited possibilities open to us, of the need to be selective, notwithstanding the stronger appeal of the rhetoric of the omnipotence which is pervasive in planning and policy fields. Particularly in complex systems, we have to value interaction as a form of analysis and use planning as a support for social practices rather than as a substitute for them. We have to understand and look for the 'intelligence of the society'. We have to include in our consideration potential intended and unintended effects.

3.4.2 Strategic Plan-Making: Connecting Knowledge Resources and Relational Resources

In several critical case accounts and reflections on planning and strategic planning, Patsy Healey proposes to stress the 'relational nature' of strategy-making, involving connecting knowledge resources and relational resources (intellectual and social capital) to generate mobilisation force (political capital) (Healey, 1998; Innes & Gruber, 2005). Such resources (capital) form in institutional sites in the governance landscape which, if a strategy develops mobilisation power, become nodes in networks from which a strategic framing discourse diffuses outwards. The strategic frame travels as an orientation, a sensibility and a focus for new debates and struggles, performing different kind of institutional work in the different arenas in which it arrives. At the same time, strategic spatial plan-making is "about building new ideas and about building processes that can carry them forward. (. . .) A social process, rather than a technical exercise, [which] seeks to interrelate the active work of individuals, within social processes (the level of agency) with the power of system forces-economic organisation, political organisation, social dynamics and natural forces (the level of structure of social relation)" (Healey, 2007, p. 198). It recognises the fact that strategic spatial plan-making, although occurring within a context of powerful structuring forces, may be used by social groups to create structure and frameworks through which to influence the flows of events that affect them (Healey, 1997, pp. 25–26). Below is an assumption about the strategic approach, based on the role of knowledge and relationality within a structured field of action, in a social, political and cultural constructivist perspective.

3.4.3 Multiple Rationalities: Dealing with Future, Legitimacy and Action

Albrechts and van den Broeck (2004), trying to bridge the gap between theoretical reflection and practical experimentation and to escape from a mechanical view of strategic planning, affirm that effective strategic planning must be able to work at four different levels. The four tracks they propose are as follows:

1. producing a long-term vision;
2. allowing immediate actions;
3. reaching the relevant stakeholders;
4. trying to reach public opinion.

"The four-track approach is based on interrelating four types of rationality: *value rationality* (the design of alternative futures), *communicative rationality* (involving a growing number of actors – private and public – in the process), *instrumental rationality* (looking for best way to solve the problems and achieve the desired future), and *strategic rationality* (a clear and explicit strategy for dealing with power relationships)" (Albrechts, 2004, p. 752; see also Albrechts et al., 1999; van den Broeck, 1987; 2001).

These four types of rationality are a great challenge to the consolidated rationality of planning, implying new ways to look at the future, to think about efficacy and action and to deal with projectuality and governance. At the same time this is a way of ordering the most relevant aspects of a strategic planning process without fixing them in a set of rigid rules. It is an approach capable of clarifying, in pragmatic terms, what we understand in theory reflecting upon the contribution of Lindblom or Healey.

3.4.4 Strategic Plans as Open Fields of Experimentation and Investigation: New Maps of Potentialities

Jean Hillier, in her recent book (2007), states that strategic spatial planning should not involve the adoption of pre-determined solutions, but might offer a 'genuine possibility' of experimentation for actants to 'internally generate and direct their own projects' in direct relevance to their own specific understandings and problematic.

Since 2005, Hillier reflects on a multiplanar theory which explores the potential of Deleuze and Guattari's concept of emergence or becoming as a creative experimentation in spatial planning practices. "These notions allow unexpected elements to come into play and things not to quite work out as expected. They allow (. . .) to see planning and planners as *experiments* enmeshed in a series of modulating networked relationships in circumstances at the same time both rigid and flexible, where outcomes are volatile; where problems are not 'solved' once and for all but are rather constantly recast, reformulated in new perspectives" (Hillier, 2005, p. 278).

She therefore proposes that strategic spatial planning be concerned with trajectories rather than specified end-points. She regards spatial planning as an experimental practice working with doubt and uncertainty, engaged with speculation as adaptation and creation rather than as proof-discovery: a speculative exercise, a sort of creative agonistic. She suggests a new definition of spatial planning along the lines of the investigation of 'virtualities' unseen in the present; the speculation about what may yet happen; the temporary inquiry into what, at a given time and place, we might yet think or do and how this might influence socially and environmentally just spatial form (Hillier, 2007). She argues for the possibility of planning to be more inclusive, democratic, open and creative, made upon improvisation, based on performance rather than on a normative/prescriptive dimension concerned with 'journeys rather than destinations', establishing conditions for the development of alternatives. She proposes a reflection on the activity of mapping practiced in strategic planning as explorations of potentials (in space–time–actors relations).

3.4.5 Governance Culture and Governance Episodes: New Limited but Practicable Paths of Sense-Making

From a convergent perspective, Healey (2007) again looks at the way in which strategic (spatial) planning is able to help specific episodes of social or institutional innovation to be absorbed into more stable governance practices and can eventually 'travel' into different contexts to re-shape the dominant governance culture.

Working through what Lindblom (1990) defines as actions of probing rather than 'planning' in a traditional way, one can find a new way to penetrate governance processes and sediment into governance culture. Making governance episodes part of a wider sense-making process, apparently weaker of the great narrations of the past, Healey (2007) offers an insight into what Lindblom calls a strategic planning approach and explains how it is possible (when dealing with complex governance problems) to introduce relevant changes and thus work on innovation starting from alterations at the margin as well as from routines.

3.4.6 Rethinking Efficacy: Governance as an Open and Complex Key Issue, Rather than a Pre-fixed Model

With Dente (2007), among many other arguments which could be raised and discussed (i.e., the relevance in a strategic approach of dealing with the issue of time intersecting long-term and short-term), we can agree that also the issue of evaluation of planning (assessing outcomes, also unexpected ones) is to be completely reframed in a strategic planning perspective.

The efficacy of strategic planning in fact has to deal with dimensions difficult to be verified and quantified, as the changes in actors' behaviours, trust, attitude to cooperation, density of network and complexity of projects and issue afforded. In this sense, since, quoting Perulli (2007), strategic planning has to deal with the capacity of identifying issues, rather than objectives to be pursued; has to produce discontinuity, rather than fostering routinary evolution; has to make out possible courses of action, rather than a generic desirable future. Hence its efficacy cannot be simply evaluated through a predefined monitoring model, inside a traditional programming convention. We have to instead develop a sort of continuous process of discussion of the core hypothesis of the plan and their operative declination; this is particularly true, states Dente, if strategic planning is able to renounce to the idea of the public actor as the main and unique actor of the plan.

According to Dente, governance is at stake and is the filtering concept for the evaluation of the efficacy of a plan. This is true when one of the issues of the plan is the difficult reconstruction of a collective actor (a strong dense coalition); it is also true when the plan has already given up with this possibility and can only count on and look for inclusive but open and hetero-direct processes of vertical and horizontal cooperation. To what extent the plan has been able to produce governance changes becomes the issue to be evaluated. But in the first the plan runs after a predefined aspiration (an ideal model); in the second the process remains open to uncertainty, and evaluation is central to feed a recursive process of probing.

This has a further consequence: in the first case it is assumed that a clear coalition, based on an assumption of reciprocal responsibilities, can play a steering collective role, more or less simple to be evaluated. In the second, which abandons from the very beginning the possibility to isolate the subjects from the situations in which they act, the steering role should be played by an actor autonomous from all the other ones involved, able to evaluate the situation and its transformation.

3.4.7 Potentialities and Transformation, Rather than Action and Outcomes

These positions are not far from those we can find in a recent book by the French philosopher and sinologue François Jullien (2004), which offers an interesting contribution defining the specificity of a strategic approach. His position is based on a 2-fold operation of distance-setting: distance from the modern conceptualisation of planned actions in relation to the eastern world and distance between the western classical thought and the modern one. Jullien, in fact, considers that there is a wide distance between the western-classic approach to the concept of strategy and the oriental (Chinese) one (more similar to the pre-classical Greek culture); looking at the first through the eyes of the second, he suggests, can help deconstructing the western approach and identify both its strengths and weaknesses.

According to Jullien, in the western classical and then modern perspective, 'efficacy' passes through a necessary process of modelling, of producing plans to deal with pre-fixed objectives. The plan precedes its application, its implementation, and has to deal with, on the one hand, the intellectual dimension of the production of the ideal form of action and, on the other hand, with the will, which defines the engagement of the individual in getting inside the reality and making the plan work. The distance between theory and practice characterises the ancient Greek classic approach which has been influencing western contemporary thought: a distance occupied and produced by occurring circumstances which deviate from theory and plans, from practice and reality, generating the same friction that one can feel walking inside water rather than on the simple ground (see Strachan & Herberg-Rothe, 2007, about the master strategist Carl von Clausewitz). Leaving behind the pre-classic *metis*, in Jullien words (the Greek word indicating the capacity to take advantage of circumstances, of seeing the situation evolving, in order to catch the favourable evolution), the classic Greek thought stands far away from the Chinese approach. This indeed, with Sun Tzu and Sun Bin, underlined the importance for the strategist to start from the situation, not one that could be modelled, but from the specific and unpredictable one inside which one happens to be thrown, trying to discover its potential and how to make use of it. In this sense the 'potential of the situation' rather than the plan (and the will of the strategist) is relevant, and circumstances cannot be regarded as just producing frictions. Thus rather than about objectives one should talk about advantages that can be taken from a situation. In spite of dealing with the 'ends-means' couple, the Chinese perspective uses a word similar to the French agencement. Since strategy can be viewed as the capacity to find all the favourable elements which can be developed in a situation in order to take advantage of it, there is no use of reasoning and acting in the light of finalities. No outcome can be expected, since the situation rather than the subject determination is central. This means also that action has to be thought in another way.

Jullien (2004) suggests using the word 'transformation' (within a process perspective), rather than action (related to a product perspective). Where occasion is central, the causal implication of the 'effect'/outcome is rejected far from the process in which it is strictly embedded. Therefore, efficacy must be indirect in relation to the attended aim. At the same time, whereas subjectivity is fading, strategy becomes indirect and modest, anti-heroic. It is not so difficult to see where Lindblom and Jullien overlap in their approaches and how the approaches to planning proposed by Albrechts, Healey and Hillier try to cope with the challenges proposed by the first one.

3.5 Reflecting Upon the Provisional Results of 'City of Cities' Project

It is too early to try to evaluate the results and outcome of this complex process. If we want to want an answer to the direct question of what changes we have been able to introduce through the strategic planning process, we will be able to indicate only initial, provisional and probably fragile results. I would like to be guided in this reflection by the four tracks proposed by Albrechts (Albrechts et al., 1999). The documents and communications used to develop and present the vision were received by the actors – from the mayors to the representatives of organised interests – with great interest, both in the content and in the perspective offered by a new orientation in a situation of rapid change.

At the same time, we have to admit that the strategy of habitability, which sought to instil a set of new ideas into governance practice failed to change the existing paradigms of the governance culture. The media in general are not attracted to planning actions and documents. The Province as a whole has not endorsed the strategy and the President continues to be more attracted by the hard mainstream 'infrastructure-and-big-projects' approach than by the soft objective of designing and implementing a multi-dimensional policy for improving habitability. This is of course linked to our ability to construct a convincing argument, capable of persuading the current leadership (Majone, 1989), but it is also due in no small part to the complex political game of symbolic politics (Edelman, 1985) in which we can play only a minor role.

So far the strategic project has been perceived as a brilliant initiative of very active *Assessore* (whose function has been re-named '*Assessore* for habitability and the strategic plan'), supported by the Polytechnic. I think the ability of the strategy to conquer the centre-stage has fallen below our expectations. This is linked to the extreme complexity of the process described by Patsy Healey, which takes place on an overcrowded and extremely fragmented arena and where simplified conventional messages always seem to have the edge in political communication.

If we look at our capacity to initiate immediate actions – the second track – we see a story of partial successes and of encouraging hopes. As I said earlier, the competition for projects and good practices achieved a great response, opening up new opportunities for this planning process. In Lindblom's terms, I see this as a promising way of using the 'intelligence of the society'; of substituting interaction for analysis; of devising a new enabling role for planning. As a consequence this approach must imply a profound change in the relationship between the public administration and the subjects; a change which emerged as a result of the competition. The problem we had is that the great energy that developed as a result of the competition was only very partially utilised. The lack of preparedness of the bureaucracy of the Province and the fear of being overwhelmed by requests for assistance and funding prevented the Province from committing the public institution to a more open interaction. Those with political and administrative responsibility decided to concentrate only on the ten winners of the competition, in our view failing to understand the nature of the demand coming from the 259 competitors who should have been recognised as discussion partners in a relevant policy process, and should have been supported in creating networks across different projects and practices and helped in creating new communication channels with the public administration.

At the same time it must be noted that many projects were developed independently from the Provincial action and that the method of the competition of projects has seen a diffusion in the planning practices of the Province.

We cannot yet say what may come of this stream of action. We have certainly seen some good developments assisted by the Province and spontaneous organisation of networking, as well as some disillusions. Even so, I do believe that this is a very promising route for planning in general. It is an opportunity to renew the field of participatory planning, engaging the community in a more proactive form of participation. This approach has conquered its legitimacy and in the last January the *Assessore* has launched a second competition of the 'City of Cities' project.

Other immediate actions are the six pilot projects proposed by the Province, which are being developed and the first implementation steps look encouraging.

All this is also indirectly connected to the third track, that of stakeholder involvement. Throughout the process we tried to establish a positive interaction with the various *Assessori*, officers of different provincial sectors, representatives of interest groups as well as other relevant actors. As stated above, some provisional results have been achieved: the cooperative effort for the preparation of the Atlas of the different sectoral policies, the collaboration in the development of the pilot projects, the partnership with the Municipality of Milan and with the Chamber of Commerce for the *Triennale* exhibition and the direct involvement of many stakeholders in the competition for projects and good practices. The problem again was how to generate a sufficient level of commitment to produce some kind of intellectual and social capital in the process (Innes et al., 1994); capital that can allow the ideas to 'travel' and to sediment into a new culture rather than being a succession of episodes, as Healey (2007) states. It is something which experts and planners can influence only to a limited extent and which depends on the general process of political communication, with a significant role for the media.

Finally, we tried to reach public opinion with information about change in the urban region, problems, opportunities and possible new perspectives. This is cited by Albrechts (2004) as a means of indirectly raising the attention of political actors for the project and also offers a way to root the ideas proposed by the strategic plan in the local community. We tried to do this mainly through the competition and through the *Triennale* event which, as stated, attracted quite a wide public considering that it was an exhibition about a planning topic. We invested relevant resources in trying to make our messages as clear as possible. This need for communication that could reach the citizens of the urban region was also important for our actions because it pushed us to translate complex concepts into non-technical language, establishing a dialogue with experts in the field of communication with whom we tried to achieve a good level of reciprocal understanding.

By looking back at this intense experience we cannot develop simple conclusions. The process was experimental, full of hopes, difficulties, disillusions and enthusiasm. We are now in the middle of a new phase in which we are working on the second competition for projects aimed at consolidating the results of the first phase, and we are bringing the *Triennale* exhibition into the territory of the Province in all the cities of the 'City of Cities'.

If we look back to this 3-years effort, we have to underline that it has been a journey of discovery in the field of uncertainties.

But the final question is: does this experience have to be interpreted as a deviation from a mainstream conception of strategic planning due to the absence of a strong leadership and to the fragmentation of powers, or could it be regarded as an appropriate approach to strategic planning in situations of growing complexity and rapid change of dynamic urban regions?

Note

1. The project team: Alessandro Balducci, Matteo Bolocan Goldstein, Paolo Bozutto, Claudio Calvaresi, Ida Castelnuovo, Bruno Dente, Paolo Fareri, Valeria Fedeli, Daniela Gambino, Marianna Giraudi, Arturo Lanzani, Antonio Longo, Fabio Manfredini, Anna Moro, Carolina Pacchi, Gabriele Pasqui, Paolo Pileri, Poala Pucci. Many materials from the Strategic Project City of Cities are available on the website http://www.cittadicitta.it.

References

Albrechts, L. (2001). From traditional land use planning to strategic spatial planning: The case of Flanders. In L. Albrechts, J. Alden, & A. da Rosa Pires (Eds.), *The changing institutional landscape of planning* (pp. 83–108). Aldershot: Ashgate.

Albrechts, L. (2004). Strategic (spatial) planning re-examined. *Environment and Planning B*, *31*(5), 743–754.

Albrechts, L., & van den Broeck, J. (2004). From discourses to acts: The case of the ROM-project in Ghent, Belgium. *Town Planning Review*, *75*(2), 127–150.

Albrechts, L., van den Broeck, J., Verachtert, K., Leroy, P., & van Tatenhove, J. (1999). *Geïntegreerd Gebiedsgericht Beleid: een Methodiek*. Research report by KU-Leuven and KUNijmegen en AMINAL, Ministry of the Flemish Community.
Amin, A., & Thrift, N. (2001). *Cities. Reinventing the urban*. Cambridge: Polity Press.
Ave, G. (2005). Play it Again Turin. Analisi del Piano Strategico di Torino come Strumento di Pianificazione della Rigenerazione Urbana. In F. Martinelli (Ed.), *La Pianificazione Strategica in Italia e in Europa. Metodologie ed Esiti a Confronto* (pp. 35–67). Milano: FrancoAngeli.
Balducci, A. (2003). Policies, plans and projects. Governing the city-region of Milan. *DISP*, *152*(1), pp. 59–70.
Balducci, A. (2004). Milano dopo la Metropoli. Ipotesi per la Costruzione di un'Agenda Pubblica. *Territorio*, *29–30*, 9–16.
Bonomi, A., & Abruzzese, A. (eds). (2004). *La Città Infinita*. Milano: Mondadori.
Bryson, J. M. (1988). *Strategic planning for public and nonprofit organisations: A guide to strengthening and sustaining organizational achievement*. San Francisco: Jossey-Bass.
Bryson, J. M., & Roering, W. D. (1987). Applying private-sector strategic planning in the public sector. *Journal of the American Planning Association*, *53*(1), 9–22.
Dematteis, G. (2006). La Città Creativa: un Sistema Territoriale Irragionevole. In G. Amato, G. Varaldo, & M. Lazzeroni (Eds.), *La Città nell'Era della Conoscenza e dell'Innovazione* (pp. 107–119). Milano: FrancoAngeli.
Dente, B. (2007). Valutare il Piano Strategico o Valutare il Governo Urbano? In T. Pugliese (Ed.), *Monitoraggio e Valutazione dei Piani Strategici*. Pianificazione Strategica. Istruzioni per l'Uso. Quaderno 1, pp. 5–10.
Edelman, M. (1985). *The symbolic uses of politics*. Urbana, IL and Champaign, IL: University of Illinois Press.
Fedeli, V., & Gastaldi, F. (Eds.). (2004). *Pratiche Strategiche di Pianificazione: Riflessioni a Partire da Nuovi Spazi Urbani in Costruzione*. Milano: FrancoAngeli.
Friend, J., & Hickling, A. (1987). *Planning under pressure the strategic choice approach*. Oxford: Butterworth-Heinemann.
Hall, P., & Pain, K. (2006). *The polycentric metropolis. Learning from mega-city regions in Europe*. London: Earthscan.
Hambleton, R. (2007). New leadership for democratic urban space. In R. Hambleton & J. S. Gross (Eds.), *Governing cities in a global Era* (pp. 163–177). New York: Palgrave.
Healey. (1997). An institutional approach to spatial planning. In P. Healey, A. Khakee, A. Motte, & B. Needham (Eds.), *Making strategic spatial plans. Innovation in Europe* (pp. 21–29). London: UCL Press.
Healey, P. (1998). Building institutional capacity through collaborative approaches to urban planning. *Environment and Planning A*, *30*(9), 1531–1546.
Healey, P. (2004). Creativity and urban governance. *Policy Studies*, *25*(2), 87–102.
Healey, P. (2007). *Urban complexity and spatial strategies: Towards a relational planning for our times*. London: Routledge.
Hillier, J. (2005). Straddling the post-structuralist abyss: Between transcendence and immanence? *Planning Theory*, *4*(3), 271–299.
Hillier, J. (2007). *Stretching beyond the horizon. A multiplanar theory of spatial planning and governance*. Ashgate: Aldershot.
Innes, J., & Gruber, J. (2005). Planning styles in conflict. *Journal of the American Planning Association*, *71*(2), 177–188.
Innes, J., Gruber, J., Neuman, M., & Thompson, R. (1994). *Coordinating growth and environmental management through consensus building*. Policy Research Program Report, University of California, Berkeley, CA.
Jullien, F. (2004). *A treatise of efficacy: Between western and Chinese thinking*. Honolulu: University of Hawaii Press.
Kunzmann, R. K. (2004). An agenda for creative governance in city regions. *DISP*, *158*(3), 5–10.
Lanzani, A. (1991). *Il Territorio al Plurale*. Milano: FrancoAngeli.

Lindblom, C. E. (1975). The sociology of planning: Thought and social interaction. In M. Bornstein (Ed.), *Economic planning east and west* (pp. 23–60). Cambridge: Ballinger.
Lindblom, C. E. (1990). *Inquiry and change: The troubled attempt to understand and shape society*. New Haven, CT: Yale University Press.
Majone, G. (1989). *Evidence, argument and persuasion in the policy process*. New Haven, CT: Yale University Press.
OECD (Organisation for Economic Cooperation and Development). (2006). *OECD territorial reviews: Milan, Italy*. OECD Policy Briefs, Paris, pp. 1–7.
Perulli, P. (2007). Una Griglia Metodologica per l'Inquadramento dei Piani Strategici. In T. Pugliese (Ed.), *Monitoraggio e Valutazione dei Piani Strategici, Pianificazione Strategica. Istruzioni per l'Uso. Quaderno* 1, pp. 27–39.
Provincia di Milano, Politecnico di Milano-DiAP. (2006). *Città di Città. Un Progetto Strategico per la Regione Urbana Milanese*. Progetto Strategico Città di Città, Milano. Retrieved 04/02/2009, from http://www.cittadicitta.it
Provincia di Milano, Politecnico di Milano-DiAP. (2007a). *10+32+217 Progetti e Azioni per l'Abitabilità: il Bando di Città di Città*. Progetto Strategico Città di Città, Milano. Retrieved 04/02/2009, from http://www.cittadicitta.it
Provincia di Milano, Politecnico di Milano-DiAP. (2007b). *6 Progetti Pilota per l'Abitabilità Promossi dalla Provincia di Milano*. Progetto strategico Città di Città, Milano. Retrieved 04/02/2009, from http://www.cittadicitta.it
Provincia di Milano, Politecnico di Milano-DiAP. (2007c). *Atlante dei Progetti e delle Azioni per l'Abitabilità della Provincia di Milano*. Progetto Strategico Città di Città, Milano. Retrieved 04/02/2009, from http://www.cittadicitta.it
Provincia di Milano, Politecnico di Milano-DiAP. (2007d). *Per la Città Abitabile: Scenari, Visioni, Idee*. Progetto Strategico Città di Città, Milano Retrieved 04/02/2009, from http://www.cittadicitta.it
Secchi, B. (2003). *Urban scenarios and policies*. In N. Portas (Ed.), *Politicas, Estratégias e Oportunitades* (pp. 275–283). Lisboa: Fondaçao Calouste Gulbenkian.
Strachan, H., & Herberg-Rothe, A. (2007). *Clausewitz in the twenty-first century*. Oxford: Oxford University Press.
van den Broeck, J. (1987). *Structuurplanning in de Praktijk: Werken op Drie Sporen* [Structure Planning in Practice: Working on Three Tracks]. Ruimtelijke Planning, 19(II.A.2.c), pp. 53–119.
van den Broeck, J. (2001). *Informal arenas and policy agreements changing institutional capacity*. Paper of the First World Planning School Congress, Shanghai.

Lindblom, C. E. (1975). The sociology of planning: Thought and social interaction. In M. Bornstein (Ed.), *Economic planning east and west* (pp. 23–60). Cambridge: Ballinger.
Lindblom, C. E. (1990). *Inquiry and change: the troubled attempt to understand and shape society*. New Haven, CT: Yale University Press.
Majone, G. (1989). *Evidence, argument and persuasion in the policy process*. New Haven, CT: Yale University Press.
[illegible]

Chapter 4
On Strategic Planning and Associated Issues

Francesco Indovina

4.1 Preamble

Several authors have remarked that the propensity – as we may call it – of local authorities to rely on 'strategic planning' originates from the interplay of various factors perceived as having a negative impact on governance capabilities.

The first factor is the speed of changes and their alleged complexity; the two aspects seem to be impossible to control (Bertuglia, Rota, & Staricco, 2004), especially by a public system more deficient than expert, better capable of administering than governing. The second factor is the budget-cutting trend in the public sector and the ensuing need for public/private partnerships to implement significant urban and territorial transformation schemes. The third factor is the ever-growing citizens' distrust of public institutions (the debate on the crisis of politics and the institutions is as endless as it is inconclusive). This attitude leads, on the one hand, to a swing to 'market' actions and, on the other, to the emergence of demand, often indeterminate, for public participation.

The right tool for unravelling these tangled rather than interwoven issues seemed to be *strategic planning*[1] seen as a kind of 'magic wand'. The schemes implemented in Italy and in the rest of Europe are well-articulated and differentiated, but often providing poor interpretations of this tool. These experiences are reported and critically analysed by an extensive body of literature (some of the most exhaustive analyses being those by Gibelli, 1996, 2003, 2007).[2]

One point often overlooked by discourse on 'strategic planning' should be highlighted from the start and will be expanded on later in this chapter: a strategy (and hence a strategic plan) can only be justified by a situation of conflict. *Urban and territorial changes* constitute a locus of conflicts in which 'contenders', using their force and attempting to neutralise that of their opponents, try to impose their own objectives (interests). Nevertheless, many 'strategic plans' fail to consider this state of things and are based on the assumption of 'collaboration' (to be achieved through

F. Indovina (✉)
Faculty of Urban and Regional Planning, IUAV University of Venice, 30135 Venice, Italy
e-mail: indovina@iuav.it

M. Cerreta et al. (eds.), *Making Strategies in Spatial Planning*,
Urban and Landscape Perspectives 9, DOI 10.1007/978-90-481-3106-8_4,

negotiation). More precisely, the most widely used term is not 'collaboration' but 'sharing': this is an important distinction but the two issues are closely inter-linked, without one the other cannot exist. The choice of whether to refer to a situation of conflict or one of collaboration (sharing) is clearly not just a matter of terminology; if it were, it would be linguistically relevant but indifferent as to substance. On the contrary, this choice concerns the interpretation of relationships within society in general and the local community in particular.

We should point out that placing the focus on sharing leaves out partiality: those interests that are not shared by all are left out or, alternatively, the aims of the plan are strictly based on the fact that all the stakeholders (and we know how problematic the definition of 'all the stakeholders' is) share its objectives and tools, thus reversing the method of formulating the objectives themselves. Indeed, the fact that shared 'strategic plans' are simpler to implement lies not perhaps in the rightness of their objectives but rather in the coincidence of interests. 'Implementability', while it should certainly not be underestimated, cannot be taken as the sole value; the formulation of sound objectives, interpreting the needs of the whole area and responding to common interest, is not necessarily tantamount to 'not' doing. This is, we might say, a strategic issue.

In the remarks that follow I will argue that, as a consequence of the above, *public governance of urban and territorial change* to achieve an adequate level of effectiveness must draw up a strategy (since it addresses a situation of conflict) which attempts to:

– respond to future uncertainty by making a reasonable medium-term proposal;
– combat public distrust of government initiatives by providing a range of real participation processes, and, especially, transparent objectives and the means to implement them, through ongoing, comprehensive dissemination of information;
– raise awareness of future opportunities for economic, social, technological and cultural development;
– incorporate (some), private interests in general public interest planning.

Whether this strategy will lead to some form of 'strategic planning' is less important than the general goal the public administration must set for itself: the governance of change. Indeed, I will argue that the focus should not be on the tool, or one of its many applications, but rather on the intended objectives. In other words – and perhaps provocatively – I will argue that it is not the strategic plan which determines the medium-term objectives of a community, but rather the strategic plan can be the tool for achieving pre-defined objectives. Obviously, talking about 'objectives' opens a wide range of questions: their formulation, definition, implementation, monitoring and so on. These issues are discussed further in this chapter, although not as exhaustively as they deserve.

The basic intention is to address the theme of the *strategic plan* by focusing on the objectives and not on the tools, on common rather than partial interests and on a general interest seen not as the sum of agreements and common needs but as responses to the demands emerging in a specific situation. Perhaps this approach is not consistent with strategic planning, but the interest in 'how' seems to me to be of secondary importance with respect to 'why'.

4.2 An Economic or a Territorial Strategy?

The *strategic plan* might appear to be largely an economic programming tool; its corporate origin can, no doubt, support this impression. But *spatial strategic planning* is a specific territorial planning tool (Gibelli, 2007) introduced in zoning and land use planning legislation by many European countries.

This fact, however, does not dispel the doubt, because no *territorial strategy* can be purely 'territorial', just as no economic strategy (e.g., for economic development) can fail to take territorial issues into account. Strategy involves not only an objective to be achieved but also expected overall effects: indeed, these expected effects are what justify the pursuit of a specific set of objectives. In other words, an objective has value not only in itself but also because of the expected results it produces (just as the possible occurrence of unexpected or perverse effects requires constant monitoring and flexibility in the tools employed).

Governance of territorial and urban changes, when not intended as mere 'administration', that is, the recording of current trends, but rather as a *strategic* type of governance that wishes to bend the trends of a given context in a precise direction to achieve specific objectives, must include economic, social, cultural and territorial contents. Hence, the crux of the matter is not whether the 'plan' should have mainly economic or territorial contents, but rather whether its strategic impact can be imputed to the action. It is from this starting point that it will be possible, from time to time and in accordance with the specific conditions of place and time, to identify that blend of economic, social, cultural, infrastructural and landscape content, locational and territorial in the broad sense, suitable for achieving the 'desired' scenario which the public administration has envisaged and adopted as a medium-term goal through stakeholder participation and empowerment processes.

4.3 The 'Appropriate' Size and Possible 'Contents'

As a consequence of both the far-reaching changes which have occurred in organisation of the territory (Indovina, 2003, 2005), increasingly tending towards the structuring of a metropolitan space ('the metropolitan archipelago'), and of the need for a 'critical mass' (both quantitatively and qualitatively) for drawing up a medium-term project, strategic design should as a rule cover a 'large' geographical area. This level offers the greatest variety of social players, economic activities, opportunities, resources, demands and needs, and above all, includes the internal and external relationships needed to determine greater quality and greater opportunities. If it were true that not all recent development is urban in the traditional sense, but has metropolitan scope, it would be essential to work at the large area level, whose size depends on the specific relationships existing in such an area. It is at this level that we can identify specific development trends, the positioning of social players, the resources available and the inevitable compromises in organisation of space. Only from this starting point can an alternative path of territorial development and organisation be traced.

Reference to the large area does not only take into account the effective *size* (mass), for positioning of the area in the international context, but may also offer efficient levels of *integration*, facilitating relationships both within the area and between the area itself and the exterior. Thus *local resources and potential can be enhanced* in a framework of internal relationships; *efficient and effective organisation of the territory* can be promoted and implemented (appropriate soil uses, environmental protection, reduction of pollution and congestion, sustainable use of resources, etc.); and all fundamental issues and essential requirements for improving quality of public life (and development) can be incorporated (Donolo, 2007).

This strategic line should comprise a coherent set of contents/objectives, including the following basic ones, provided here only as an example:

- *equity*: counteracting both the existing or future imbalances between the various zones in the area (territorial justice), and social differentiations which can be mitigated through appropriate provision of welfare services and redistribution of resources;
- *densification*: appropriate active (infrastructure) and passive tools (constraints, limits, etc.) are needed to reduce land degradation processes and abnormal soil consumption;
- *control of physical and environmental resources*: seen as the rational use of resources, including energy planning, water consumption and so on;
- *local development*: social capital and local resources may, if recognised, become opportunities for local development projects;
- *spread of innovation*: favouring the creation of facilities and connections between scientific research and the business world, favouring technology transfer and business and social spin-offs;
- *improved awareness and education*: local development, the spread of innovation, but also the sustainable use of resources which requires improved cultural awareness and education of individuals and higher quality vocational training at all levels;
- *infrastructure*: appropriate organisation of the territory and achievement of the aforesaid objectives require a widespread, well-articulated infrastructure network which, however, must not be wasteful of resources and territory;
- *solution of local conflicts*: situations of conflict between *local public interest* and a *higher-level public interest* (Indovina, 2007) can be better managed by regional planning which can bring into play a range of methods for achieving the general interests of the whole area.

The combination of the above contents/objectives, or the degree of importance attributed to each element with respect to the others, will depend on the context, area prospects, current trends and the general goals pursued.

4.4 Conflicts and Strategy

The social, individual and institutional players acting and operating in a given area have a *common interest*, on the one hand, in ensuring that the territory functions to the best of its capabilities and possibilities but, on the other, in pursuing their specific objectives they often oppose this common interest and the interests of the other players.

It is not so much a case of schizophrenic behaviour, as of a condition that emerges during the implementation process. Additionally, the common interest presents different facets according to the social conditions of each player. Basically, the pursuit of a common interest is both a necessity and an obstacle, which the players, on the one hand, constantly call for and on the other constantly attempt to violate.

The dynamics of any area are determined by 'social practices' (Indovina, 1997), that is, by the actions taken by single or associated individuals to achieve their aims. These social practices determine the dynamics and innovation of a specific area. However, such practices are often in contrast with one another; they affirm individual and partial (hence non-shared) interests; they generate geographical and social imbalances; they lead to unsound territorial organisation and they produce obstacles to development. In short, these practices contain both positive and negative elements: they tend to 'bend' any regulatory framework through a process of self-organisation, they make the 'strongest' to emerge as the winner in every contrast or conflict and they undermine equity by privileging a sort of social Darwinism. Social practices tend to produce significant negative externalities but at the same time, they produce dynamism and innovation.

Needless to say, organisation of the city cannot consist of the sum total of social practices; it is a collective 'product' in the sense that it must provide guarantees for all stakeholders. It must satisfy expectations of equity, improve efficiency and effectiveness of organisations and contribute to strengthening social relationships, integration, dialogue, equal opportunities and development.

One might correctly argue that the 'conditions' which the city should guarantee pertain to *urban ideology*, because in actual fact the city (which is the spatial projection of social structure) cannot be equitable organised and is strongly marked by social discrimination (which manifests itself in the social organisation of space). The reality of the city, as generated by socio-economic and market mechanisms, collides with urban ideology.[3] While the real city may be considered a 'deviation' from the role society (ideologically) assigns to it, we must also be aware of the fact that urban ideology creates demands and expectations among social groups and citizens. Not surprisingly, all public schemes launched in and for the city declare the aim of affirming those elements of urban condition which we may classify as ideological. However, these elements can be achieved only partially, because of the contrast between social practices and their partiality and, above all, because of the emergence of conflicts.

We may summarise the above argument as follows:

- the collective body of individuals and institutions recognises the need to affirm a *common interest*, a form of city and territorial organisation guaranteeing (this) common or general interest, in order to facilitate achievement of the objectives of individual players;
- in striving to achieve their aims, individual players seek to bend the constraints, the rules and the very organisation which guarantees common or general interest. They also enter into competition with the other stakeholders and leverage on their power to meet their ends;
- the role of 'guardian' of public interest is assigned to the public institutions. Public institutions are provided with their own form of 'power';
- the power of the public institutions, used for affirming common and general interest, is not opposed by a private interest but rather by a number of diverse private interests. Basically, the public/private dichotomy oversimplifies a far more complex reality.

It is only in this sphere of conflicts, of opposing forces, of new and dissolved alliances, that strategy may come into play (Cecchini & Indovina, 1992) even in a reality that presents both conflicting and collaborative relationships,[4] a combination which constitutes a factor of dynamism and transformation of social organisation (Busino, 1978).

In this context, strategic thinking means the science of action: it contemplates conflict and the use of force, but it does not rule out collaboration.[5] In the sense considered here, which is not military, strategy targets the achievement of an objective which may not be shared by some social groups and is therefore opposed by them. Hence the need to draw up a strategy emerges.

4.5 Before Strategy: Defining the Objective

The organisation of space is not an autonomous form of the configuration of reality, but rather constitutes a *projection* of social organisation in space. The structures (social relations of production), which mark societies over time give shape to a specific spatial organisation, but this takes place in a 'built' territory (history, fixed social capital, prior forms of urbanisation, etc.), which dialectically represents an element of *resistance* to change of the territory, but is also the basis for such change. Any specific spatial configuration, as observed at a given point in time, incorporates the tensions, social and cultural change, innovations (technological and organisational), the temper and the deep and surface movements of the resident community, and it reveals how all these factors have shaped the territory.

This awareness does not imply powerlessness of collective and public action, but makes explicit the material which such action must deal with. The urban condition, whatever this may mean, is a collective construction comprising: the values of those living in a community, social and cultural relations; attempts to build a desired future; as well as handling of contradictions and contrasts, rather than the sum of particular interests. While social practices oppose the collective dimension to enhance particular interests, *governance* consists of the *ongoing reconstruction* of the collective dimension.

However, this is not *conservative* reconstruction but rather – taking into account the innovations and dynamisms induced by social practices – the search for higher-level balance that meets the expectations and hopes of social players. In other words, the aim is to define a desirable future. It is in the definition of this future that the *intentionality* of the 'policy' comes into play; a future which bends the reality of a given situation, which also modifies its trends, in order to achieve desirable targets. The construction of this future is a public function, therefore in order to avoid self-referentiality, lack of realism and unsoundness, it must be based on *knowledge, science and participation*. This means obtaining comprehensive knowledge on the changes under way, the processes these changes have induced or may induce, trends and the inter-dependencies between the reality concerned and 'external' dynamics (regional, national or international). A knowledge that instead of simply outlining the 'state of facts' seeks to identify change trends (in economic, territorial, social and cultural configuration). A number of scientific and methodological tools make it possible, on the basis of this knowledge, to design, what we might call, the *probable future*. The tools available enable us to make not just a simple future projection of quantifiable trends, but also to assess qualitative elements, highlight relationships between the various trends, identify points of friction and so on. In the absence of actions aimed at modifying current trends and processes, the *probable future* basically outlines the likely future conditions (economic, social, cultural and territorial), of liveability and the quality of life of that community.

In contrast to this *probable future* stands a *possible future*, an objective aimed at change (now for the future), which seems more desirable than mere waiting for the realisation of trends. This of course raises the issue of how this image of the future should be built.[6]

A fundamental role is played by *public intentionality*, that is, by the government's interpretation of the possible future and of the prospects which seem to hold more promise for society than the realisation of the existing trends (not only is this activity legitimate, it is a prime responsibility and is indeed what is expected of the elected leaders).

A key element required to dispel any doubt of self-referentiality and a collective foundation for this future scenario is *participation*, which can play a fundamental role in both the proposal and verification stages. Within this framework 'conflicts' should be considered as the expression of dissatisfaction and demands. Widespread participation in the overall scenario building should be ensured; the social partners should have their say – in short, the focus should be on the greatest possible degree of participation and contribution.

An important condition of the participation process is that the debate on the possible future should, as far as possible, be free from the 'particular' interests of the various stakeholders. Citizens, social partners, organisations and so on should be called upon to discuss, contribute to and define an imagined future as a better and more desirable scenario.

While participation can occur spontaneously (e.g., through conflict), it must also be organised. Although organised participation loses the element of spontaneity (the direct expression of dissatisfaction), it does constitute a formidable *listening* tool. This process should be interactive, and here a fundamental role is played by the public authority's 'team' which has the task of: (1) recording and re-formulating the elements arising from policy planning and the participation process, (2) identifying the positive and negative elements of current trends and (3) highlighting any contradictions which might emerge from the diverse demands and the various alternative prospects. This interactive work leads to the identification of some 'cornerstones', that is, general policy objectives; they include sustainability (Ferlaino, 2005), social equity (Fregolent & Indovina, 2002), the achievement of better and more widespread knowledge, development and so on. The choice of these cornerstones is among the most significant contributions of policy-makers.

Depending on the context, such general goals may differ and may be subjected to debate (in this case, they should be based on strong grounds and shared by the majority of society). Specific models and techniques have been developed to improve participation (Rizzi, 2004).

When building participation, an important aspect is the attitude of policy-makers; it would be a serious mistake – to be carefully avoided – to consider participation as a tool for building 'political consensus'. Although such an objective is explicitly excluded from desired outcomes, in actual fact the search for political consensus is an implicit objective.[7] What needs to be sought with great care is *support for a project of change and its joint design*, which is the aim of participation in this context. Different conditions and experiences, various future expectations, the 'dreams' (*I have a dream*), compared and assessed against current trends, possibilities and available resources, contribute to generating a process through which the whole community, via the 'common learning' process, builds a shared future scenario (i.e., shared by the majority of the community).

The process of defining the *possible* (and hoped for) *future*, relatively widely shared, will be marked by *realism* (in the sense that it takes into account the real possibilities of the context) and *innovation* (insofar as it corrects current trends). It will define a development pathway which, on the one hand, targets some shared political objectives (sustainability, equity, etc.), and, on the other, attempts to govern the existing trends to achieve an improved situation. Lastly, this process will be characterised by consensus (insofar as it is shared), and by *opportunity* because it creates favourable conditions for each individual to achieve his or her specific objectives in the framework of a common project.

By comparing the *probable future* with the *possible future*, we can gauge the gap between the two, and hence measure the 'effort' needed to move from the former to the latter. Here 'effort' means many things – not just financial resources but also political resources, the capacity of mobilising public opinion, organisational resources and so on. This comparison also outlines the scope of governmental action, which will require an appropriate *strategy* to achieve the overall scheme.

4.6 Strategy and Stratagem

At this point it becomes imperative for the public administration to formulate a strategy for the drafting of a *strategic plan* – not a plan defining objectives but rather a plan identifying *players, forces, interests, policies and actions* – to set in motion the realisation of the *possible future*. Admittedly, the approach I propose in this chapter departs from the standard practice (and perhaps also from the theory) of strategic planning as applied in many contexts. However, the poor outcome of many strategic plans and the partial success of others, often in contrast with an overall development vision, seem to warrant the proposing of different approaches.

The public authority cannot act alone in designing a *possible future*, due to chronic lack of financial resources and due to the fact that the broad nature of the objectives requires involvement and commitment of other public and private parties. However, this in no way diminishes its role but rather increases its governance responsibility and orients its action.

The contribution of the other players should not be limited to the executive phase but should also include analysis of issues, definition of the plan and identification of the appropriate tools, policies and actions. With a positive combination of government and governance.

This approach is based on the following requirements:

- the public authority must be the *director* of the overall plan, acting as guarantor of its soundness and credibility;
- in construction of the *strategic plan* each player is assigned specific tasks and takes on specific commitments;
- the private players involved, while pursuing their own objectives, also *contribute to the achievement of general objectives*, since their own specific objectives fall within the general plan;
- the strategic plan arises from a process of governance which *must be devoid of any under-the-counter deals*.

I argue that the public administration, as a depository of the objectives identified by consensus, is the party which defines the strategy, by means of: (1) joint and separate concertation and negotiation forums or other tools and (2) setting up the time milestones, subjects and concrete initiatives, thus giving life to the intended future. This however, does not mean *imposition* on the part of policy-makers, on the contrary, discussion may in many aspects be open, but with clear objectives. Indeed, any self-referentiality on the part of policy-makers will have been dispelled by the prior process of participation allowing definition of the *possible future*.

Thus, the public administration should draw up its own *strategy* while preparing a *strategic plan*. In this process, we cannot simply make a distinction between public and private; indeed, the situation is much more complex, not only due to the different objectives pursued by the various 'private' players, but also because only the interests of some of these private players will be fully met by the envisaged future, while others will be wholly or partly sacrificed. Furthermore, other public institutions may have objectives in partial contradiction with those of the public administration promoting the plan.

Hence the theme is to construct a *strategy* leading to the full definition of a *strategic plan* which clearly identifies actions, timelines and players. This is certainly not a simple operation, and the task is often aggravated by a shortfall in the required professional skills in the public administration.

The structure of this strategy should also include specific stratagems to achieve compromises, actions, breakthroughs and so on. A useful example is the first of the 36 Chinese stratagems *the most skilled in warfare directs his opponent, but never permits his opponent to direct him* (Casacchia, 1990), which, in our case, applies to the directing role to be played by the public administration.

Of course, in this case the term 'war' is replaced by *negotiation* which also has strategic rules (Schelling, 1960), as summarised hereunder:

- do not focus on negotiation themes whose results can be profitable for all parties – take them for granted. Do not play down the difficulty of finding agreement on these themes, but assume that the opposing party too is well aware of the situation;
- focus the greatest attention on agreements characterised by dissymmetry, that is, whose outcome is better for one of the parties than for the other. Bear in mind, however, that this situation is determined by the fact that any agreement is better than no agreement at all. An excessively broad range of possibilities generates instability which may, in turn, lead to failure to reach an agreement. An agreement may be reached through 'voluntary sacrifice of the freedom of choice', that is, imposing constraints on oneself (the public administration in this case) can reduce the adversary's freedom of choice;
- when the result depends on the development of the negotiation, it can be useful to clearly explain specific objectives (through 'public statements'), to make it quite clear which are not open to compromise;
- the distribution of costs and benefits should be clearly identified. Clarity on this point facilitates balanced choices;
- the 'threat' constitutes an important tool – a party threatens to do something it actually has no interest in doing since it would be damaging to both parties. This is a tool to be used with caution, indeed a threat is made only if one is convinced that it will produce the desired results; basically, it is the threat itself which achieves the aim, not its fulfilment;
- the 'promise' is on the other hand a positive tool, particularly when the final action is outside the control of one of the two contenders.

The approach proposed in this section can be summarised as follows:

- the public administration (policy-maker) is a *guarantor* of a project for the future;
- the achievement of this vision is facilitated by the drawing up of a *strategic plan*;
- the public administration defines a *strategy* for drafting the strategic plan.

This might appear as a somewhat 'Baroque' construction and, as already said, as a variation on the more consolidated model of strategic planning. Indeed, I would argue that the main purpose of the strategic plan is to identify not general objectives but rather methods (strategic methods) to achieve interventions, actions and policies able to shape and realise pre-defined objectives through a process that combines *political intentionality*, *knowledge of the situation* and *participation*. This approach, as already stated, departs from more traditional forms of strategic planning which – bluntly stated – tend to diminish the role of the public authority and favour the objectives of the stronger players (perhaps also because their objectives are easier to achieve). The point made in this chapter is that the traditional approach should be reversed by refocusing on the role of public authorities in the organisation of cities and territories and in the quality of life of their communities.

Notes

1. On the 'production' of strategic plans see Spaziante and Pugliese (2003) and Martinelli (2005).
2. See also Savino (2003), who remarks on strategic planning as part of his analysis of local area governance.
3. This theme was explored in depth in the 1960s. See the classic Castell (1974) and Bolognini (1981).
4. We can confidently reject the *cliché* that all relationships within society are of an antagonistic nature, and also the opposite contention that all relationships are collaborative, and that any exceptions to this rule are simply anomalies to be eradicated. Indeed, not only are both types of relationship found in society, but also conflicting relationships may not always be removed by employing force (Cecchini & Indovina, 1989).
5. For an analysis of collaboration, see Gambetta (1989); for collaboration within a situation of conflict see Rapoport (1960) and Cecchini (1989).
6. We could also consider three futures, that is, probable, desirable and possible, where the latter is a 'realistic' version of the desirable. However, this subdivision does not seem to provide significant advantages.
7. It is not unusual, for instance, for the electoral defeat of a party which promoted and supported a participation process to appear as being incomprehensible. This bafflement is summarised by the question: "How could this happen? We fostered citizens' participation to give them a say in decision-making, and yet they voted against us!".

References

Bertuglia, C. S., Rota, F. S., & Staricco, L. (2004). *Pianificazione Strategica e Sostenibilità Urbana*. Milano: FrancoAngeli.

Bolognini, M. (1981). *Spazio Urbano e Potere. Politica e Ideologia della Città*. Milano: FrancoAngeli.

Busino, G. (1978). Conflitto. In VV.AA. *Enciclopedia Einaudi* (Vol. 3). Torino: Einaudi.
Casacchia, G. (Ed.). (1990). *I 36 Stratagemmi*. Napoli: Guida.
Castells, M. (1974). *La Questione Urbana*. Padova: Marsilio.
Cecchini, A. (1989). Minaccia e Negoziato tra Simulazione e Gioco. In A. Cecchini & F. Indovina (Eds.), *Simulazione*. Milano: FrancoAngeli.
Cecchini, A., & Indovina, F. (1989). *Simulazione*. Milano: FrancoAngeli.
Cecchini, A., & Indovina, F. (1992). Il Pensiero Strategico e le Tecniche. In A. Cecchini & F. Indovina (Eds.), *Strategie per un Futuro Possibile*. Milano: FrancoAngeli.
Donolo, C. (2007). *Sostenere lo Sviluppo*. Milano: Mondadori.
Ferlaino F. (Ed). (2005). *La Sostenibilità Ambientale del Territorio*. Torino: UTET Università.
Fregolent L., Indovina F. (Eds.). (2002). *Un Futuro Amico. Sostenibilità ed Equità*. Milano: FrancoAngeli.
Gambetta, D. (1989). *Le Strategie della Fiducia*. Torino: Einaudi.
Gibelli, M. C. (1996). Tre Famiglie di Piani Strategici: verso un Modello Reticolare e Visionario. In F. Curti & M. C. Gibelli (Eds.), *Pianificazione Strategica e Gestione dello Sviluppo Urbano*. Firenze: Alinea.
Gibelli, M. C. (2003). Flessibilità e Regole nella Pianificazione Strategica: Buone Pratiche alla Prova in Ambito Internazionale. In A. Spaziante & T. Pugliese (Eds.), *Pianificazione Strategica per le Città: Riflessioni dalle Pratiche*. Milano: FrancoAngeli.
Gibelli, M. C. (2007). Piano Strategico e Pianificazione Strategica: un'Integrazione Necessaria. *Archivio di Studi Urbani e Regionali*, *89*, 211–221.
Indovina, F. (1997). Nuove Condizioni ed Esigenze per il Governo Urbano. In C. S. Bertuglia & F. Vaio (Eds), *La Città e le sue Esigenze*. Milano: FrancoAngeli.
Indovina, F. (2003). *La Metropolizzazione del Territorio. Nuove Gerarchie Territoriali*. Economia e Società Regionale. *Oltre il Ponte*, *3–4*, 46–85.
Indovina, F. (2005). La Nuova Dimensione Urbana. L'Arcipelago Metropolitano. In M. Marcelloni (Ed.), *Questioni della Città Contemporanea*. Milano: FrancoAngeli.
Indovina, F. (2007). Spazi e Luoghi Contesi. *Contesti*, *1*, 13–23.
Martinelli F. (Ed.). (2005). *La Pianificazione Strategica in Italia e in Europa*. Milano: FrancoAngeli.
Rapoport, A. (1960). *Fights, games and debates*. Ann Arbor, MI: Michigan University Press.
Rizzi, P. (2004). *Giochi di Città*. Bari: La Meridiana.
Savino, M. (2003). Verso Nuove Forme di Governo del Territorio. Alcuni Possibili Percorsi. In M. Savino (Ed.), *Nuove Forme di Governo del Territorio. Temi, Casi, Problemi*. Milano: FrancoAngeli.
Schelling, T. C. (1960). *The strategy of conflict*. Cambridge: Harvard University Press.
Spaziante, A., & Pugliese, T. (Eds.). (2003). *Pianificazione Strategica per le Città: Riflessioni dalle Pratiche*. Milano: FrancoAngeli.

Chapter 5
Notes on Strategic Processes in Land Use Planning

Luigi Mazza

5.1 Introduction: Land Use and Mobility Perspective

The available planning knowledge is unfortunately poor or, at least, relatively poor as it is not adequately consolidated and shared. This implies the need for defining themes and words for any talk (these are the reasons why I'm not, in most cases, prepared to talk about planning since the issue is too general to deserve a discussion).

Therefore I want to make it clear that I assume a land use and mobility perspective (for brevity, I use the word planning to mean land use and mobility planning). It must be understood that what I say from this perspective should not be assumed as necessarily meaningful in other planning perspectives.

My assumption is not irrelevant for at least four basic reasons which are the main characteristics of a land use and mobility planning process.

1. Land use planning is about land, a limited and scarce resource; other forms of planning involve other types of resources which are generally reproducible and not limited. This difference implies different problems and planning approaches.
2. Land use planning involves property, which is a cornerstone of our political, social and economic system. This implies that land use planning, on many occasions, has to be 'continuous'; it has to consider the whole land and not a part of it (land is a continuous resource, and administrative borders rarely have a functional meaning). Again it does not apply to other forms of planning, or not in the same way.
3. Land use planning allocates and controls a particular resource, property rights. The resource is generally controlled only by the state, while other forms of planning allocate and control resources which are allocated and controlled by both public and private agencies.

L. Mazza (✉)
Department of Architecture and Planning, Polytechnic of Milan, 20133 Milan, Italy
e-mail: luigi.mazza@polimi.it

M. Cerreta et al. (eds.), *Making Strategies in Spatial Planning*,
Urban and Landscape Perspectives 9, DOI 10.1007/978-90-481-3106-8_5,

4. Last, but certainly not least, land use planning decisions are long-term decisions; the public decision-maker may not be in charge when the decision becomes effective and, more important, it is usually very difficult, if not impossible, to change or cancel the future consequences of a decision. For this reason land use planning decisions should require a large majority agreement.

The first characteristic reminds us that any development process is necessarily a competitive one: many interests are usually competing for the same land. More often than not, the conflict of interests cannot be solved by the market because the urban market is largely a product of planning choices, therefore it is expected that a solution to the conflict will be achieved through political intervention, usually a plan. The second and the fourth characteristics join the third one in stressing the political nature of the planning process and the need for political interventions to get the process working.

5.2 Strategic, Communicative and Political Behaviour

Strategic thinking is generally based on the pair winner/loser and it is aimed at achieving success. Strategic thinking and acting is usually defined in opposition to the communicative thinking and acting, which implies negotiation and compromise between conflicting interests. Focused on the result, a strategic behaviour is always selective.

A strategic actor – no matter whether individual or collective – considers the context as a resource to be used to maximise the success and to reduce risks and the unexpected effects of the action. A strategic actor is not expected to locate his/her goal in a system of collective aims which can reduce his freedom of action. And he is not keen on a cooperative behaviour; his preferred behaviour is a competitive one. Strategic actors – no matter whether individual or collective – pursue goals and take decisions on the basis of different and sometimes conflicting sets of values, different allocation of resources and different risks related to unexpected effects. Despite these differences, strategic actors' choices interact and are often in conflict due to these differences.

The clash between strategic and communicative behaviour is or may be resolved by political behaviour, and particularly by the political exchange aimed to reach an agreement between competing interests. Political thinking and acting are based on the conjunction of the dual rationalities of strategic and communicative thinking and acting. In other words, political acting is based on the conjunction of communicating and decision-making. In particular, political rationality consists in the capability of giving a reasoned account and evaluation of the consequences of the interruption or the impossibility of a dialogue. A political behaviour risks making a decision with which only some of the involved interests agree, and despite this, it assumes the responsibility of developing the process.

The question is: when and why is the risk rightly taken?

5.3 Planning Strategic Processes

Because of the nature of the resource which is contended by the competitive interests, the process of land use development is governed by a mechanism of expansion and exclusion which is typical of capitalist processes. Any land use transformation that is, land use expansion, produces a land use exclusion. In other words, the land use development process is always, and necessarily, producing winners and losers. It follows that the land use development process is always based on a strategic rationality, even if and when this is not explicit. But a planning strategic process is never based only on a strategic thinking and acting. Because of its political nature, a land use development process cannot escape a communicative rationality and the consequent search for an adequate consensus. Even if we call it a strategic process, a planning process is a political one in which strategic and communicative rationality participate in the effort to achieve the desired aims. A strategic planning process is always a political process; it is a voluntary one.

From the land use and mobility perspective, when we talk about a strategy we mean a political and technical process aimed, in principle, at the community development. A strategic process pursues the community development through selected projects. The success of the strategy can be evaluated with reference to the specific results of each project. The comprehensive goal of community development is pursued through a few basic activities:

(a) designing a political vision;
(b) building a coalition supporting the vision;
(c) selecting some specific projects consistent with the vision;
(d) involving actors capable to implement the projects.

Where (a), (b), (d) are mainly political activities with a relatively poor technical content, (c) is a political and technical activity.

The four activities confirm that, even if a strong interaction between political and technical actors is developed, in practice, a strategic process has a major political nature and content, and it is basically a political process. The four activities are not the temporal and logical phases of a process; they in fact interact continuously. They constitute the four elements of a circular process: to design a vision a coalition is needed, but a vision is necessary to coalesce a coalition. A political vision is necessary for involving interests in the building of a coalition, but interests intervening in the formation of a coalition will participate in redefining the political vision.

There is not a starting point in the process because the politicians proposing the first draft of the vision are not strangers to the interests; they represent most of them and are often part of them. Again the selection of the projects and policies necessary to implement the vision is not completely dependent on the vision itself, because some projects will need the vision to be described in detail and so at least partially modified. Thus the vision is certainly the term of reference for the selection of the projects but there is a feedback from the projects: the projects help to understand which part of the vision is too difficult or impossible to implement and/or may suggest other ways to select new goals which are consistent with the aims of the vision. Finally, the actors forming the coalition are not necessarily the same actors prepared to implement the selected projects. Involvement of the implementers may produce a redesign of the projects and a partial reformulation of the vision.

The circularity of the process is possible because a strategy is made by a collection of projects which are not components of one temporal chain. However, within the circularity the process is phased. In the first phase the vision is general in two ways: (1) 'comprehensive' and (2) 'open' (not well defined). In the second phase, during the coalition building, the vision becomes selective, focused on aims and comprehensive about means. Gradually, the process selects projects and policies which concern their content, location and scheduling. From this standpoint we can define the process as general, comprehensive and open, selective about aims and comprehensive about means.

5.4 Visions and Myths

A political or strategic vision is a political programme aimed at community development, that is, a future community that we assume to be better than the present one. Community is an ambiguous term. It is used here in a very general sense – national, regional and local community – and without any specific philosophical and political implications. Community – and not, for instance, locality – means that the majority of people involved in the strategic process share the assumptions of the chosen political vision. Once again we stress that the strategic process implies at least a good majority consensus.

A political vision is a planning resource, which is usually rooted in 'myths', where the myth is understood as a social construct – a fusion of ('religious') values and political ('ideological') manifestos. Values and manifestos are the product of national and international cultures and histories. The myth points to the existing fears and hopes and devises an overall answer to erase fears and realise hopes. The shorthand of the myth avoids any reference to practical difficulties that might complicate the simple question and answer. The myth thereby became an appealing part-truth, attractive for its simplicity. The function of myths is (a) to condense a complicated intellectual message into a shorthand format which can readily be translated into a political vision, a strategic process and action and (b) to organise and make a political vision cohesive.

Political visions and strategies are *bets* on the future and sometimes against the future; the myth is the core, the marrow of the bet. Due to its religious/ideological content, a myth may become a dreadfully dangerous tool (everybody knows the disasters produced by the translation of some 'neocon' myths in economic and political visions and war strategies). But myths do not necessarily produce bad effects, they may produce good ones too; that is, joining Europe was a powerful myth designed as a national strategy for reducing part of the Italian public debt, and it was crucial to gain acceptance of the Italian people despite of its high cost for most families.

Myths may assume a technical form, which is only one expression of their common religious/ideological nature. For instance, in the land use planning, field anti-urbanism, low density and neighbourhood units are examples of myths that inspired planning visions during the last century. Most land use planning has been and is still dominated by these myths. In many countries public ownership of the land, as an indispensable condition for good planning, has been another powerful myth. Sustainable development has been at the basis of most planning visions in the last decades. It may be a useful myth in the northern European countries, but not very helpful in the very dense southern ones.

Myths expressed by a technical form may be as powerful as the political ones, but they can be only one component of a political vision aimed at community development. A powerful political vision is necessarily based on a powerful myth.

5.5 Visions, Strategies and Plans

The political visions, produced by the different tiers of the strategic process and by the different localities, have a common basis in the shared myths which are rooted in national and international cultures and histories. As long as political visions are consistent with the myths, they form a chain which connects them vertically and horizontally. The visions are mutually reinforcing and supporting through the chain. And the more a political vision is consistent with the basic shared myths, the more it is locally reliable and powerful. Even if we assume a circular relationship between the visions of different tiers and cooperative relationships between visions of different regions, and a progressive vertical and horizontal interaction between visions, the strategic character of the visions implies their hierarchical organisation in a system which has its centre in the nation and is articulated through the regions to the localities.

If we agree with the previous paragraph, two important questions arise:

1. Can a local vision be meaningful and effective even if a superior vision is not available, if there is no chain?
2. Can a local strategic process be developed even if the local vision is weak?

To design a local vision is always possible and it may be a useful political exercise to understand strengths and weaknesses, fears and hopes of the locality. But if there is no superior vision to link and interact with, it is doubtful that the local vision be effective. In this case a vision may be a claim for and a contribution to the design of a regional and national vision. Again it is always possible to act strategically at the local level even when a superior vision is not available. However, in general, when lacking a shared vision the strategic process may more easily become subservient to particular interests.

A coalition, the selected projects and the spatial policies are the final products of a strategic process. To implement projects and policies a further step is needed: development rights must be allocated through the land use plan. The development projects, called in by the strategic process, must be located in the legal space of the plan and approved to be implemented. The plan is the rule which mediates the interests in a general system of land use and mobility compatibilities.

If a true strategic process is developed, the land use plan becomes the tool used by the locality to register and adjust the outcome of the strategic process to the land. The relationship between the projects selected by the strategic process and the land use plan is not a one-way relation, but an interactive one. The land use plan represents the context in which the projects have to be implemented. An adjustment of the projects to the context is necessary before their approval and implementation. The whole development process is then the result of two parallel circular processes: the strategic one in which a vision is defined, a coalition built and some projects selected; a (land use) planning one in which the selected projects interact with the land use plan.

The most important contribution of the plan is offering a comprehensive representation of the context with which the selected projects can interact. The resistance of the context to the change that the projects want to introduce is proportional to the weakness of the vision and of the strategic process. The resistance of the context is weak or null while facing a powerful strategic process, in particular, because the land use planning knowledge is poor, and therefore, it may not always be possible to give a good and complete account of the expected impacts produced by the implementation of the projects.

Finally, to stress the differences between strategic processes and land use and mobility plans, it is possible to summarise their main characteristics by distributing them according to four main dimensions: (1) structure and power, (2) resources and tools, (3) products and (4) aims (Fig. 5.1).

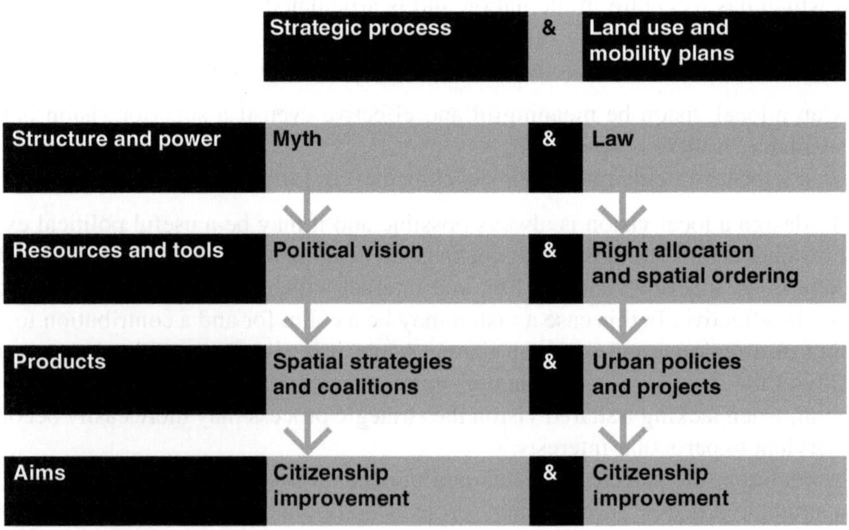

Fig. 5.1 The differences between strategic processes and land use mobility plans

While the structure of a strategic process and its persuasive power are granted by myths, the structure of plans and their normative power are conferred by the planning law.

The main resources and tools of a strategic process are visions based on myths; the main resources and tools of a plan are the allocation of development rights and the definition of spatial ordering and control, both of which are conferred by the planning law. The main products of a strategic process are coalitions and spatial strategies; opportunities for public and private development are the products of plans. While the general aim of a strategic process is community development, the general aim of a plan is citizenship improvement. But citizenship may be considered a general evaluation standard of both strategic processes and land use plans. An evaluation standard is a matter of a political choice, for instance, a synthetic indicator of community development might be improvement of the average income or improvement of the social capital and so on. It seems that a more satisfying standard would be the flourishing of the citizenship, the political and social citizenship.

References

Albrechts, L., Healey, P., & Kunzmann, K. R. (2003). Strategic spatial planning and regional governance in Europe. *Journal of American Planning Association*, *69*(2), 113–129.

Blowers, A. (2002). Political modernisation and the environmental question: The case for strategic planning. *Local Government Studies*, *28*(2), 69–87.

Faludi, A., & van der Valk, A. (1994). *Rule and order. Dutch planning doctrine in the twentieth century*. Dordrecht: Kluwer Academic Publishers.

Healey, P. (2007). *Urban complexity and spatial strategies*. London: Routledge.

Healey, P., Khakee, A., Motte, A., & Needham, B. (Eds.). (1997). *Making strategic spatial plans*. London: UCL Press.

Hillier, J. (1995). *Planning rituals: Rites or wrongs*? Paper presented at the 9th AESOP conference, 16–19th August 1995, Glasgow, UK.

Innes, J. E. (1986). *Myths in policy processes: The importance of the implicit debate*. Berkeley, CA: IURD.

Innes, J. E. (1987). *Myths and the definition of policy problems. An exploration of homeownership and public-private Partnership*. Berkeley, CA: IURD.

Mazza, L. (1995). Order and change, rule and strategy. In C. S. Bertuglia, G. Bianchi, & A. Mela (Eds.), *The city and its sciences* (pp. 559–576). Heidelberg: Springer.

Mazza, L. (1999). *The specific domains of planning*. *European Planning Studies*, *7*(5), 557–561.

Mazza, L. (2002). Technical knowledge and planning actions. *Planning Theory*, *1*(1), 11–26.

Mazza, L. (2004). *Piano, Progetti, Strategie*. Milano: FrancoAngeli.

Needham, B. (1997). Planning strategies and planning methods. In P. Healey, A. Khakee, A. Motte, & B. Needham (Eds.), *Making strategic spatial plans* (pp. 173–190). London: UCL Press.

Needham, B. (2000). Spatial planning as a design discipline: A paradigm for Western Europe. *Environment and Planning B*, *27*(3), 437–453.

Vigar, G., Healey, P., Hull, A., & Davoudi, S. (Eds.). (2000). *Planning governance and spatial strategy in Britain*. London: Macmillan.

Young, I. (1990). *Justice and the politics of difference*. Princeton, NJ: Princeton University Press.

Chapter 6
Post-structural Complexity: Strategic Navigation in an Ocean of Theory and Practice

Jean Hillier

6.1 Introduction

As my embarkation point for the voyage, that is, this chapter, I take a quotation from Salman Rushdie's *Haroun and the Sea of Stories* (1990, p. 72): "When the boy, Haroun, looked into the sea of stories: [h]e looked into the water and saw that it was made up of a thousand thousand and one different currents, each one a different color, weaving in and out of one another like a liquid tapestry of breathtaking complexity; and Iff explained that those were the Streams of Story. (. . .) And because the stories were held here in fluid form, they retained the ability to change, to become new versions of themselves, to join up with other stories and so become yet other stories."

In the 'Ocean of Stories', that is, planning theory and practice we see human and non-human stories flowing, interconnecting, congealing and transforming the molecular and molar lines of trajectories. It is generally accepted by academics and practitioners that a city is an 'endless kaleidoscope of possible viewpoints' or landmarks; a "mobile panorama of interacting events" (Cooper, 2005, p. 1693). We are beginning to regard cities, human and non-human actants not as 'things-in-themselves' but as complex, multiple and mutable elements of connections and disconnections, relations and transitions.

The recent introduction of a range of concepts, including complexity, multiplicity, emergence, becoming, assemblage and so on, represents a relatively new and important shift in thinking of theorisation and, by extension, of methodology in planning (Venn, 2006).

I offer a multiple, relational approach of dynamic complexity to understanding contingencies of place and actant behaviours, based predominantly on the work of Gilles Deleuze and Félix Guattari. If spatial planning is concerned with "discovering the options people have as to how to live" (Thrift, 1996, p. 8), then it is concerned

---

J. Hillier (✉)
School of Architecture, Planning and Landscape, Newcastle University,
Newcastle Upon Tyne NE1 7RU, UK
e-mail: j.s.hillier@ncl.ac.uk

M. Cerreta et al. (eds.), *Making Strategies in Spatial Planning*,
Urban and Landscape Perspectives 9, DOI 10.1007/978-90-481-3106-8_6,

with understanding the world in terms of practical effectivity rather than of classificatory representation – not the *what*, but the pragmatic Deleuzean *how*: not so much 'what does it mean?' but 'how does it work?' (Deleuze & Guattari, 1984, pp. 109, 129).

Rather than seeing the flux of movements in societies as sets of things with stable qualities and thinking of themselves as people who might act upon these things, planning theorists and practitioners might instead view movement as productive, a Deleuzoguattarian open whole traversing and connecting across space and time.

I am drawn to Michel Foucault's ideas of immanence and to Gilles Deleuze and Félix Guattari's ideas of becoming or moving beyond. These notions allow unexpected elements to come into play and things not to quite work out as anticipated. They allow us to see planning, governance, planners and other agents of governance as *experiments* or *speculations* entangled in a series of modulating networked relationships in circumstances both rigid and flexible, where outcomes are volatile, and problems are not 'solved' once and for all, but are rather constantly recast and reformulated in new perspectives. Questions become issues of problematisation rather than of neat solutions. Speculation is to be viewed as creation rather than as scientistic proof-discovery. I regard experimentation as a violation of prescribed conventions; a transgression of boundaries, in which genres are blurred and jumbled. Speculation is a tentative method of knowing, working within an ideology of doubt and uncertainty: of what might become.

The material in this chapter is a temporary fixity of my ongoing "gropings in the dark, experimentation, modes of intuition" (Deleuze & Guattari, 1987, p. 461), entangled in the oceans of complex post-structuralism or post-structuralist complexity. It reflects my theoretical genealogy through work by Habermas, Foucault, Laclau and Mouffe and Lacan and its recent rupture with my discovery of the potential for creative transformation offered by Deleuze and Guattari.

I regard this chapter as a particular "direction in motion" (Deleuze & Guattari, 1987, p. 21); a multiplicity connected to other multiplicities, yet possessing a kind of stability or coding of information. I offer a Deleuzoguattarian-inspired approach to planning as speculation and experimentation, before outlining a multiplanar theory of planning as strategic navigation and some initial thoughts about the method: how might the theory translate into strategic spatial planning practice. I non-conclude that spatial planning theory and practice will inevitably be anexact stories where the unexpected and the aleatory lurk.

6.2 Strategic Spatial Planning as Strategic Navigation: Multiplanar Speculation and Experimentation

"[A] productive encounter with chaos" (O'Sullivan, 2006, p. 62).

Imagine. . .Several people are on a raft in the middle of the ocean. But this is *not* a shipwreck. They have come together for some reason (it could be a race, a dare or such). It is a makeshift raft made up of flotsam and jetsam, with a makeshift paddle. The people do not know where they are very precisely and they are out of sight of land, trying to make headway in an ocean of varying currents, with varying waves, wind speeds and directions. Their overall objective – or trajectory – is to reach land.

How might they go about getting there? I argue that they would probably attempt this through a series of short-term projects, such as:

- trying to work out where they are. They might think through questions such as: where do they think they came from last? How might they have got here from there? What can they remember? What landmarks did they pass?
- trying to work out future potentialities. Questions might include: what are the elements involved and what are the relations or connections (and disjunctions) between them? Which relations are likely to be more powerful? With what implications? For example:
  - *the condition of the raft and the paddle* – these are very makeshift and need constant repair/patching up; elements fall off; chemicals in a steel container are leaking and eating away the rope binding the raft together; – the ocean and its currents – the people may anticipate the general direction of the currents but they cannot predict them;
  - *the weather* – what might the effects of sunshine or storms be? Is it possible to see stars and the position of constellations? Or is it too cloudy to see the night sky?
  - *the strength of the people on board the raft* – this relates to availability of food, water, heat, body mass, physical and mental strength and so on;
  - *desires of the people on board* – to survive? to be a hero? to remain adrift just long enough to attract sponsorship for a book deal with potential film rights?
  - *chance* (the aleatory), including hazards such as icebergs, huge containers lost from ships;
  - *hope* – of seeing a ship which stops; that the blur on the horizon turns out to be land.

Having worked out the relations between these (and other) elements, can the people on board tweak any of them so that the outcomes become favourable? There is a need to negotiate between philosophical or ethical ideals and the practical necessity of getting things done. What experimentation might be productive? For instance, ditching a sick person to save food and water for the others? Dumping the chemicals container to save the rope, although the container gives the raft extra buoyancy? Making the raft look aesthetically attractive?

The people will need to 'live' together, with flexibility and adaptability; a situation of creative experimentation. Depending on circumstances and what seems to work (or not), they will probably change their means (perhaps making a sail, ditching a container), the direction they go in (possibly someone thinks they can see land far off to the left) and perhaps even their goals. Of course, with several people on the raft, they probably will not agree on the direction in which they want to go or the actions they should take to get there.

I refer to the above as 'strategic navigation'.[1] I believe that it resonates with practices of strategic (spatial) planning. It also resonates with the work of the French scholars, Gilles Deleuze and Félix Guattari, whose theorising promotes pragmatic, speculative experimentation. Deleuze, in particular, focuses on the potentialities of the multiplicity of forces which could be activated, rather than on transcendent questions of the 'good' or on a negativity of what is absent. Trajectories, movement and transformation are vital. 'Becoming' is a movement between things, disrupting meanings, understandings and ways of being. Concepts are fluid, folding across and into each other, not always harmoniously, and often in agonistic dissonance where differences come into contact. Becoming, then, is "supremely pragmatic" (Massumi, 1992, p. 100). Deleuze actually calls his philosophy a 'pragmatics', as he aims to lever open new spaces and make new connections of lines between elements. To me, spatial planning represents an issue of becoming. It evolves, it functions and it adapts, somewhat chaotically, always pragmatically, concerned with what can be done, how new things, new foldings and connections can be made experimentally, yet still in contact with reality (Hillier, 2007). Spatial planning attempts to embrace a future that is not characterised by the continuity of the present, nor by the repetition of the past, but by a difference that can never be fully grasped (May, 2005). I identify a need for spatial planning to accommodate fluidity and immanence and to have some form of temporary fixity. I propose that strategic spatial planning be concerned with trajectories rather than specified end-points. In regarding spatial planning as an experimental practice working with doubt and uncertainty, engaged with speculation as adaptation and creation rather than as scientistic proof-discovery – a speculative exercise, a sort of creative agonistic. I suggest a new definition of spatial planning as strategic navigation along the lines of the investigation of 'virtualities' unseen in the present; the speculation about what may yet happen; the temporary inquiry into what at a given time and place we might yet think or do and how this might influence socially and environmentally just spatial form (Hillier, 2007).

6.2.1 Multiple Planes

In theorising strategic spatial planning, I adopt the ontological conceptualisation of planes used by Deleuze and Guattari.[2] For ease of association with strategic planning practice, I term the first type of Deleuzoguattarian plane, a plane of immanence (1994), and the second type a plane of organisation (1987) (Fig. 6.1).

The broad plane of immanence is defined not by what it contains, but "rather by the forces that intersect it and the things it can do" (Kaufman, 1998, p. 6). It is the temporary product of a mapping of forces (see below). As Kaufman (1998, p. 6) continues, such mapping "is at once the act of charting out a pathway and the opening of that pathway to the event of the chance encounter".

The plane is an object of construction; a practice (Bonta & Protevi, 2004, p. 62), which maps and records performance of actants' desires: "a disorganised flux that allows itself to be coded" (Colebrook, 2002, p. 114). The plane is open to "new connections, creative and novel becomings that will give it new patterns and triggers of behaviour" (Bonta & Protevi, 2004, pp. 62–63). The 'key move' is to construct a plane by collaborative experimentation – to work experimentally together on my

Fig. 6.1 Schematic descriptors of the planes of immanence and transcendence (Hillier, 2007, p. 243)

Plan of immanence Consistency	**Plane of organisation** Transcendence
Becomings/emergence	**Transcendence**
Open-ended trajectories	**Closed goals**
Multiplicities of meshworks	**Hierarchical relations of power**
Chance/aleatory	**Planned development**
Smooth space (with some striation)	**Striated space** (with some smoothing)
Unstructured	**Structured**
Dynamism of unformed elements	**Stability of judgement and identity**
Flux and fluidity	**Inertia of sluggish movement**
Power to	**Power over**

metaphorical raft in order to reach land. The plane of immanence is a praxis that leaves the ends of each line of knowledge open to extension (Skott-Myhre, 2005); not something closed or the end of a process. A plane (long-term strategic plan or trajectory), of foresight, of creative transformation, of what might be. Rather it "functions like a sieve over chaos" (Boundas, 2005, p. 273), implying a sort of "groping experimentation" (Deleuze & Guattari, 1994, p. 41), of multiplicities of concepts, many of which never come to be as originally intended. The plane of organisation, in contrast, is a transcendent plan or blueprint with certain goals for development. These goals are predetermined standards (such as land use regulations or a design guide) to which things are submitted in judgement and ordered by the forms of representation (whether applications meet the standard criteria, etc.). Local area action plans, design briefs and detailed projects are typical planes of organisation. They tend to be relatively local or micro-scale, short-term and content specific. They facilitate small movements or changes along the dynamic, open trajectories of planes of immanence. The planes of immanence and organisation exist simultaneously and are interleaved; a multitude of layers that are sometimes fairly closely knit together and sometimes more separate. We, as actants, inhabit both planes at the same time. In the words of Deleuze and Guattari (1987, p. 213), "every politics is simultaneously a macropolitics and a micropolitics." I argue for the broad trajectories/visions of strategic spatial planning to be background plan(e)s of immanence and for more specific local/short-term plans and projects such as foregrounded plan(e)s of organisation:

- several (or perhaps one collectively preferred) trajectories or 'visions' of the longer-term future, including concepts towards which actants desire to move, such as sustainability (planes of immanence);
- shorter-term, location-specific detailed plans and projects with collaboratively determined tangible goals, for example, for mainstreet regeneration, provision of affordable housing and so on (planes of organisation).

6.3 Theorising Multiplanar Cartographies for Strategic Navigation

> It's always about going from one place to another and how you get there (Lepage, 2003, p. 153).

In this section, I attempt to translate the broad picture of multiplanar theory into a practice of strategic navigation. For Deleuze, relations are vital to the active construction of existence. It is the contingent "circumstances, actions, and passions" (Deleuze & Parnet, 2002, p. 56) of life which provide for the specific forms of relations between different elements. Relations are endowed with a positive reality as they are not derived from the elements themselves. Relations are not subordinated to the essence of things. Rather, they come into being via practice.

The challenge for strategic navigation is to think relationally, following the relationalities and intensities which cut across objects, events and us as theorists and researchers.

Deleuze and Guattari (1987, p. 146) describe their cartography or 'pragmatics' as comprising four circular components:

1. the generative component – the *tracing* of concrete mixed semiotics and pointing towards the potentiality of what might emerge;
2. the transformational component – making a transformational *map* of the regimes and their possibilities for translation and creation;
3. the *diagrammatic* component of the relational forces that are in play 'either as potentialities or as effective emergences';
4. the machinic component – the study of assemblages and outlines of programmes to figure out of what new assemblages might emerge?

Analysis does not only trace relational connections, conjunctions and disjunctions (Deleuze & Guattari, 1984; Deleuze, 1990) between elements in an assemblage, but it also maps their potential transformations. What events might transpire from the relations between discourses, texts, practices, laws, affects and silences?

Deleuze and Guattari (1987, p. 146 above) suggest a pragmatic method of making a tracing; then making a transformational map of possibilities for translation and creation; then, making diagrams of the forces that could play in each case, either as effective emergences or potentialities. Inclusive, democratic discussion could develop negotiated trajectories and Deleuzean fabulations (strategic spatial plans on the plane of immanence), as well as major project plans or local action plans (on the plane of organisation), in a form of strategic spatial planning as strategic navigation.

There is a "pivotal opposition" (Bosteels, 2001, p. 894) between tracing or 'interpreting' something in retrospect and mapping trajectories through diagrams to anticipate whether relations "can serve as indicators of new universes" (Guattari, 1986, p. 102, cited in Bosteels, 2001, p. 895). To trace, or interpret, entails looking back, often from above, in a systematic manner. To map involves discovery and perception of indicators or landmarks.

Landmarks are useful for orientation purposes as something to head towards. An analytic cartography thus involves *both* the deductive interpretation, especially of ruptures and discontinuities, of 'symptoms' of an actual situation *and* the invention of new heterogeneous, experimental assemblages and pragmatic diagrams, "a furtive glance sideways into an undecidable future of desire" (Bosteels, 2001, p. 895).

6.3.1 Tracing

Tracing involves discovering relations between elements in assemblages. To trace is to describe and to analyse the diversity of relations, the modalities of co-ordination, the discourses, the emotions, affects and so on, and how they were mobilised to shape actants' frames, representations and behaviours. Deleuze and Guattari's pragmatism is agonistic, referring to the role of relational difference and conflict in creative transformation. Special attention should, therefore, be paid to tensions or strife between different ways of connecting, controlling and framing issues. Assemblages are continuously subject to change as relationships fold and unfold, compose and decompose in the play of internal antagonisms and agonisms. Conflicts tend to arise over the relations which control framing, and also about which entities and issues are included in the connections (which are present) and which are excluded (absent). The assemblage is thus not independent of and does not precede the traced relationships: "it is nothing other than the occurrence of these relationships" (Eriksson, 2005, p. 601).

In order to trace relationships, I introduce three main sets of 'variables': Foucault's (1980, 1984) notion of the *dispositif* and its elements of power, knowledge and subjectivity, analysed, in particular, not only through discourse, but also through materiality or visibility; together with Deleuze and Guattari's, two axes of materiality/expressivity and territorialisation (for more detail, see Hillier, 2010).

6.3.2 Mapping

Deleuze and Guattari (1987, p. 12) urge us to "make a map, not a tracing". What distinguishes a map from a tracing is that a map is oriented towards experimentation. Deleuzoguattarian maps are concerned with creative potential. Creative mapping of connections and potentialities pays attention, therefore, not only to affect and trajectories of future becomings, but also to the already-delineated tracings of *dispositifs*, representations and signification: the beliefs and habits which express actants' desires.

Mapping in strategic spatial planning practice would entail attempting to select and to facilitate potentially 'good' encounters and to avoid 'bad' ones. This is a pragmatic exercise in which strategic planners would attempt to map relational forces and their ethologies of potential connections, conjunctions and disjunctions; their possible trajectories, bifurcations and mutations. In other words, strategic planners would attempt to diagnose what might become (Bergen, 2006, p. 109). This is, of course, impossible. Nothing eventuates precisely as anticipated. "Becoming is directional rather than intentional" (Massumi, 1992, p. 95). The aleatory (or chance) is often a powerful force.

So what might strategic planners do? Mapping (strategic navigation) is not a process of standing back and describing, but of entering the relations between elements and 'tweaking' (Massumi, 2002, p. 207) as many as possible in order to get a sense of what may emerge: "pragmatic tweaking: a hands-on experimentation in contextual connectivity" (Massumi, 2002, p. 243).

Strategic navigation offers a process whereby strategic planners may be able to identify and engage the virtual events immanent within their worlds. Such a process would involve 'teasing out the proliferating inter-connections' between elements entering into 'the play of virtual differences' (Bogue, 2007, pp. 9–10), experimenting with them and mapping potential tensions and conflicts. As Bogue (2007, p. 10) describes, this is "both a process of exploring and hence constructing connections among differences, and a process of undoing connections in an effort to form new ones": a practical 'thinking otherwise' in an experimental activation of the potential of the virtual. A 'what might happen if . . .?' approach, not so much to predict, but to be alert to as-yet unknown potentialities (Deleuze, 1988, pp. 1–2).

6.4 Non-conclusion

> There is too much out there: nothing has to be there, so many things that can be (May, 2005, p. 62).

The future is outside control, conceptually and behaviourally. "There is no transcendence guiding the present, giving form to a particular future that has to happen this way and not others" (May, 2005, p. 63). Spatial planning is a kind of creative agonistic between presence and absence, manifest and latent and the general and the particular. It is about learning something new and providing the opportunities for the emergence of 'people-to-come' and the 'not-yet', not pre-determined or pre-identified by a 'rational space or an adequate place' (Rajchman, 1998, p. 31). I regard planning as strategic navigation: speculative and creative, yet structured, experimentation in the spatial – Deleuzoguattarian fabulation.

In fabulation the subject (geographical area, people or policy area) is constructed as a site of oscillation between reality and the virtual, which intersect in a state of transformation or a becoming. Deleuze (1989) suggests that becoming should be expressed as a collective will; "a collaborative process of invention" (Bogue, 2006, p. 212). Becoming and its fabulation belong "to a people, to a community, to a minority whose expression they practice and set free" (Deleuze, 1989, p. 153). Fabulation, then, is the discourse of minorities: a collective but non-unifying articulation of differences which "harmonizes difference through interpenetration" (Follet, 1998, p. 34) to produce what Follet terms "common thought" (1998, p. 34).[3] Common thought is the outcome of a process in which differences are neither suppressed nor superseded, but in which they are integrated into a 'whole'. Moreover, "the strength of this whole lies precisely in the preservation and interrelation of difference" (Holland, 2006, p. 197), a form of agonistic pragmatism.

Strategic spatial plans are inevitably fables, political fictions or 'visions' (Deleuze, 1997), which 'speak the possibles' (after Boundas, 2006, p. 24). To think or fabulate a field of possibility means "arranging it according to some concept (...) thereby constructing a temporary and virtual arrangement according to causal, logical and temporal relations. Such thinking is always a response to some particular set of circumstances" (Stagoll, 2005, p. 205). For instance, Deleuze and Guattari (1987, p. 251) suggest that tentative criteria may be developed from practical experience and judgement in order to anticipate potential becomings. However, the range of potentialities that can become possibilities that can become actualised is constrained by "an ordering and filtering system" (Due, 2007, p. 9) which imposes a determinate structure on the socio-economic-political processes with which thinking, foresighting and fabulation performances are entangled. This 'ordering and filtering system' (i.e., institutional structures) is Deleuzoguattarian, and it may block creative transformation. Through mechanisms of organisation, signification and subjectification,[4] powerful entities with a desire for constancy and stability can dogmatically halt lines of flight and block fluidity (May, 2005).

Improvisation is important in forms of strategic planning practices. These practices would be performative rather than strictly normative/prescriptive, concerned with 'journeys rather than destinations' and with establishing the conditions for the development of alternatives. I advocate fabulation, potentially an inclusive, democratic 'what might happen if...?' approach which allows disparate points of view to co-exist; which has a concern for indeterminate essences rather than ordered ones; for emergent properties rather than fixed ones; and for intuition and uncertainty, multiplicity and complexity rather than systematic predictabilities. Strategic spatial planning as strategic navigation is a performance of risk-taking, of not being in total control and of transcending the technicalities of planning practice to create an "open reading frame for the emergence of unprecedented events" (Rheinberger, 1997, p. 31).

Notes

1. I am indebted to Catherine Wilkinson for introducing me to the work of Richard Hames (especially 2007), who uses the term 'strategic navigation'. Hames is clearly thinking along similar lines to myself, but he also puts theory into practice in the field of organisational management (http://www.richardhames.com).
2. Deleuze and Guattari confusingly use different terms for the planes in their 1987 and 1994 work. In 1994, the plane of consistency (1987) is referred to as the plane of immanence, while the plane of organisation (1987) becomes the plane of transcendence.
3. This concept of common thought includes more similarities with Mouffe's (2005) notion of 'conflictual consensus' than with a Habermasian consensus produced through communicative action: "the essential feature of common thought is not that it is held in common, but that it has been produced in common" (Follet, 1998, p. 34).
4. Often referred to as subjection: subjectification involves the subjective identification of others and the accordance of a subject-position to others.

Acknowledgements Material in this chapter presents an abridged and amended version of Hillier (2010).

References

Bergen, V. (2006). La Politique Comme Posture de Tout Agencement. In M. Antonioli, P. A. Chardel, & H. Regnauld (Eds.), *Gilles Deleuze, Félix Guattari et le politique* (pp. 103–114). Paris: Éditions du Sandre.

Bogue, R. (2006). Fabulation, narration and the people to come. In C. Boundas (Ed.), *Deleuze and philosophy* (pp. 202–223). Edinburgh: Edinburgh University Press.

Bogue, R. (2007). *Deleuze's way: Essays in transverse ethics and aesthetics*. Aldershot: Ashgate.

Bonta, M., & Protevi, J. (2004). *Deleuze and geophilosophy: A guide and glossary*. Edinburgh: Edinburgh University Press.

Bosteels, B. (2001). From text to territory: Félix Guattari's cartographies of the unconscious. In G. Genosko (Ed.), *Deleuze and Guattari: Critical assessments of leading philosophers* (Vol. 2, pp. 881–910). New York: Routledge.

Boundas, C. (2005). The art of begetting monsters: The Unnatural Nuptials of Deleuze and Kant. In S. Daniel (Ed.), *Current continental theory and modern philosophy* (pp. 254–279). Evanston, IL: Northwestern University Press.

Boundas, C. (2006). What difference does Deleuze's difference make? In C. Boundas (Ed.), *Deleuze and philosophy* (pp. 3–28). Edinburgh: Edinburgh University Press.

Colebrook, C. (2002). *Understanding Deleuze*. Sydney: Allen and Unwin.

Cooper, R. (2005). Relationality. *Organization Studies, 26*(11), 1689–1710.

Deleuze, G. (1988) *Nouveau Millénaire*. Défis littéraire, n. 257. Accessed September 29, 2006, from http://www.france-mail-forum.de/index2b.html#Deleuze

Deleuze, G. (1989) [1985]. *Cinema 2: The time-image* (H. Tomlinson & R. Galeta, Trans.). Minneapolis, MN: University of Minnesota Press.

Deleuze, G. (1990) [1969]. *The logic of sense* (M. Lester & C. Stivale, Trans.). London: Athlone.

Deleuze, G. (1997). Desire and pleasure. In A. Davidson (Ed.), *Foucault and his interlocutors* (D. W. Smit, Trans.) (pp. 183–192). Chicago: University of Chicago Press.

Deleuze, G., & Guattari, F. (1984) [1972]. *Anti-Oedipus: Capitalism and schizophrenia* (R. Hurley, M. Seem, & H. Lane, Trans.). London: Athlone Press.

Deleuze, G., & Guattari, F. (1987) [1980]. *A thousand plateaus: Capitalism and schizophrenia* (B. Massumi, Trans.). London: Athlone Press.

Deleuze, G., & Guattari, F. (1994) [1991]. *What is philosophy?* (H. Tomlinson & G. Burchill, Trans.). London: Verso.

Deleuze, G., & Parnet, C. (2002) [1977]. *Dialogues II* (H. B. Habberjam, Trans.). New York: Continuum.

Due, R. (2007). *Deleuze*. Cambridge: Polity Press.

Eriksson, K. (2005). *Foucault, Deleuze, and the ontology of networks. The European Legacy*, *10*(6), 595–610.

Follet, M. P. (1998). *The new state: Group organization and the solution of popular government*. Pennsylvania, PA: Pennsylvania State University Press.

Foucault, M. (1980). The eye of power. In C. Gordon (Ed.), *Power/knowledge: Selected interviews and other writings 1972–1977* (C. Gordon, L. Marshall, J. Mepham, & K. Soper, Trans.) (pp. 146–165). New York: Pantheon.

Foucault, M. (1984). *Polemics, politics and problematizations. Interview with Paul Rabinow* (L. Davis, Trans.). Accessed February 2, 2007, from http://foucault.info/foucault/interview.html

Guattari, F. (1986). *Les Années d'Hiver 1980–1985*. Paris: Bernard Barrault.

Hames, R. D. (2007). *The five literacies of global leadership*. San Francisco: Jossey-Bass.

Hillier, J. (2007). *Stretching beyond the Horizon: A multiplanar theory of spatial planning and governance*. Aldershot: Ashgate.

Hillier, J. (2010). Poststructural complexity: Strategic navigation in an ocean of theoretical and practice stories. In J. Hillier & P. Healey (Eds.), *Conceptual challenges for planning theory*. Aldershot: Ashgate (forthcoming).

Holland, E. (2006). Nomad citizenship and global democracy. In M. Fuglsang & B. M. Sørensen (Eds.), *Deleuze and the social* (pp. 191–206). Edinburgh: Edinburgh University Press.

Kaufman, E. (1998). Introduction. In E. Kaufman & K. J. Heller (Eds.), *Deleuze and Guattari: New mappings in politics, philosophy and culture* (pp. 3–19). Minneapolis, MN: University of Minnesota Press.

Lepage, R. (2003). Interview with Aleksandar Dundjerovic. In A. Dundjerovic (Ed.), *The cinema of Robert Lepage: The poetics of memory* (pp. 147–157). London: Wallflower Press.

Massumi, B. (1992). *A user's guide to 'capitalism and schizophrenia': Deviations from Deleuze and Guattari*. Cambridge: MIT Press.

Massumi, B. (2002). *Parables for the virtual*. Durham: Duke University Press.

May, T. (2005). *Gilles Deleuze: An introduction*. Cambridge: Cambridge University Press.

Mouffe, C. (2005). *On the political*. London: Routledge.

O'Sullivan, S. (2006). *Art encounters Deleuze and Guattari*. Basingstoke: Palgrave Macmillan.

Rajchman, J. (1998). *Constructions*. Cambridge: MIT Press.

Rheinberger, H. J. (1997). *Towards a history of epistemic things: Synthesising proteins in the test tube*. Stanford, CA: Stanford University Press.

Rushdie, S. (1990). *Haroun and the sea of stories*. London: Granta.

Skott-Myhre, H. (2005). Towards a minoritarian psychology of immanence and a psychotherapy of flight: Political meditations on the society of control. *Parallax*, *11*(2), 44–59.

Stagoll, C. (2005). Event. In A. Parr (Ed.), *The Deleuze dictionary* (pp. 87–89). Edinburgh: Edinburgh University Press.

Thrift, N. (1996). *Spatial formations*. London: Sage.

Venn, C. (2006). Cultural theory, biopolitics, and the question of power. *Theory, Culture and Society*, *24*(3), 111–124.

Part II
Exploring Phenomena

At one time, practice was considered an application of theory, a consequence; at other times, it had an opposite sense and it was thought to inspire theory, to be indispensable for the creation of future theoretical forms. In any event, their relationship was understood in terms of a process of totalisation. For us, however, the question is seen in a different light. The relationships between theory and practice are far more partial and fragmentary. On one side, a theory is always local and related to a limited field, and it is applied in another sphere, more or less distant from it. The relationship which holds in the application of a theory is never one of resemblance. Moreover, from the moment a theory moves into its proper domain, it begins to encounter obstacles, walls and blockages which require its relay by another type of discourse (it is through this other discourse that it eventually passes to a different domain). Practice is a set of relays from one theoretical point to another, and theory is a relay from one practice to another. No theory can develop without eventually encountering a wall, and practice is necessary for piercing this wall.

Gilles, D. (1980). Intellectuals and power: A conversation between Michel Foucault and Gilles Deleuze. In C. Gordon (Ed.), *Power/knowledge: Selected interviews and other writings, 1972–1977* (p. 206). Brighton: Harvester.

Chapter 7
Spatial Planning, Urban Policy and the Search for Integration: The Example of a Medium-Sized City

Rob Atkinson

7.1 Introduction

Across Europe there has been considerable discussion and debate in academic and policy circles over the role of cities in economic, social and environmental development. Compared to even 20 years ago, cities are now viewed in a much more positive light; rather than being seen as a source of problems, they are now widely seen as the 'motors of economic development' in the European, national and regional economy (CEC, 1997, 1998a, 2005, 2007a; Kelly, 2006; Core Cities Working Group, 2004; ODPM, 2006a, 2006b, 2006c). The associated notion of urban competitiveness has gained considerable prominence, not least in relation to the knowledge economy.

Linked to this is the recognition that 'quality of life and place' also plays an important role in the attraction and retention of workers central to that economy (Florida, 2000, 2002). At the same time, at least in terms of rhetoric, there has also been a growing recognition that urban economic development must be part of and integrated with policies that address social cohesion and environmental problems. All of these policy areas are brought together under the rubric of sustainable development, which requires a long-term, strategic and integrated approach to thinking, policy and action. Increasingly, spatial planning is being seen as one of the key ways to develop a strategic and integrated approach, create more spatially targeted and effective initiatives and thereby make better use of limited resources.

The main focus of the chapter is on attempts to facilitate integrated urban development through the spatial planning approach currently being developed in the United Kingdom and elsewhere in Europe. In the first part of the chapter, I will discuss some of the wider European debates, before going on to discuss the recent development of spatial planning in the United Kingdom, how it has become articulated with a new regional agenda, with a focus on the South West of England. Finally, I will turn to Plymouth and the regeneration of a particular neighbourhood in the city. By adopting this approach, the chapter will illustrate and discuss

R. Atkinson (✉)
Cities Research Centre, University of the West of England, Bristol, BS16 1QY, UK
e-mail: rob.atkinson@uwe.ac.uk

M. Cerreta et al. (eds.), *Making Strategies in Spatial Planning*,
Urban and Landscape Perspectives 9, DOI 10.1007/978-90-481-3106-8_7,

some of the overarching issues pertinent to the new spatial planning approach from the European, through the national, to the regional, the city and finally the neighbourhood level.

7.2 Spatial Planning and Urban Policy at the European Level

In this section, I will briefly discuss some of the key issues/debates at European level, which have helped frame (Faludi, 2003a) the debates within the United Kingdom, and spatial planning,[1] mainly urban regeneration policy,[2] in the South West of England and Plymouth. Within the European Union (EU), spatial planning is synonymous with the *European Spatial Development Perspective* (ESDP), embodied in a non-binding inter-governmental document (CSD, 1999), agreed upon between Member States in 1999. The indeterminate status of this document and the lack of any specific legal competence to justify actions of the European Community in this sphere mean that the ESDP has had something of a chequered history which often makes it difficult to trace direct relationships between it and particular policies and outcomes. Nevertheless, one should not underestimate its impact at European and national levels through its influence on the Structural Funds and particular initiatives such as INTERREG and ESPON. Given the extensive coverage of the ESDP's evolution (Williams, 1996, 1999, 2000; Faludi, 1997a, 1997b, 2003a, 2003b, 2005), I do not intend to discuss it in detail, merely to note the main points relevant to this chapter. In particular, I would highlight three basic goals that the ESDP identifies:

1. economic and social cohesion;
2. sustainable development;
3. balanced competitiveness of the European territory (CSD, 1999, p. 10).

It is argued that these goals need to be "(. . .) pursued in combination, with attention also being paid to how they interact" (*ibid*, p. 11), and that this should take the form of balanced and polycentric development within a framework of competition and cooperation.

The ESDP also stresses the importance of integration and coordination of the activities of the EU, Member States, regions and localities. If the new challenges facing the EU are not to lead to greater divisions – the notion of vertical, horizontal and territorial integration is central to the ESDP and its vision of Europe.

The elaboration of an 'EU urban policy' was initially promoted by a series of documents published by the Commission of the European Communities during the 1990s (CEC, 1997, 1998a) and led to a greater focus on urban issues at EU level. Indeed, since 1998, the majority of EU Presidencies have had an urban theme as part of their work programme and have held urban fora/conferences. This has been both a strength and a weakness. A strength in the sense that it has helped raise the profile of urban issues and keep them on the European agenda, but a weakness in that even when successive Presidencies have had an urban theme, each has had a somewhat different focus and emphasis, and there has been no real sense of continuity between Presidencies. As a result, there has been a sense of 'policy drift'.[3]

Moreover, the importance of cities has been enhanced by the increased acceptance, amongst politicians and policy-makers, that cities are Europe's 'motors of economic development/growth'; this function has become closely associated with the Lisbon–Gothenburg Strategy (see for instance CEC, 2005, 2006a, 2006b, 2007b, 2007c), thereby strengthening the position of urban areas in the thinking of many key actors at all levels, from European to local. The issue of an EU urban policy has been discussed elsewhere (Atkinson, 2001, 2002, 2006; Parkinson, 2005), however the key, interdependent, objectives are:

- strengthening economic prosperity and employment in towns and cities;
- promoting equality, social inclusion and regeneration in urban areas;
- protecting and improving the urban environment, towards local and global sustainability;
- contributing to good governance and local empowerment (CEC 1998a, pp. 5–6).

As with the ESDP, there is a strong emphasis on the need to ensure that actions taken at EU, national, regional and local levels are vertically, horizontally and territorially integrated. Furthermore, there has been a growing recognition that the EU's sectoral policies have important impacts on urban areas and their development, and that these policies should take into account their 'spatial impact' and 'urban dimension' (CEC, 1998b, 2007b, 2007c).

Attempts have been made to bring together the urban and spatial agenda within the framework of the EU (e.g., the *Lille Agenda*) (CSD, 2000), but they have met with limited success. However, perhaps more has been achieved by linking the concept of polycentricity with the increasingly popular concept of the city region. The renewed popularity of the latter reflects the role increasingly attributed to the city region in the development of a 'knowledge-based' (regional) economy and the apparent need to ensure that a certain quality of life is available in order to attract and retain key knowledge workers (Florida, 2000, 2002).

By bringing together the city region and polycentricity, it is possible to conceive how a 'core city' may symbiotically exist within an interrelated network of smaller towns and cities (and rural areas), in which there is a complementary distribution of functions that support and/or provide a range of key facilities and services essential to economic development within the polycentric city region, and thereby enhance the competitive position of the region.

Finally, we need to highlight two other issues that have figured prominently in the spatial and urban agendas at both EU and national levels – governance and stakeholder engagement. With regard to the focus of this chapter, we simply need to note the importance that has been placed on these two issues (Atkinson, 2002) in terms of both spatial planning and urban policy.

There is a general recognition that 'good governance' is essential to both activities and that there is a need to improve urban and regional governance to both facilitate and integrate, in institutional and policy terms, the developments outlined above, and to ensure that citizens (and other stakeholders) are actively engaged in them from an early stage in the policy process. The emphasis on stakeholders not only relates in part to the need to enhance the legitimacy of institutions and policy and its outcomes but also reflects a belief that stakeholders' involvement improves the effectiveness of policy interventions.

7.3 Spatial Planning and Urban Policy in the United Kingdom

In the United Kingdom, until recently, the emphasis has been on the more traditional forms of land use planning introduced by the *Town and Country Planning Act* (1947); the more expansive and ambitious spatial planning discourse has been less influential in the discourse and practice of planning in the United Kingdom.[4] However, over the last decade, as the ideas associated with spatial planning have been debated and elaborated at European level, they have become more widely disseminated and accepted among academics, professionals, politicians and policy-makers.

As a result, in the United Kingdom, this spatial planning discourse has begun to exercise some influence over planning policy and practice (Shaw & Sykes, 2003, 2005, 2006, 2007) and, in particular, helped frame government policy and how planners see their role and practices. In addition, it has been gradually brought into contact with urban policy discourse and practice as part of an attempt to create a more integrated approach to urban and regional development. Spatial planning also complements New Labour's emphasis on joined-up thinking, policy and action and the attempt to stimulate economic growth and competitiveness, whilst simultaneously addressing issues of social exclusion, social segregation and inter-regional disparities.

What is also interesting is the new emphasis given to the regional dimension and regional planning since 1997. It is important to recognise that traditionally 'the region' has had a rather weak presence within the United Kingdom, particularly in England; whilst 'regions' have existed for many years, they have largely done so for central administrative convenience. As a result they lacked any genuine material reality in terms of institutional/organisational forms, and those regional bodies that did exist tended to have little or no powers and responsibilities.

Prior to the 1997 election, the Labour Party had begun to investigate the possibilities for greater regional decentralisation and autonomy within England (see the RPC Report, 1996). Following on from Labour's election victory in 1997, it was decided that eight Regional Development Agencies (RDAs) be set up.

The RDAs set up in 1999 primarily had an economic focus with a remit to improve the competitiveness of the English regions and reduce disparities between the economically weakest regions, located in the North and South West, and the South East of England. Each RDA is responsible for drawing up the Regional Economic Strategy (RES) and addressing a range of other issues. Currently, RDAs are shadowed, and subject to a degree of scrutiny, by a non-elected Regional Assembly (RA), which is made up of representatives from various regional stakeholders.[5] Gradually, the RAs have taken on a range of other roles including preparation of the Regional Spatial Strategy (RSS). As part of these developments, there has also been a greater emphasis on 'joining-up' economic, transport, environmental and social (including urban) policies at the regional and sub-regional level. The RES and RSS are intended to be complementary documents that support each other. Nevertheless, a degree of uncertainty remains over how the RES and RSS relate to one another and which has precedence.[6] Informing this process is a 'vision' for the long-term development of each region and a strategy to implement that vision over a 20-year period.

These changes have given regional planning a new prominence (Marshall, 2004), although whether this amounts to real devolution and a further 'hollowing out' of the central state is open to debate. As Marshall (2004) argues: "The central government remains the dominant policy-maker, even though regions and non-state agents are allowed some space to think about their policy alternatives. The form is different, but the power shift is actually (. . .) from more shared governance arrangements (. . .) to greater central government control" (p. 468).

It is in this context that spatial planning has assumed an increased, and some would argue, central role in the development and realisation of 'regional visions'. Governments' endorsement of spatial planning was signalled in Planning Policy Guidance (PPG) 11 (DETR, 2000) and was part of a wider shake up of the planning system (for more detail see Nadin, 2006, 2007, 2003, 2005; Marshall, 2004). In particular, this approach was designed to deliver the government's economic agenda, sustainable development agenda and its Sustainable Communities strategy (ODPM, 2003, 2005a) at the regional and sub-regional level.

Spatial planning was defined in the following terms in Planning Policy Statement (PPS) 1: "The new system of regional spatial strategies and local development documents should take a spatial planning approach. Spatial planning goes beyond traditional land use planning to bring together and integrate policies for the development and use of land with other policies and programmes which influence the nature of places and how they can function" (ODPM, 2005b, p. 12).

PPS 11 on RSS (ODPM, 2004a) further elaborated the role and status of spatial planning, making it clear that it should address both regional and sub-regional issues in an integrated manner. A sub-regional approach should include: "(. . .) functional relationships between settlements (. . .) within the area affected by the same strategic planning issue or issues which may well differ from administrative boundaries. This could include consideration, for example, of how the strategic planning system can assist not only in creating and sustaining the economic competitiveness of a city or a cluster of towns but in spreading the benefits of a prosperous city to the wider region (the concept of a 'city-region'[7])" (*ibid*, p. 4).

The new planning system also included Local Development Frameworks (LDFs) elaborated at local authority level but consistent with and framed by regional strategies. The role of LDFs is set out in PPS 12 (ODPM, 2004b); this emphasises the requirement for local authorities to adopt a spatial planning approach and the need to ensure that the LDF "take account of the principles and characteristics of other relevant strategies and programmes (. . .)" (*ibid*, p. 4). At the heart of the LDF is a core strategy that sets "(. . .) out the key elements of the planning framework for the area. . . a spatial vision and strategic objectives (. . .) a spatial strategy; core policies; and a monitoring and implementation framework with clear objectives for achieving delivery" (*ibid*, p. 7). Within this framework, more detailed Action Area Plans should be developed that "(. . .) provide the planning framework for areas where significant change or conservation is needed" (*ibid*, p. 9). The emphasis is very much on implementation/delivery – the LDF objectives should include:

– planning growth areas;
– stimulating regeneration;
– focusing on the delivery of area-based regeneration initiatives (*ibid*, p. 9).

The intention is that local planning bodies work closely with Local Strategic Partnerships (LSPs) and take into account Communities' Strategies (CSs) (on LSPs and CSs see Atkinson, 2007). The emphasis is, once again, on horizontal, vertical and territorial integration and on the need to develop appropriate structures of governance. In the development of all of these strategies and frameworks, there is a clear intention that relevant stakeholders should be involved in the planning process from an early stage. As part of its guidance, government has issued lengthy lists of stakeholders who should be involved. Although given the accompanying indicative timetables outlined for this overall process concerns remain over the capacity of many of the stakeholders identified to participate in an effective and timely manner.

Clearly, these developments reflect the influence of wider European debates on spatial planning, although as several authors have noted (Nadin, 2007; Shaw & Sykes, 2005), post-2004 explicit references to the ESDP seem to have declined somewhat. Nevertheless, ideas and 'ways of seeing' associated with the ESDP continue to exercise an influence, even if they are often used in a highly selective manner to support particular strategies and policies by those involved in spatial planning. However, the degree to which a genuinely integrated and coordinated regional strategy, along with associated policies and delivery mechanisms, has been developed remains questionable (Treasury, 2007).

7.4 Spatial Planning in the South West of England

This section outlines some of the key issues in the South West and describes the overall strategy developed to address them.

As can clearly be seen from the map (Fig. 7.1), the South West is a somewhat peripheral region spread over a considerable distance, this peripherality is emphasised by poor intra-regional transport connections and the lack of good links from much of the region to the rest of the United Kingdom and Europe. In economic terms, the region's economy "(. . .) is shifting in the direction of a knowledge economy. . . although. . . growth in this sector is low (. . .) [and it] (. . .) seems to be public sector driven (. . .)" (SWRA, 2004a, p. 2).

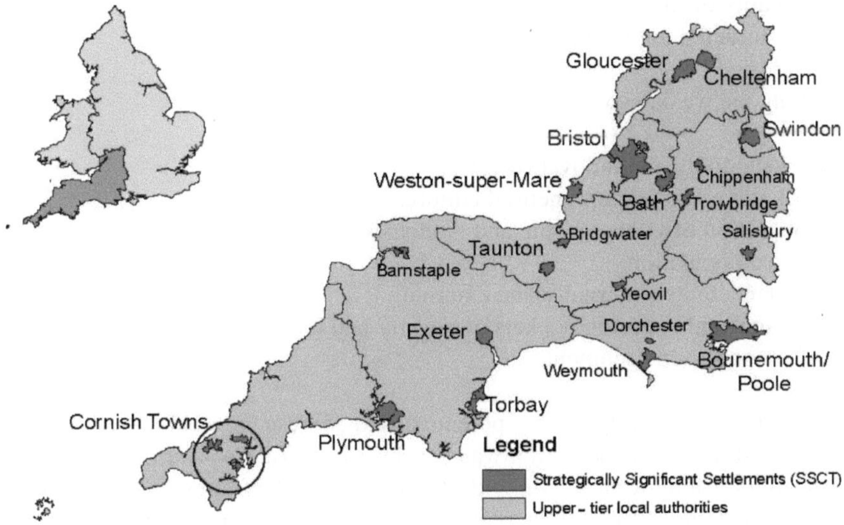

Fig. 7.1 The map of the South West

Moreover, there are weak intra-regional relationships. The Draft Regional Spatial Strategy (SWRA, 2006, p. 20) notes that "The eastern part of the region, particularly Swindon and South East Dorset and increasingly Gloucestershire and the West of England, are now well within the functional 'City Region' of London." Furthermore, the region has a dispersed population of around 5 million (SWRA, 2006) and is the most rural of the English regions (SWRDA, 2002). Approximately 35% of the population live in settlements of less than 10,000 people and it has a higher proportion of very small villages than any other English region (SWRA, 2006, p. 2). The urban areas are "(. . .) comparatively small in the main ranging from Taunton with a population of just over 50,000 people to Bristol with just over 400,000. The [11] PUAs [Principal Urban Areas] accommodate 46% of the region's population, and have 55% of the region's jobs (rising to 66% within the urban travel to work areas). However, these proportions are lower than the national average reflecting the importance of smaller towns and rural economies (. . .)" (SWRDA, 2002, p. 3).

This fragmentation is further exacerbated by any clear sense of regional identity[8] that has prevented the region from identifying and articulating a coherent set of regional priorities and needs on the national level.

There is a general agreement that the South West has a rather complex mixture of 'strengths and weaknesses' (SWRA, 2004a, 2006); among the former are:

– growing economy;
– high employment rate;
– expanding population;
– attractive natural environment;

– high quality of life;
– people attracted to the region both to tour and to live.

Among the weaknesses are:

– low productivity and innovation rates;
– lack of a dynamic entrepreneurial culture;
– out-migration of young people and graduates;
– dispersed population;
– poor connectivity (internally and externally);
– entrenched and enduring pockets of poverty and inequality;
– pressure on the environment.

In order to both build on these potentials and address the weaknesses, an overall vision and strategy for the region was developed in *Just Connect* (SWRA, 2004b), which offers a long-term vision and "(...) a clear set of aims and objectives for joining up and integrating all the region's strategies" (*ibid*, p. 1). Spatial planning is seen as central to the region's strategy as it will "(...) allow for strategic responses across sub-regions where necessary to complement existing activity" (SWRDA, 2004c, p. 4). Thus a series of nested regional and sub-regional strategies have been developed to realise the aims of *Just Connect.* One of these strategies is focussed on the city of Plymouth.

In the geographical and spatial analyses produced to inform this vision and associated strategies (SWRDA, 2002, 2004c), urban areas are assigned a key role. The 11 PUAs are central to the region's development, and the majority of the region's future development is to be focussed on these areas, with Plymouth designated as one of the four sub-regional centres (SWRDA, 2002). These 'core cities' are seen as crucial to the regions' competitiveness; they increase productivity and enable the development of a knowledge economy. Nevertheless, it is these cities, particularly Bristol and Plymouth, which contain some of the regions' most deeply entrenched and persistent problems of poverty and deprivation, and addressing these remains a key issue for the region.

7.5 Spatial Planning and Urban Regeneration in Plymouth

Plymouth is a medium-sized city.[9] It is the second largest city in the South West and has a population of approximately 240,000, although over the last 30 years, the city has been losing population; for instance, between 1991 and 2001, it lost 1.1% of its population (City of Plymouth, 2004, p. 10). To this extent, the city deviates from the majority of similarly categorised cities across the EU.

In terms of its socio-economic profile, the city displays the usual variations found in any British city, containing areas that are among the most affluent in the South West and several that are in the worst 10% nationally.[10] In the South West the local authority of Plymouth is ranked third for multiple deprivation.

Some brief statistics provide an overview of the city's economic/employment structure (City of Plymouth, 2004; all data are for 2001 unless otherwise stated):

- distribution, hotels and restaurants – 19.9% of employment (GB 24.1%);
- public administration, education and health – 29.3% (GB 24.1%);
- manufacturing – 14.8% (GB 14%);
- banking, finance and insurance – 16% (GB 19.5%);
- armed forces – 6.6% (GB 0.7%);
- unemployment in 2003 – 2.3% (1.5% in the South West and 2.4% in England);
- self-employment – 8.8% (14.9% in the South West and 12.4% in England);
- (people working in) managerial and professional occupations – 18.1% (24.9% in the South West and 26.5% in England);
- (employed in) sales/customer services – 11.2% (8.1% in the South West and 7.7% in England);
- (in) elementary occupations – 14.2% (12.2% in the South West and 11.8% in England);
- population aged 16–74 with a degree or higher – 13.5% (18.8% in the South West and 19.9% in England).

This paints a picture of a city with an economic structure that is not orientated towards high value-added sectors such as the knowledge economy, one that lacks high-value occupations and is over-reliant on the public sector and on low-skilled and low value-added forms of employment.

Perhaps the key point to note about the city's economy is that traditionally it has been dominated by defence related industries and associated employment. The decline of the Royal Navy Dockyard as a source of employment has had a major impact on the city over the last 30 years. In 1965, the Dockyard employed some 24,000 people, but by 1981, this had declined to 15,000 and by 1997 to 3,500, and the decline has continued to the present day. The city has low levels of entrepreneurship and self-employment, lacks a culture of innovation in business, has low rates of start up and survival amongst firms and lacks formally recognised basic and key skills. Together, these factors play a major role in explaining the high rates of unemployment and deprivation in parts of the city. Within the city, this situation has been recognised for some years. A range of initiatives have been launched through the development of a series of partnerships and they have achieved some success in their attempts to create a more varied and modern economic/employment structure.

To address these issues, and in common with the South West as a whole, Plymouth has developed its own 20-year vision for the future of Plymouth (known as the Mackay Vision – produced by MBM Arquitectes and AZ Studio) (Plymouth2020 2003). The Mackay Vision places considerable emphasis on high-quality urban design, investment in infrastructure, the development of the city's extensive waterfront and the regeneration of the city centre, along with other areas, within a strategic and integrated framework for the city as a whole. The city's long-term development strategy is embodied in a series of core strategy documents (City of Plymouth, 2004, 2007[11]). These documents contain a long-term vision for the city's development and an integrated and coordinated strategy to achieve that vision. In addition, the Core Strategy provides "(. . .) broad guidance on the scale and distribution of development and the provision of supporting infrastructure. It contains 'higher level' policies for delivering the spatial vision, guiding broad patterns of development and constraint" (City of Plymouth, 2007, p. 2).

The Core Strategy seeks to locate itself in a wider context, making vague references to the European context and more specific links to national and regional policies/strategies. Following on from the RSS, one of the key aims is for the city to become "(. . .) the economic hub of the far South-West (. . .)" (*ibid*, p. 4). Building on the Mackay Vision, redevelopment of the waterfront area and the city centre is seen as crucial to the city's strategy. This is to be accompanied by new investment in infrastructure, a more varied and modern employment structure and improvements in the quality of life of all the city's citizens. These developments are also an integral part of the proposal to address problems of multiple deprivation in the city. The Core Strategy identifies "(. . .) ten priority areas within the city because of their opportunities for change, city wide importance, or urgent need for regeneration" (*ibid*, p. 24), and for each of these areas an Area Action Plan is outlined, one of which covers the Devonport area.

In practice, Plymouth's council has found it difficult to bring together the various strands of its activities in an integrated and coordinated manner. Too many of the council's departments have been under-performing and resolving their own internal problems was a necessary first step, before addressing wider issues across the city's governance system. As a result the council provided weak strategic leadership within the city for much of the 1990s and the first half of the 2000s. More recently, following on from a series of reforms within the council, and a greater emphasis on a strategic approach, this situation appears to have improved, although the extent to which these changes have actually been translated into delivery/action remains debateable.

The city's LSP, Plymouth2020, is also charged with bringing together and integrating the activities of all of the disparate organisations and partnerships that constitute the governance system of Plymouth.[12] Among the tasks allocated to the LSP is drawing up a city-wide Neighbourhood Renewal Strategy (NRS) which, amongst other things, aims to "Bring together and co-ordinate policy development and action at neighbourhood level (. . .) [put in place] (. . .) appropriate mechanisms (. . .) to ensure effective delivery and improvements for people living in these neighbourhoods; (. . .)" (Plymouth2020 2002, p. 5). Unfortunately, Plymouth2020 lacked the powers and resources to carry out this task and has only very slowly begun to have an impact on the governance system and service delivery mechanisms in the city.[13]

Both Plymouth City Council (PCC) and Plymouth2020 have strategic leadership roles in the city and are expected to create a more integrated and strategic approach to governance and the city's problems. However, they have been allocated overlapping tasks and the division of responsibilities and powers between them is unclear; a situation that typifies the position in many other English cities. The reality is that the city council and other public service agencies control far more resources than the LSP, and the latter depends on them to 'bend' their service provision to support the LSPs strategy.

7.6 Regeneration in Devonport. The New Deal for Communities

Devonport is about 1.5 km to the west of the city centre and has a population of approximately 4,800 and is located, principally, in the St Peter ward, with part in the Keyham ward (the split is roughly 80:20 in favour of St Peter). The area exhibits high levels of multiple deprivation. The area is dominated by part of the Royal Navy Dockyard (the South Yard), much of which is used primarily for storage or lies redundant. It is this site that represents one of the key areas for redevelopment in both the area and Plymouth. Devonport is made up of three quite distinct communities. The South Yard divides the north and south of the area, making communications between the two difficult, with main roads further fragmenting the area. It is dominated by large areas of high density post-war social housing (80% social housing, 20% owner-occupied) and has been subject to a number of previous, largely unsuccessful, regeneration initiatives, including: urban renewal schemes, Estate Action Programme, Single Regeneration Budget and Housing Partnership Scheme. When the New Deal for Communities (NDC)[14] was launched in 2001, the area was also included in a Health Action Zone, Employment Zone and Education Action Zone, all of which had projects running in the area.

At the heart of the NDC process, in common with all contemporary regeneration initiatives, is a partnership board, Devonport Regeneration Community Partnership (DRCP), involving the local community, public sector organisations, private sector representatives and other relevant interests. Its role is to develop and implement a long-term strategy to renew the area. The DRCP is a key mechanism for bringing together all the different, relevant organisations whose services and actions impact on the area and its problems. This is particularly important given that the level of finance (approximately £50millon over 10 years) allocated to the initiative is miniscule compared to the problems to be dealt with. Public (and private) sector agencies involved in the partnership are expected to 'bend' their resources and service provision to support the NDC's actions. The partnership is in part a means to 'lever-in' additional resources and coordinate the actions of the NDC with other agencies.[15]

Government guidance required all NDCs to address the following through an integrated long-term strategy:

- worklessness;
- crime;
- health;
- education;
- housing and the physical environment.

In common with all NDCs, DRCP drew up a key strategy document identifying the area's problems and how it intended to address them. Among the key problems were:

- the population of the area was relatively young and the proportion of single parents particularly high – double the national average;
- in terms of housing, the owner occupied sector was particularly small and the local authority rented sector, although declining, particularly large;
- a poor physical environment;
- worklessness was a significant problem in the area in 2003, the worklessness rate was more than twice that for Plymouth and around three times the national figure;
- poor health, including a high incidence of low birth weight, and a range of other health problems (including high rates of mental illness, heart disease, drug use and smoking rates);
- people in the area lacked the type of skills and qualifications recognised in the labour market, and educational attainment was low;
- high crime rates and fear of crime.

At the core of the strategy, to address these problems, was 'breaking the poverty cycle'. Poverty cycle was seen as being at the heart of the area's problems, and addressing it constituted the key driver of change. The analysis carried out suggested that in the past, the lack of employment led to long-term unemployment, welfare dependency (combined with 'work' in the informal economy) and poverty. This fed through to young children, creating low expectations even before they entered school, which in turn produced a situation of poor attainment at all levels in education that in turn led to long-term unemployment, welfare dependency and poverty. This also led to the area being stigmatised. Breaking this cycle involves not only training people and getting them back into work but also intervening in education at an early age to raise both expectations and performance in school, which in turn will lead to increased opportunities in the labour market. It was hoped that this would create a virtuous circle. By developing supporting initiatives to address housing, health, crime and community development, an integrated and self-reinforcing series of measures were put into place over a period of several years.

As well as further developing its own delivery plan by 2004, DRCP had, after extensive consultation with the community, drawn up the Devonport Development Framework (DDF) which represents the key plan for the physical redevelopment of the area. It is important to note that the DDF is additional to the NDC framework; however, it represents a significant means of delivering physical change seen as vital to DRCPs strategy. Therefore it has a crucial role in transforming the area up to 2016 and beyond. The DDF is an extremely complex and ambitious series of interlinked projects. It relies upon the Ministry of Defence releasing land from the Navy Dockyard for development. Fortunately this has happened, if rather slowly. The DDF has involved the setting up of an additional partnership organisation, linked to the NDC, to develop and deliver it and this in turn has required a complex, and uncertain, series of financial relationships to fund the associated projects.

The DDF represents a form of neighbourhood spatial planning designed not only to physically regenerate the area but also to integrate these developments with the wider NDC process and ensure the two are mutually reinforcing. Moreover, as part of its own LDF process, PCC has produced a Devonport Area Action Plan (City of Plymouth, 2006) that is largely based on work already done by the NDC. This further legitimates the NDC's planning and incorporates it into the formal processes of local spatial planning. Nevertheless, doubts remain over the extent to which the DRCP's objectives are fully compatible with PCC's Core Strategy as the latter has a more economic focus than the former.

There is no doubt that the DRCP has made significant progress since its inception in 2001. However, no one involved in the partnership believes that by the end of its 10-year life, it will have succeed in transforming the area and achieving the objectives set out in the original delivery plan. Indeed, most of those involved talk in terms of a generation (i.e., 20–30 years), being required for real change to take place in the area. It is not unreasonable to suggest that the first 6 years of the NDC have also been a learning process for all concerned, not least in terms of how to develop an integrated strategy, put in place the required delivery mechanisms and how to work together. Even after 6 years, much remains to be done in terms of these key processes, and the actual outcomes produced often remain difficult to measure. The development of an increasingly strategic approach by the Board of DRCP and the positive involvement of its partners are important strengths for the future. While these are encouraging developments, they are not ends in themselves.

Despite the progress made, there remains a need to establish more and improved collaborative working relationships across the city and at sub-regional level in order to better support partnership development and mainstreaming and to ensure that the latter becomes firmly entrenched within agencies. In common with other English cities, Plymouth has a bewilderingly complex array of partnerships and delivery organisation, in which lines of responsibility and accountability are often unclear (Atkinson, 2003, 2005, 2007; Stewart, 2003). This situation frequently impedes action and makes it difficult to focus resources.

The extent to which a spatial planning approach has been able to help create a more vertically, horizontally and territorially integrated approach within Plymouth remains questionable. Furthermore, residents in Devonport are, quite understandably, primarily concerned with the area's development rather than that of the city as a whole and there is, by no means, a clear conception about how, if at all, they see the area fitting in to the Mackay Vision or PCC's Core Strategy, let alone regional strategies. Moreover, some of the wider implications of DRCP's achievements, developments underway and those planned for the future, have yet to become apparent, and it is quite possible that their impact could fundamentally change Devonport, including its population structure, in a way that may well impact negatively on the existing population (Atkinson, 2003).

7.7 Conclusions

As noted earlier, Faludi (2003a) argues that the 'discourse of spatial planning' has helped frame the way in which issues relevant to the organisation of space and development are understood. But we should bear in mind that it is only one influence among a number, and as I have pointed out, the 'spatial planning discourse' is protean and subject to a variety of interpretations. As Shaw and Sykes (2003, 2005) have noted, in the United Kingdom spatial planning's emphasis on integration complements New Labour's emphasis on joining-up and the New Managerialism that has come to dominate the public sector (Newman, 2001). Furthermore, as Healey (2007) has argued, the framing dimension represents only one dimension of spatial planning and its role in integration. Other dimensions include coordination between policies/strategies and bringing multiple actors together to share knowledge and experience (Healey, 2007).

In reality of course, these three dimensions tend to become intertwined, but it is useful to bear these distinctions in mind. Above all we should recognise that spatial planning can be co-opted, in a highly selective manner, to serve a range of different, even competing and conflicting, ideological, political and policy agendas.

Nevertheless, because of the perceived benefits of integration, a great deal of time and resources have been put into developing integrative strategies to bring together and focus the fragments of governance and policy systems. An over-reliance on this integrative approach brings with it the danger of 'joining-up' functioning as a 'technical/managerial fix'; this implies that deep-rooted socio-economic and political problems can be addressed (and solved) solely by technical/administrative means, without addressing the causes of those problems and thereby effectively depoliticising issues associated with distribution, inequality and deprivation. Moreover, this runs the risk of spatial planning simply becoming one among many (public) management techniques that can be used to address deficits in governance and policy as well as wider societal problems. So we need to be cautious when considering spatial planning and not limit ourselves to the 'integrationist' dimension or the political rationales of those who seek to use it.

Even if we take the 'integrationist' perspective at face value, we need to ask to what extent has the turn to spatial planning produced a more integrated and coordinated approach to urban and regional policy? At European level, for some time, it has been recognised that the degree of integration between policies is at best poor and often totally lacking, and the situation is, if anything, worse when it comes to considering and understanding the spatial (and urban) impacts of EU policies (CEC, 1998b). The problem is that despite the framing dimension of a spatial planning discourse its impact on thinking and action (through policy coordination) has been relatively limited within the Commission and even within Directorate General for Regional Policy. Nor are national governments any better in this respect.

As we have seen, with regard to the South West the RES is drawn up by the SWRDA, while the RSS is drawn up by the SWRA, and there is a degree of dissonance between the two. This is not something confined to this region; indeed, the issue has so concerned central government that the RAs in England are to be phased out from 2010, and RDAs will become responsible for a single integrated regional strategy. This change seems to reflect a concern that this division of responsibilities leads to a 'lack of alignment' between the two strategies (i.e., a policy coordination deficit and possibly also a framing deficit). The conclusion reached by a Treasury review is "(. . .) that there should be a single integrated regional strategy which sets out the economic, social and environmental objectives for each region" (Treasury, 2007, p. 92). Whilst this may sound eminently reasonable on paper, I would certainly entertain strong reservations over RDAs being the single responsible body for this new integrated regional strategy, given their overwhelming economic focus and the lack of concern, often shown, for social and environmental issues/problems. This may well amount to another example of increased central control by other means, suggesting that RAs have not been as 'on message' as central government would have wished. It seems likely that economic development will be the primary strategic driver that serves as the organising principle of regional development and integration. As a result (democratic) accountability to the region's population may be further reduced from its current, inadequate level.

At local level, the relationships among the multiple organisations that exist as part of the new architecture of local governance have produced a confused, and frequently confusing, situation, in which lines of accountability and responsibility for policy are often unclear. The situation in Plymouth is typical of this, where the plethora of delivery organisations, partnerships and strategies make it difficult to achieve any real coherence in either organisational or policy terms. This has been exacerbated by a central government, where different central departments issue incompatible targets that too often have blocked the development of an integrated approach. For example, there is a lack of an integrated and coordinated approach at central level that structures (and fragments) action at local level. Government is currently seeking to remedy this situation (Atkinson, 2007), but the situation has, in part, been created by New Labour's lack of trust of sub-national government and a politically driven obsession with micro-management. Spatial planning fits rather uncomfortably into this situation, as there has traditionally been a tendency not to think spatially among many service delivery organisations. To date, spatial planning has been unable to contribute to clarifying the confusion outlined above.

Of course, as hinted in the previous paragraph, much of the current debate around integration comes under the rubric of governance, or perhaps more correctly 'good governance'. As I have argued elsewhere (Atkinson, 2002), governance was developed as an analytical concept to help understand the changes in the structure, organisation and operation of government and governments' relations with society that had taken place within contemporary Western societies (Lefèvre, 1998). Essentially, analysts were reacting to a restructuring of the state in which the market appeared to be taking on a greater role, along with other groups in civil society (e.g., NGOs), in the delivery of services. Researchers observed that the system of government and policy was becoming more fragmented and difficult to manage, thus the need for integration. Parallel, and linked to these changes, were demands by groups in civil society for greater participation, other than through traditional party-based representative forms, in decision-making in the policy process and service delivery – much of this came under the rubric of empowerment.

Of course we must bear in mind that the degree of integration 'pre-governance' was never particularly extensive, but there is no doubt that recent developments have exacerbated the task of achieving organisational and policy integration. Nevertheless, some analysts have argued that these changes have not reduced the need for government and that "(. . .) governance may generate a need for new forms of government" (Jordan, Rudiger & Zito, 2005, p. 493) to provide goals, oversee and manage it. However, this is a very different approach to the normative connotations, frequently attached to the notion of (good) governance by politicians, policy advisors and so on. The problem is that much organisational and policy integration largely exists in documents rather than in terms of action situations that are permeated by competition and conflict. Indeed, it may well be that competition and conflicts are the dominant leitmotif, and that spatial planning is simply used to provide a veneer of integrationist rhetoric and legitimacy. Nor is the search for integration without costs in terms of time spent, resources invested in coordinating mechanisms and the attempts to bring about difficult change in the 'silo mentality' that dominates most policy systems. To date, this search has only produced limited benefits.

In my view, it would be wise to bear in mind Kunzmann's (1998) salutary warning that the search for the 'holy grail' of integration and coordination is in fact a chimera because: "(. . .) actors, ministries, institutions or agencies just do not wish to be coordinated, for whatever real or strategic reason, be it simple disagreement on goals, more subtle envy and greed, or just for power reasons" (*ibid*, p. 101).

From this perspective, we should not see spatial planning as some shining sword that will cut the Gordian knot of incoherence and magically integrate policies and actions in particular spaces and places, and somehow merge them into a nested interlocking hierarchy of policies stretching from the European, through the national, to regional, local and neighbourhood level. As we have seen in the English case, there is incoherence at all levels from the national to the local, and perhaps the best spatial planning can do is to help expose this and suggest alternative ways of addressing problems.

Notes

1. As Faludi (2003a) points out, the term increasingly used is 'spatial development policy' (*ibid*, p. 2). However, given that the term, and arguably the practice of, spatial planning are currently in vogue in the United Kingdom, I will continue to use the term in relation to the EU.
2. The idea of an EU urban policy is also highly contentious. As with spatial planning, the EU lacks a legal competence to develop and implement an urban policy. Nevertheless, the *Third Report on Economic and Social Cohesion* (CEC, 2004) explicitly points to the importance of EU 'urban policy', which amounts to a de facto acknowledgement that EU policies have important implications for the future development of cities and that there is, at the very least, an implicit EU urban policy – or what we might term, an 'urban agenda' (Atkinson, 2001). Moreover, if we take the EU 15, whilst most would claim to have an urban policy of some sort, it is not clear that all have what Parkinson, Bianchini, Dawson, Evans and Harding (1992) term an 'explicit urban policy' (overviews of 'older' Member State urban policies are

in van den Berg, Braun, & van der Meer, 1998, 2007). If we turn to new Member States, then none of them has an urban policy (van Kempen, Vermeulen, & Baan, 2005). Nevertheless, as Atkinson and Eckardt (2004, p. 63) argue "(. . .) it is possible to recognise that broadly similar [urban] issues are being addressed by a range of policies that do share certain similarities and which increasingly appear to form the 'new conventional wisdom' of urban regeneration across much of Europe." There is no doubt in my mind that the EU and its programmes, particularly through the Structural Funds and Community Initiatives such as URBAN, have played an important role in this process.

3. Since the Dutch Presidency of 2004 and the agreement of the Rotterdam *Urban Acquis*, there does seem to have been more coherence about the Urban Agenda and Presidencies. Efforts have been made to enhance and develop the agenda through the *Bristol Accord* (UK Presidency, 2005) and, more recently, under the German Presidency, through the *Leipzig Charter* (German Presidency, 2007a). Indeed, one of the accompanying explanatory notes boldly states "The Leipzig Charter will create a foundation for a new urban policy in Europe" (German, 2007b, p. 1).
4. It is important to bear in mind that the United Kingdom is made up of four countries and that, while there is a broad similarity in policies between them, there are often differences. Moreover, the devolution processes instigated by New Labour have accentuated some of these differences. The focus of this chapter is on England, so readers are advised not to assume that what is said about spatial planning and urban policy in England neatly transfers to Scotland, Wales and Northern Ireland (With regard to spatial planning many aspects of this are addressed in Tewder-Jones & Allmendinger, 2006, particularly in Part II of the book; on Scottish urban policy see Turok, 2004). For overviews of (English) Urban Policy see Imrie and Raco (2003) and Johnstone and Whitehead, 2004; on planning see Cullingworth and Nadin (2006).
5. In July 2007, as part of a review conducted by the Treasury (Treasury, 2007), it was decided that from 2010 RAs would be phased out and their scrutiny role reallocated to local authorities. At the time of writing, it is unclear how these new arrangements will operate.
6. In addition to RDAs and RAs, Government Offices for the Regions (GoRs) also exist. They represent the 'regional arm' of central departments and are responsible for overseeing a range of social policies within each region. The division of responsibilities between RDAs, RAs and GoRs is not altogether clear, nor does the division help develop a strategic approach. In my opinion RDAs and GoRs should have been merged in 1999.
7. In 2006, these developments were reinforced by the *White Paper on Local Government* (DCLG, 2006, Chapter 4) that argued the case for realigning some local authority boundaries with functional urban economic areas and stressed the need for stronger strategic (political) leadership in city-regions to meet the economic challenges that cities face (on/for these debates see Atkinson, 2007; Lloyd & Peel, 2006).
8. While accepting that any notion of regional identity is somewhat nebulous, the South West does seem to have peculiar problems that arise, in part, from administrative decisions over which areas to include in it. For instance, Dorset is classified as part of the South West, yet is more closely aligned with the South East of England. There are also inherent problems related to the history and culture of the 'region'. Cornwall, in the far south west of the region, has its own officially recognised language (although it is spoken by only a handful of people) and distinct identity, and it is more likely to compete with the neighbouring county of Devon than cooperate with it. Other problems relate to the predominantly rural nature of the region and the poor relations between the regions' main urban areas and surrounding rural areas. Perhaps the 'best' example of this is the city of Bristol and surrounding local authorities in the north east of the region which have, to say the least, a history of poor relations with little evidence of cross-boundary cooperation. Such divisions and tensions make a mockery of any notion of a regional identity and impede the development of coherent regional strategies.
9. I am using the Urban Audit definition of what constitutes a medium- (or mid-) sized city, that is, one with a population between 50,000 and 250,000. Plymouth's population is towards

the upper limit of that range. While between 70% and 80% of the EU's population lives in cities, the *State of European Cities* report (CEC, 2007a, p. 2) notes that "(...) the majority of urban population growth has occurred and is occurring in Europe's medium and small-sized cities."

10. For instance based on the Index of Multiple Deprivation 2000, the St Peter ward (in which most of the Devonport NDC is located) is among the worst 5% in England; two other wards are in the worst 10%; four are in the worst 15% and two in the worst 20%.
11. The City Growth Strategy (City of Plymouth, 2004) is described in the following terms: Plymouth City Growth Strategy is a new agenda for business and economic growth, based on a model developed in the United States. The Strategy covers the whole of the City of Plymouth. It identifies and seeks to develop business clusters – local concentrations of interconnected firms and institutions in related industries – that gain competitive advantages from being located close together (*ibid*, p. 2).
12. LSPs exist across the country and central government has charged them with rationalising and streamlining the local governance system in their areas in order to develop a 'joined-up' strategy within the local authority area and facilitate the development of policy and delivery.
13. For the first 3 or 4 years of its existence relations between the DRCP and Plymouth2020 could not be described as bad, simply because they were largely non-existent. Things have improved more recently, but this has been a slow process and even today it is doubtful if one can talk about a clear and integrated governance system in the city, or an integrated NRS that can effectively deliver its objectives. This does not reflect a lack of willingness on the part of the LSP, simply a lack of resources and powers to carry out its role, a situation that has been reproduced across England (Atkinson, 2005, 2007; Geddes, 2006).
14. There are 39 NDCs in England, each lasts for 10 years, with each area receiving around £50 million over its lifetime; in many ways they are the test-bed of Labours' urban policy. They embody key ideas about developing and implementing a long-term integrated strategy that requires close coordination with other public service providers, the private sector and the voluntary and community sectors. The lessons learnt from NDC about 'what works' and how best to deliver services are intended to be mainstreamed by other public service providers, thus they represent, in theory at least, part of a wider attempt to shake up the whole public sector and how it operates (for general discussions of overall NDC progress see Lawless, 2004, 2006).
15. It was several years before many public sector organisations realised this. Some thought that because the NDC area was receiving additional finance they could scale back their involvement, while others thought they could substitute NDC resources for their own. Initially, few of the 'outside agencies' realised that NDC (and central government) required them to rethink how they operated and find new ways of delivering services to the area (and to other deprived areas). This created considerable problems for some agencies; for instance, parts of PCC, the Local Education Authority and the Primary Care Trust were all slow to take on board these implications. Other partners such as the Police and Job Centre Plus realised the implications very quickly and began to reconfigure their service provision to the area; they were positive and enthusiastic partners almost from the beginning of the NDC. However, NDC demands for additional resources posed serious problems for these agencies; PCC's housing department pointed out that if they were to fully respond to the NDC request for additional investment in housing, it would use up their housing capital budget for several years. In addition, during the early years of the NDC relations between the partnership and public service agencies, particularly PCC, were not good as the local community blamed them for their situation and, initially, wanted to run the regeneration programme without outside interference. However, over time there has been a growing realisation by community representatives of DRCP of the need to work with these agencies, and, combined with a more proactive stance by the agencies, better working relationships and outcomes have gradually developed.

Acknowledgements My thanks to Dr Ian Smith for providing the map of the South West.

References

Atkinson, R. (2001). The emerging 'urban agenda' and the European spatial development perspective: Towards and EU urban policy? *European Planning Studies*, *9*(3), 385–406.

Atkinson, R. (2002). The white paper on European governance: Implications for urban policy. *European Planning Studies*, *10*(6), pp. 781–792.

Atkinson, R. (2003). Addressing social exclusion through community involvement in urban regeneration. In R. Imrie & M. Raco (Eds.), *Urban renaissance? urban policy, community, citizenship and rights*. Bristol: Policy Press.

Atkinson, R. (2005). New actors in the field of urban regeneration: The development of area-based initiatives and multi-sectoral partnerships. In K. R. Gupta (Ed.), *Urban development debates in the new millennium. Studies in Revisited theories and redefined praxes* (Vol. II). New Delhi: Atlantic Books.

Atkinson, R. (2006, November 30). *EU urban policy, European urban policies and the neighbourhood: An overview of concepts, programmes and strategies*. Paper presented at a European Congress Urban Renewal Strategies of European Neighbours – Examples, Munich.

Atkinson, R. (2007). Under construction. The city-region and the neighbourhood: New actors in a system of multi-level governance? In I. Smith, E. Lepine, & M. Taylor (Eds.), *Disadvantaged by where you live? Neighbourhood governance in contemporary urban policy*. Bristol: Policy Press.

Atkinson, R., & Eckardt, F. (2004). Urban policies in Europe: The development of a new conventional wisdom? In F. Eckardt & P. Kreisl (Eds.), *City images and urban regeneration*. Frankfurt am Main: Peter Lang.

CEC (Commission of the European Communities). (1997). *Towards an urban agenda in the European union*. Communication from the Commission, Commission of the European Communities, Bruxelles.

CEC (Commission of the European Communities). (1998a). *Sustainable urban development in the European Union: A framework for action*. Communication from the Commission, Commission of the European Communities, Bruxelles.

CEC (Commission of the European Communities). (1998b). *Report on community policies and spatial planning*. Working document of the Commission Services, Commission of the European Communities, Bruxelles.

CEC (Commission of the European Communities). (2004). *A new partnership for cohesion. Convergence, competitiveness, cooperation*. Third Report on Economic and Social Cohesion, Commission of the European Communities, Bruxelles.

CEC (Commission of the European Communities). (2005). *Cities and the Lisbon agenda: Assessing the performance of cities*. Commission of the European Communities, Directorate General Regional Policy, Bruxelles.

CEC (Commission of the European Communities). (2006a). The Territorial State and Perspectives of the European Union Document. Towards a Stronger European Territorial Cohesion in the Light of the Lisbon and Gothenburg Ambitions (First Draft). Commission of the European Communities, Bruxelles.

CEC (Commission of the European Communities). (2006b). *Communication from the commission. Regions for change*. Commission of the European Communities, Bruxelles.

CEC (Commission of the European Communities). (2007a). *State of European cities report*. Commission of the European Communities. Bruxelles: DG Regio.

CEC (Commission of the European Communities). (2007b). *The urban dimension in community policies for the period 2007–2013. Part 1*. Commission of the European Communities, Bruxelles.

CEC (Commission of the European Communities). (2007c). *The urban dimension in community policies for the period 2007–2013. Part 2*. Commission of the European Communities, Bruxelles.

CSD (Committee on Spatial Development). (1999). *European spatial development perspective: Towards balanced and sustainable development of the territory of the EU (ESDP)*. Luxembourg: European Communities.
CSD (Committee on Spatial Development). (2000). *Proposal for a multi-annual programme of co-operation in urban affairs within the European Union*. Report by the Committee on Spatial Development, Marseilles, 6 October, 2000.
City of Plymouth. (2004). *Core growth strategy*. Plymouth: Plymouth City Council.
City of Plymouth. (2006). *Devonport area action plan 2006–2021*. Plymouth: Plymouth City Council.
City of Plymouth. (2007). *Core strategy*. Plymouth: Plymouth City Council.
Core Cities Working Group. (2004). *Our cities are back. Competitive cities make prosperous regions and sustainable communities*. Third Report of the Core Cities Working Group, ODPM, London.
Cullingworth, J. B., & Nadin, V. (2006). *Town and country planning in the UK*. London: Routledge.
DCLG (Department of Communities and Local Government). (2006). *Stronger and prosperous communities. The local government white paper*. London: HMSO.
DETR (Department of the Environment Transport and the Regions). (2000). *Planning policy guidance note 11: Regional planning*. London: DETR.
Faludi, A. (1997a). A roving band of planners. *Built Environment*, *2*(4), 281–287.
Faludi, A. (1997b). European spatial development policy in 'Maastricht II'? *European Planning Studies*, *5*(4), 535–543.
Faludi, A. (2003a). The application of the European spatial development perspective. Introduction to the special issue. *Town Planning Review*, *74*(1), 1–9.
Faludi, A. (2003b). Unfinished business. European spatial planning in the 2000s. *Town Planning Review*, *74*(1), 121–140.
Faludi, A. (2005). Territorial cohesion: An unidentified political objective. Introduction to the special issue. *Town Planning Review*, *76*(1), 1–13.
Florida, R. (2000). *Competing in the age of talent: Quality of place and the new economy*. Report prepared for the K Mellon Foundation, Heinz Endowments and Sustainable Pittsburgh, Pittsburgh.
Florida, R. (2002). *The rise of the creative class*. New York: Basic Books.
Geddes, M. (2006). Partnership and the limits to local governance in England: Institutionalist analysis and neoliberalism. *International Journal of Urban and Regional Research*, *30*(1), 76–97.
German Presidency. (2007a). *Leipzig charter on sustainable European Cities*. Document prepared by the German Presidency of the European Union, Leipzig.
German Presidency. (2007b). *Renaissance der Städte*. Document prepared by the German Presidency of the European Union for the Informelles Ministertreffen, Leipzig.
Healey, P. (2007). Territory, integration and spatial planning. In M. Tewdwr-Jones & A. Allmendinger (Eds.), *Territory, identity and spatial planning. Spatial governance in a fragmented nation*. London: Routledge.
Imrie, R., Raco, M. (Eds.). (2003). *Urban renaissance? urban policy, community, citizenship and right*. Bristol: Policy Press.
Johnstone, C., & Whitehead, M. (Eds.). (2004). *New horizons in British urban policy*. Aldershot: Ashgate.
Jordan, A., Rudiger, K. -W. W., & Zito, A. (2005). The rise of 'new' policy instruments in comparative perspective: Has governance eclipsed government? *Political Studies*, *53*(3), 477–496.
Kelly, R. (2006). *Speech to core cities summit*. Speech by Ruth Kelly, Secretary of State for Communities and Local Government to the Core Cities Summit, Bristol.
Kunzmann, K. R. (1998). Planning for spatial equity in Europe. *International Planning Studies*, *3*(1), 101–120.
Lawless, P. (2004). Locating and explaining area-based initiatives: New deal for communities in England. *Environment and Planning C*, *22*(3), 383–399.

Lawless, P. (2006). Area-based urban interventions: Rationale and outcomes: The new deal for communities programme in England. *Urban Studies*, *43*(11), 1991–2011.

Lefèvre, C. (1998). Metropolitan government and governance in western countries: A critical review. *International Journal of Urban and Regional Research*, *22*(2), 9–25.

Lloyd, G., & Peel, D. (2006). City-regionalism: The social construction of an idea in practice. In M. Tewdwr-Jones & P. Allmendinger (Eds.), *Territory, identity and spatial planning. Spatial governance in a fragmented nation*. London: Routledge.

Marshall, T. (2004). Regional planning in England. Progress and pressures since 1997. *Town Planning Review*, *75*(4), 447–472.

Nadin, V. (2006). The role and scope of spatial planning. Literature review. In J. Baker, L. Talljard, J. Clayden, M. Baker, V. Nadin, & D. Shaw (reporters), *Spatial planning in practice: Supporting the reform of spatial planning*. London: Report prepared for the Department for Communities and Local Government.

Nadin, V. (2007). The emergence of the spatial planning approach in England. *Planning, Practice and Research*, *23*(1), 1–27.

Newman, J. (2001). *Modernising governance, new labour, policy and society*. London: Sage.

ODPM (Office of the Deputy Prime Minister). (2003). *Sustainable communities: Building for the future*. London: ODPM.

ODPM (Office of the Deputy Prime Minister). (2004a). *Planning policy statement 11: Regional spatial strategies*. London: ODPM.

ODPM (Office of the Deputy Prime Minister). (2004b). *Planning policy statement 12: Local development frameworks*. London: ODPM.

ODPM (Office of the Deputy Prime Minister). (2005a). *Sustainable communities: People, places and prosperity*. London: ODPM.

ODPM (Office of the Deputy Prime Minister). (2005b). *Planning policy statement 1: Delivering sustainable development*. London: ODPM.

ODPM (Office of the Deputy Prime Minister). (2006a). *State of the English cities* (Vol. I). London: ODPM.

ODPM (Office of the Deputy Prime Minister). (2006b). *State of the English cities* (Vol. II). London: ODPM.

ODPM (Office of the Deputy Prime Minister). (2006c). *A framework for city-regions*. London: ODPM.

Parkinson, M. (2005). Urban policy in Europe. Where have we been and where are we going? In E. Anatalovsky, J. Dangschat, & M. Parkinson (Eds.), *European metropolitan governance. Cities in Europe-Europe in the cities*. Wien: Node Research.

Parkinson, M., Bianchini, F., Dawson, J., Evans, R., & Harding, A. (1992). *Urbanization and the function of cities in the European community*. Luxembourg: Office for Official Publications of the European Communities.

Plymouth2020. (2002). *Neighbourhood renewal strategy 2002–2007*. Plymouth: Plymouth2020.

Plymouth2020. (2003). *A vision for plymouth*. Prepared by MBM Arquitectes and AZ Urban Studio, Plymouth2020, Plymouth.

UK Presidency. (2005). *Conclusions of ministerial informal meeting on sustainable communities in Europe* [known as Bristol Accord]. UK Presidency, Bristol 6–7th December 2005, ODPM, London.

RPC (Regional Policy Commission). (1996). *Renewing the regions: Strategies for regional economic development*. Sheffield: Sheffield Hallam University.

SWRA (South West Regional Assembly). (2004a). *A geographical analysis of the south west*. London: *A report for the south west regional assembly*. Prepared by the Local Futures Group.

SWRA (South West Regional Assembly). (2004b). *Just connect. An integrated regional strategy for the south west 2004–2026*. South West Regional Assembly, Taunton.

SWRA (South West Regional Assembly). (2006). *The draft regional spatial strategy for the south west 2006–2026*. Taunton: South West Regional Assembly.

SWRDA (South West of England Regional Development Agency). (2002). *South west urban economic study*. Final report. Roger Tym and Partners, Truro.

SWRDA (South West of England Regional Development Agency) . (2004c). *Spatial dynamics final report*. Bristol: Prepared by DTZ Pieda Consulting.
Shaw, D., & Sykes, O. (2003). Investigating the application of the European spatial development perspective (ESDP) to regional planning in the United Kingdom. *Town Planning Review*, *74*(1), 31–50.
Shaw, D., & Sykes, O. (2005). European spatial development policy and evolving forms of territorial mobilisation in the United Kingdom. *Planning, Practice and Research*, *20*(2), 183–199.
Stewart, M. (2003). Towards collaborative capacity. In M. Boddy (Ed.), *Urban transformation and urban governance*. Bristol: Policy Press.
Tewder-Jones, M., & Allmendinger, P. (Eds.). (2006). *Territory, identity and spatial planning. Spatial governance in a fragmented nation*. London: Routledge.
Treasury, H. M. (2007). *Review of sub-national economic development and regeneration*. London: H M Treasury.
Turok, I. (2004). Scottish urban policy: Continuity, change and uncertainty post-devolution. In C. Johnstone & M. Whitehead (Eds.), *New horizons in British urban policy: Perspectives on new labours urban renaissance*. Aldershot: Ashgate.
van den Berg, L., Braun, E., & van der Meer, J. (Eds.). (1998). *National urban policies in the European Union*. Aldershot: Ashgate.
van den Berg, L., Braun, E., & van der Meer, J. (Eds.). (2007). *National policy responses to urban challenges in Europe*. Aldershot: Ashgate.
van Kempen, R., Vermeulen, M., & Baan, A. (Eds.). (2005). *Urban issues and urban policies in the new EU Countries*. Aldershot: Ashgate.
Williams, R. H. (1996). *European Union spatial policy and planning*. London: Paul Chapman.
Williams, R. H. (1999). Constructing the European spatial development perspective: Consensus without a competence. *Regional Studies*, *33*(8), 793–797.
Williams, R. H. (2000). Constructing the European spatial development perspective. For whom? *European Planning Studies*, *8*(3), 357–365.

Chapter 8
Strategic Planning and Urban Governance: Effectiveness and Legitimacy

Panagiotis Getimis

8.1 Planning as a Political Process in the Framework of Transforming Governance

Since the 1990s, planning as a political process has changed profoundly. Planning is no longer considered a state function governed by strict hierarchies accompanied by explicit competencies. Central, regional and local governments are no longer the only stakeholders involved in the process, neither is land use regulation the only policy area affected.

Today a plethora of elected and non-elected, governmental, quasigovernmental and private sector actors and institutions from all spatial scales, voice their interests in new systems of local governance. These multi-actor and multi-level systems of local governance which are emerging to combat the lack of horizontal and vertical integration in traditional planning processes, it is believed, will help policy-making become more flexible, adaptable and holistic in approach. Resulting policies benefit from an enhanced sense of ownership and the planning process becomes more sustainable through greater participation. Special emphasis is given also to environmental issues, which have to be integrated into all sectoral policies including spatial planning.

Tewdwr-Jones (2002, p. 278) characterises planning as having undergone a "transformation from an end product into a strategic enabling of means-based activity within a much broader framework of governance" driving spatial agendas and resulting in customised policies.

In the first part of this chapter, I will examine the transition from government to governance. Considering government failure as a lack of effectiveness and legitimacy, the shift from 'government' to 'governance' is explained as a re-orientation away from 'hierarchies' towards 'heterarchies'. The main questions posed refer both to the opportunities and risks that may be derived from governance arrangements and to the prerequisites for the avoidance of governance failure. It is important for

P. Getimis (✉)
Panteion University of Political and Social Sciences, Athens 17671, Greece
e-mail: getimisp@gmail.com

M. Cerreta et al. (eds.), *Making Strategies in Spatial Planning*,
Urban and Landscape Perspectives 9, DOI 10.1007/978-90-481-3106-8_8,

policy-makers to be aware of the co-existence and complementarities of governance modes, avoiding risks and enhancing opportunities for participatory governance, thus ensuring both effectiveness and legitimacy.

The second part will look into the Europeanisation of domestic politics and the main principles of strategic planning and participation within the framework of urban governance. Europeanisation is understood not as a linear and homogeneous adaptation of the domestic institutional structures to an ideal type of norms and regulations, but as an interactive process of political and institutional changes in which territorial specificity plays a crucial role. Institutional innovation and learning processes differ from country to country and the principles of partnership and strategic planning (e.g., *European Spatial Development Perspective* – ESDP, CSD, 1999) for territorial cohesion have different outcomes in different localities.

The third part of the chapter discusses the transformation of planning and the shift from the traditional/conventional planning policies to contemporary planning policies which enable multi-level and multi-actor governance arrangements. However, this shift from 'government' to 'governance' creates not only opportunities but also risks, which planners must be aware of in their quest for strategic, collaborative and sustainable planning, in order to avoid new problems.

The last part offers an insight into strategic planning and the diversity of small- and medium-sized cities in Europe. Based on the data provided by a specific study on small- and medium-sized cities in European countries (ESPON 1.4.1, 2006) different definitions and typologies are presented, and three important aspects concerning the dilemmas and perspectives of strategic planning are highlighted: (1) principles, (2) territorial specificity and (3) alternatives.

In the end presents some conclusions regarding the transformation of planning as a political process in light of the transition from government to governance.

8.2 From 'Government' to 'Governance'

8.2.1 Government Failure: Lack of Effectiveness and Legitimacy

Haus, Heinelt and Stewart (2005) consider effectiveness and legitimacy to be the criteria for evaluating government success or failure. Effectiveness they define as the 'governing capacity' of the government to solve problems by reflecting on its options, arriving at strategies for addressing these problems and having the ability to follow these strategies in their political actions. Legitimacy refers to the acceptance, ownership and justifiability of the decision and implementation processes and the policy objectives themselves. Legitimacy is closely linked to democratic self-government and participation. The principle forms of democratic legitimation are presented in Table 8.1.

Input-legitimation through participation relates to the possibility to voice one's opinions and have these opinions considered in the formulation of policy. Throughput-legitimation means that with transparent institutions and processes,

Table 8.1 Different forms of democratic legitimation (Haus et al., 2005, p. 15)

Forms of democratic legitimation	Principle	Criteria	Phenomena of crisis
Input-legitimation	Participation	Consent	Decrease of voter turnout, etc.
Throughput-legitimation	Transparency	Accountability	Opaque institutions, etc.
Output-legitimation	Effectiveness	Problem-solving	Policy failure, etc.

understanding of the policy-making process and actor accountability are enhanced, rendering implemented policies more legitimate. Finally, output legitimacy relates first and foremost to the legitimation of policies based on the involvement of the necessary actors and the use of available information to make informed decisions.

8.2.2 Definitions and Contents: From Government to Governance

Different forms of democracy and democratic reform score differently with regard to the legitimation 'principles' in the second column of the above Table 8.1. However, none is successful in all the above forms of legitimation and effectiveness and in the resolution of market failures. Hence a discourse advocating the shift from government to governance emerges. "The literature on governance rejects the dichotomy of 'state' vs. 'market' and re-examines the interrelations between civil society, state and market, arguing that the boundaries have become blurred. The shift from 'government' to 'governance' signifies a re-orientation away from the hierarchy of the state and the institutions whose role is to promote conventional forms of political representation (party system, electoral participation, majoritarian principle), to heterarchy. In a heterarchy a highly diverse range of actors with different interests, power and histories, pursue their goals through participation in cooperative forms of action, and joint decision-making processes. In this sense, 'governance' places emphasis on the conditions enabling 'civic cooperation', formal and 'informal arrangements', 'networking and coordination of efforts' and 'alliances/coalitions' between different interest groups in concrete policy domains in a multi-level framework. These prerequisites refer to the tasks and objectives of 'mutual understanding', 'negotiation and bargaining', 'institutional capacity', 'trust' and 'social capital'" (Getimis & Georgandas, 2001, p. 2).

8.2.3 Governance Opportunities

Getimis and Kafkalas (2002, pp. 157–158) consider that the emergence of new forms of governance presents five main opportunities.

1. *Widening the forms of representation – legitimacy gains.* Given the growing crisis of the political institutions and the democratic deficit at all levels of political representation, new forms of governance, based on arguing and bargaining, broaden legitimacy through the involvement of new types of actors (e.g., committees, new bodies) and through new forms of interest intermediation. It should be mentioned however, that empirical examples suggest that in so far as the broadening of legitimacy is concerned, more often than not, the old government structures have been maintained and the new forms of governance have been simply added upon them.
2. *Broadening participation – effectiveness gains.* The new governance arrangements provide for empowerment and access to holders, with or without legal entitlements, and thus may lead to effective policy outcomes (i.e., effectiveness), which cannot be derived from conventional forms of government. The new cooperative partnerships, oriented on common tasks, go beyond legalistic rights, supporting cooperation and widening forms of participation (e.g., at the European, national and local level). However, it is not always clear whether effectiveness comes as a result of more participation per se or because participatory governance triggers the reconsideration of certain failures of command and control policies. In any case, some real progress can be detected.
3. *Continuous learning and improvement – knowledge gains.* New governance arrangements give new opportunities for permanent learning to the different actors involved, regardless of the success or failure of the policy outcome. Different actors, with different histories and power, test their knowledge, arguments and powers and learn from each other in the new forms of participation. Again, empirical examples cannot provide conclusive evidence on whether learning processes correspond to the new governance arrangements or to the combination of other factors at work.
4. *Early conflict resolution – consensus gains.* Participatory governance emerges as a means of conflict resolution. One could argue that the aim is to avoid a conflict resolution by courts. This can be achieved through early integration of specific actors with their respective interest from the beginning (i.e., in the phase of development and implementation of the policy instrument). This opportunity is linked to the rules of selection and the empowerment/disempowerment of those holders who participate in, who are excluded from or who 'opt out' of the new governance arrangements.
5. *Institutional, organisational and technological restructuring – innovation gains.* Participatory governance seems to trigger organisational restructuring, sometimes as a direct response to failures in the application of command and control policies ('hierarchies'). For example, the turn towards participatory governance can lead to institutional and organisational innovations bypassing structures dedicated to respond to top-down hierarchical decision-making. This role becomes even more important whenever it is coupled with broader societal objectives such as the pursuit of sustainability or consensus building.

8.2.4 Governance Risks

We are turning to governance as the solution (Getimis & Georgandas, 2001, p. 3) after the crisis of the welfare state (top-down), in the 1970s, the subsequent turn to market forces of the neoliberal political project of global deregulation, which peaked in the 1990s, and the market's failure to solve developmental, social and ecological problems (externalities). However, we should also anticipate and address possible governance failures.

1. *Ineffectiveness – 'eye wash effect'*. In this case all decisions are already taken, and the involvement of certain actors has a purely public relations or marketing purpose. The aim could be to obtain information for a better negotiating position or to gain knowledge about new technologies, for which the enterprise would otherwise have to pay. Thus, increased participation does not necessarily lead to the achievement of certain policy goals, such as sustainability, which may simply be ignored or added to lists of goals without intent or commitment. This effect allows the new governance structure to become an instrument of shifting responsibilities rather than committing all actors to the pursuit of specific policies (Getimis & Kafkalas, 2002, p. 169). Jessop (2002) labels this governance risk as 'noise' or a 'talking shop'.
2. *Non-accountability – transparency – legitimacy loss*. This is associated with the diffusion and probably dilution of responsibilities within ad hoc governance agreements where unequal partners participate in a policy process with an uneven distribution of costs and benefits. This dilution of responsibility makes the participants non-accountable in both political and legal terms. Non-accountability feeds the temptation to pursue targets that no actor acting on their own could support. This leads us to the increased danger of the reproduction of the uneven distribution of power among the participants, entering the process based on different forms of legitimacy and power (e.g., legal entitlement on the one hand and de facto power on the other). This should be compared with and weighed against the performance of existing government structures (Getimis & Kafkalas, 2002, p. 168).
3. *Governance overestimation*. There is an underestimation of the strengths of existing normative frameworks of hierarchies (e.g., political representation, party system and majority) and an overestimation of the potential of another value system of heterarchy (negotiation, bargaining, commitment to dialogue, networking, etc.). This leads to a shift from general rules and legal perspectives (political and civil rights) to partial rules and holder claims ('citizenship' vs. 'holdership') (Getimis & Georgandas, 2001, pp. 2–5).
4. *Compartmentalisation of policy – fragmentation – comprehensiveness loss*. Although governance arrangements reduce the general problems of democratic participation and the democratic deficit through structured participation procedures and problem-solving in concrete policy domains, this is done without reference to the broader political and socio-economic context. It is a participation and democracy 'à la carte'. Participation procedures of coalition partners

take place in fragmented policy areas (fragmentation), while cooperation and networking among actors are conceived and analysed on the basis of selective incentives and tasks (selectiveness) (Getimis & Georgandas, 2001, pp. 2–5). Inconsistencies may thus be multiplied and synergies undermined between particular policies that become apparent in their parallel pursuit within the same territory without any *ex ante*, ongoing or ex post assessment of their combined impact upon the territory (Getimis & Kafkalas, 2002, p. 169).

5. *Instrumentalisation – substantial rationality loss*. The emphasis on problem-solving and the 'effectiveness' of policy outcomes, combined with the dominance of a technocratic rational, may underestimate important aspects of political legitimacy and social justice. The danger lies in the overestimation of the internal and external functionality in the policy process and the dominance of technocratic knowledge (e.g., 'managerial' assessment of policy outcome, benchmarking, etc.), at the cost of democratic participation and the empowerment of civil society. Empirical cases support this but not in a systematic way, while countervailing tendencies have also been recorded, for example, in the combination of managerial trends with sustainability objectives (Getimis & Kafkalas, 2002, p. 169).

In order to avoid governance failure the following advice is offered.

1. *Co-existence and complementarity of coordination modes*. If all modes of economic and political coordination (government, market and governance) are prone to failure, successful policy-making may depend on the complementing of market, state and network modes of governance (Getimis & Kafkalas, 2002, p. 157) and "on the capacity to switch modes of coordination as the limits of any one mode become evident (...) [or] meta-governance" (Jessop, 2002, p. 52). The interest in governance becomes, in fact, a search for the appropriate combination of markets, hierarchies and networks that will collectively provide the steering and control capacities (Getimis & Kafkalas, 2002, p. 157).
2. *Reflexive learning*. By encouraging 'self-reflection, self-regulation and self-correction' learning will be facilitated, side-stepping the risks outlined above. The reflexive process will further facilitate the selection of the optimal mode or mix of coordination (market, government, governance) (Jessop, 2002, p. 55).
3. *Participatory governance*. It is based on the complementarity between political leadership and community involvement. Forms of participatory governance achieving a good balance and complementarity between leadership and democratic participation can enhance legitimacy and effectiveness. "Leadership may solve some of the problems related with community involvement through a participatory management of policy networks and by ensuring their public accountability. Community involvement on the other hand can bring dispersed knowledge and awareness of negative externalities in decision-making and implementation processes and can shed public light on proceedings in representative and administrative bodies" (Haus et al., 2005, p. 23).

8.2.5 Some Conclusions

Clearly, the understanding and application of local governance as a solution to the failures of alternative modes of coordination pose a number of challenges.

Against the types of risk outlined above we should think of both: (1) the possible benefits stemming from the mobilisation of many, until now underused or isolated, individual and institutional resources and (2) the achievement of consensus through deliberation and active participation with freedom of entry for an increasing percentage of the population. In order to increase the possibility of a positive outcome we have to reconsider the important aspects of democracy, participation, political legitimacy and social justice, not only in fragmented and specific policy fields, but in all policy-making frameworks (in which the state still plays a key role), and at all levels (especially at the global level, where despite the proliferation of many political and economic institutions the lack of democratic representation remains a crucial issue). It should be mentioned however, that the situation is characterised by the rather low probability of success, despite the fact that the meaning of success itself becomes conditional upon the achievement of the fragmented partial targets of each particular governance agreement (Getimis & Kafkalas, 2002, p. 169).

The inconclusive effects of governance point towards the importance of the flexible coexistence of old and new forms of government and governance including state administrative hierarchies, market-led solutions and participatory governance initiatives. This argument is equivalent to a plea for the selective re-regulation of particular stages in policy-making and policy implementation in order to achieve the optimum combination of effectiveness and legitimacy through participatory governance (Getimis & Kafkalas, 2002, p. 170).

The prudent combination of different coordination modes, including hierarchies and the market, will allow the pursuit of both effectiveness and legitimacy between which, as many have concluded, there is a trade-off.

8.3 Europeanisation and Domestic Politics: Urban Governance, Partnership and Strategic Planning

8.3.1 Different Aspects of Europeanisation

By the term Europeanisation we refer to a set of processes through which the EU political, social and economic dynamics become part of the logic of domestic discourses, identities, political structures and public policies (Radaelli, 2000). For the needs of this chapter, the dynamics of Europeanisation will be confined to the domain of political structures and policies.

Broad Europeanisation changes can be discerned in two domains of urban politics (Getimis & Grigoriadou, 2004, pp. 3–7). The first is related to the transition of traditional urban government towards urban governance focusing on new horizontal partnerships, networking and community involvement in policy formulation and decision-making. In particular, divisions and conflicts between different politico-administrative units and between public and private actors have to be eliminated for urban resources to be mobilised and the potential access to EU funding to be utilised (Benz & Eberlein, 1998). The result is the empowerment of politics at the local level and their transformation from nationalised and hierarchical forms towards more negotiated and independent practices in a manner that involves the urban society and a wide range of interest groups (Peter, 2000).

The second one concerns the reorientation of urban policy away from fragmented actions of arbitrary development towards integrated, strategic, local action plans and initiatives for sustainable development policies, which contribute to the improvement of the quality of life in cities and the preservation and enhancement of the urban environment. Strategic, sustainable, urban development very often implies a commitment to a shared vision of urban change requiring a combination of resources from different sectors (public, private and community).

The promotion of sustainable urban development and the implementation of the partnership principle are two complementary, mutually reinforcing goals of EU policies aiming at successful urban governance. The former seeks the protection and improvement of the urban environment so as to improve the quality of life, safeguard human health and protect local and global eco-systems. This is achieved through the encouragement of partnership building. In particular, the establishment of good urban governance entails the vertical integration of activities at different levels of government and the better horizontal integration at the local level among the concerned organisations and citizens. In accordance with EU policies, partnership building emerges as a crucial factor for improving the quality of life in cities and for managing the urban environments in more sustainable ways (CEC, 2001a). For example, the programmes Urban and Life for the Development and Implementation of Community Environmental and Urban Regeneration Policies have had a catalytic effect on urban policies and partnership formation.

Furthermore, the *White Paper on European Governance* (CEC, 2001b) is indicative of the importance the European Union (EU) places on community involvement as integral part of good governance. In this paper, the European Commission strongly argues that broad citizens' participation should be ensured throughout the policy chain from design to implementation. Consequently, the White Paper's proposals are underpinned by two good governance principles: openness and participation (Knodt, 2002). However, the *implementation* of Agenda 21 has already introduced the principle of citizens' participation in the EU political agenda. Many of the European Community's programmes and policies have been based on the principle of the active involvement of the concerned groups throughout the relevant procedures. Consequently, civil society has been given specific mechanisms for participating in the development and implementation of Community policies (CEC, 1997a).

According to the EU, a number of interdependent factors explain the importance of the implementation of these principles:

1. the establishment of a more balanced European urban system as a precondition for economic and social cohesion;

2. the reinforcement of the cities constituting the drivers of the European economy;
3. the achievement of the new EU commitments and obligations vis-à-vis the global environment[1];
4. the resolution of complex and interrelated urban problems and the maximisation of urban potential, which are both undermined by the predominance of traditional sectoral approaches and the fragmentation of powers and responsibilities among various levels of government (CEC, 1997b).

The Structural Funds constitute the main funding mechanism for urban partnerships promoting sustainability in the EU. More specifically, partnership is one of the key principles underlying the Structural Funds. From the 1988 reform on structural policy, which introduced the principle of partnership as an institutional basis for implementation, to the recent 1999 reform, the definition of partnership has been broadened. In the 1988 reform, partnership in line with the principle of subsidiarity was defined as close consultation for the pursuit of common goals between the Commission, the concerned member states and the competent authorities, which are familiarised with the problems of disadvantaged regions. In subsequent reforms, a broader approach of partnership was adopted to ensure the involvement of all the concerned partners such as economic and social partners and environmental and non-governmental agencies. Subsequently, the 1999 regulation abandoned the 1988 decentralised approach to partnership for a wider approach that addresses all concerned bodies (Bache, 2000; Bollen, 2000).

Although partnership formation is a substantial prerequisite for the implementation of the Structural Funds, a recent report funded by the European Commission (Kelleher, Batterbury, & Stern, 1999) underlines the existence of significant variations and differences in the implementation of the partnership principle among the member states. In particular, this report indicates that where member states have little experience in partnership formation, the EU requirements have often 'kick-started' processes of partnership building. Regarding the composition of these partnerships, it is argued that the role of social partners and NGOs has often been limited. To explain these variations, a number of factors have been proposed. Of particular importance are the national institutional and cultural traditions, the well-established corporatist models and prior experience in partnerships.

8.3.2 Europeanisation and Institutional Innovation

The inherent ambiguity in the concept of 'Europeanisation' is reflected in the different and often controversial theoretical approaches (Getimis, 2003, pp. 81–83). Intergovernmental approaches stress that Europeanisation enhances the role and power of nation states vis-à-vis supranational and sub-national political actors (Moravcsik, 1995). On the other hand, neo-institutionalist approaches to European integration argue that supranational European policy provides new opportunities and resources to sub-national actors ('sub-national mobilization'), and this in turn leads to the gradual weakening of the nation state.

A third approach, which accepts neither the 'hollowing out' of the state nor its strengthening, argues that EU policy is produced by a complex web of policy networks of actors ('organized feedback loops') in a multi-level policy arena (Heinelt & Smith, 1996; Hooghe, 1996; Marks, 1993; Staeck, 1996). However, these networks are not highly stabilised and integrated but are characterised by a variety of differentiations: (a) the new internal organisational differentiations of collective actors; (b) the differentiation (sequentialisation) of decisions on different levels; as well as (c) the functional differentiation between a decision-making arena on the EU level and implementation arenas in the member states/regions (Heinelt, Lang, Malek, & Reissert, 2001).

European-level regional policy, based on the structural funds and aiming at socio-economic and territorial cohesion and European integration, is a very important policy area.

In this context it is important to clarify to what extent European regional policy, besides its positive redistribution effect, promotes institutional innovation based on the 'partnership' principle at the sub-national, regional level. How do new policy networks emerge and what is the degree of fragmentation or coherence? How important is local/regional embeddedness with regard to institutional capacity, and socio-political and cultural specificity of the region, in the success or failure of regional institutional innovation in the different member states?

European regional policy constitutes a rather enduring and long-standing challenge for the administrative and institutional structures of the member states. At the same time it provides opportunities for institution building and network creation at the national and sub-national levels, even if the pre-existing institutional capacity is poor (e.g., in many Objective 1 regions) (Paraskevopoulos, 2001; Paraskevopoulos, Getimis, & Rees, 2006).

It is generally accepted that 'the principle of partnership has enabled local elected representatives, social and economic organisations, non-governmental organisations and associations to be more involved in decision-making. However, apart from the formal respect for the obligation, the extent of partnership in practice has differed greatly' (CEC, 2001c; Kelleher et al., 1999).

8.3.3 Multi-level and Multi-actor Governance Arrangements. The Need for Loose Coupling Mechanisms

It has been argued that European policy in a multi-level governance system faces a risk of fragmentation and isolation of sectoral or territorial policies and needs to build further coherence mechanisms for 'loose coupling' of the policy networks' structures and arenas (Benz, 2000; Heinelt, 1996). This multi-level governance approach has gained wide acceptance in the academic debate since it provides fruitful understanding of the political integration of Europe at all levels (local, regional, national, European).

Europeanisation is not conceived as an 'homogeneous' and 'cohesive' top-down process, derived as an 'independent' (external) variable, that affects domestic institutions. It is rather an interactive and conflicting process of creating fragmented/differentiated policy structures with loose, coupling coherence mechanisms within the framework of an emerging system of multi-level governance, in which different European, national and sub-national actors in competition and/or cooperation share their power (Getimis, 2003, pp. 81–83).

8.3.4 Territorial Specificity and Local Embeddedness

The different political structures of each member state operate as a filter, which refracts Europeanisation pressures in different directions and styles. European policy impacts differ by area, because domestic responses to EU policies have varied considerably across policies and countries (Knill, Heritier and Borzel cited in Getimis & Grigoriadou, 2004, pp. 3–7). The regions' responses to the opportunities offered by European policy vary depending on their institutional capacity and endogenous potential (Getimis, 2003; Keating & Jones, 1995).

The territorial specificity depends upon and varies according to the specific socio-economic local development and the concrete political and institutional context. This is reflected by the fact that the strengths and weaknesses of all policy schemes and initiatives designed at higher sectoral and territorial levels become visible at the local level. But there are additional reasons for the relative importance of the local level for the introduction of the new forms of governance. On the one hand, markets become increasingly global and, in any case, they correspond to the exchange of products without any particular consideration of the social and political conditions under which these products are produced. On the other hand, hierarchies in the form of either nation states or the various intergovernmental schemes retain an administrative character that represents variations of the subsidiarity principle that assigns controversial but hierarchically determined functions to each territorial level. As such, the local context may be viewed as the testing ground not only for the effectiveness of the new forms of participatory governance but also for the success or failure of the hierarchical regulatory policies. As the interrelations at the European and national level have changed, the national and regional and/or local scale has been altered accordingly, in some specific areas maintaining the command and control approach of the past, and in other cases integrating and adapting new and old forms of governance to their structural individualities. The local level emerges as the most appropriate for the implementation of the new ideas of participatory governance, involving the networking of actors and the participation of a variety of holders in specific types of partnerships, initiatives and policy networks. In this respect the status and the quality of actors is instructive (Getimis & Kafkalas, 2002, pp. 160–161).

The integration assumptions leading to the study of the vertical relations between regions, nation states and the EU as well as to their transformation have recently been complemented by studies focusing more systematically on the horizontal changes of domestic policy processes resulting from the impact of EU policy. According to these approaches, significant importance has been attributed to the degree of acceptance or resistance to change from domestic urban political institutions and structures, illustrating the importance of domestic factors in adapting to European principles and funding conditions (Bache, 2000; Borzel, 1999; Paraskevopoulos et al., 2006).

8.3.5 Learning Process

There are differences with regard to the political influence of the state government vis-à-vis the sub-national level in programming, implementation, monitoring and evaluation of European policies. Especially in unitary states such as Greece, Ireland and France, national government dominates the regional policy process: from negotiations with the Commission to the programming and implementation of regional development plans and operational programmes. Sub-national authorities have only limited political influence, however, they gradually gain important benefits through institution building and learning at the regional level (Getimis, 2003, pp. 81–83).

Besides the aforementioned differences, there are others, with regard to centre–local relations in each country, that relate to the administrative styles and to the dominant models of interest intermediation among local, regional, national and European levels of governance: confrontational and/or consensus-oriented. These factors determine the substance of formal network building (like the I, II and III Monitoring Committees of the Community Support Framework), which are established at the regional level in all member states. The implementation of the 'partnership' principle in countries with a tradition in negotiation, bargaining and social dialogue, either through institutional arrangements (e.g., Germany, bureaucratic and negotiating administration), or through non-institutionalised processes (e.g., United Kingdom, Ireland, dissention and flexible negotiation), demonstrates extensive and successful network and institution building, where public and private actors cooperate with mutual understanding and trust. On the contrary, in countries which lack consensus-oriented governance through negotiations, formal networks and ad hoc cooperation of actors are cultivated, aiming primarily at fragmentary benefits of the European regional programmes. Under these conditions, these network structures are susceptible to central influence and control and they are unable to build a permanent and comprehensive web of locally embedded institutions (Getimis, 2003, pp. 81–83).

However, even in these cases, there is evidence of a slow learning process in which different actors from the public and the private sector and non-governmental organisations (NGOs) test their knowledge, rights and negotiating power and learn from each other (Paraskevopoulos et al., 2006). The Greek experience, starting from the programming and implementation of the Integrated Mediterranean Programmes (1987) and passing through two/three Community Support Frameworks (CSF I 1989–1994; CSF II 1994–2000; CSF III 2000–2006), is a characteristic case of this positive impact that the 'Europeanisation' of regional policy had on the existing institutional and administrative edifice (Getimis, 2003, pp. 81–83).

Although pre-existing features of hierarchical and clientelistic relations, a confrontational mode of interest intermediation and a weak civil society hindered extended institution and network building, significant progress can be ascertained in the last 15 years. Important factors contributing to this change are the institutional decentralisation reform at the regional level (1987), the strengthening of the political legitimacy and efficiency of the Local Government (I and II tier 1984, 1994), the institutional and financial incentives towards public–private partnership and the motivation of network building and institutional learning through education and training policy (Getimis, 2003, pp. 81–83; Paraskevopoulos et al., 2006).

8.3.6 Territorial Cohesion and the ESDP as a Strategy

The concept of territorial cohesion is a key concept for integrated and holistic solutions to different territorial problems and geographical inequalities. It integrates diverse values in a wide range of territories (typologies): social inclusion and equity, parity of access, innovation, competitiveness, entrepreneurship, protection of natural and cultural resources, partnership and cooperation. The complementarity of these values is, however, not given. It is always a difficult goal to achieve and it requires new forms of multi-level governance. These are based on principles of participation, negotiation and partnership between all actors involved, strengthening institutional capacity (especially at the regional level) and increasing democratic participation. The building of public–private partnership networks and the increase of political legitimacy are crucial for territorial cohesion (Getimis, 2005).

More precisely, the concept of territorial cohesion is reflected in the ESDP. The ESDP is built on the acknowledgement that the achievement of the fundamental goals of the EU requires taking into account the territorial dimension, though spatial planning, regional planning and geography. This first major contribution to this '(new) way of thinking'[2] derives from an integrated view of the EU fundamental goals, illustrated by the 'triangle of sustainability': economy, society and environment. The triangle suggests the balance and complementarity between the goals.

According to the ESDP, these three fundamental goals must be considered together, pursued simultaneously in all regions, and their interactions must be taken into account. Such a vision is closely linked to the concept of territorial cohesion.

The Commission expects that, although regional disparities have grown after the 2004 Enlargement, cohesion policies and especially the territorial cooperation policies will strengthen economic growth, productivity and competitiveness in the middle-term and will promote a new spatial transformation towards a more polycentric structure of the European territory. Such policies are expected to increase the wealth for all the European citizens, since the satisfaction of social needs can be fulfilled mainly through the economic growth and development, spatial integration and cohesion of the European territory. Such a concept helps to broaden public participation procedures and increase political legitimacy that safeguards social stability (e.g., through social integration policies in less developed areas or in declining neighbourhoods within big agglomerations in the core of Europe).

The ESDP as an 'intergovernmental' consultation and negotiation process is an example of Europeanisation having different impacts on spatial planning traditions in the different member states. It is clear that the aim was not to prepare a European Master Plan, nor did the Commission have the competency or role to impose binding regulations and directives. The ESDP is a strategy about new principles and discourses concerning the European territory, generating structural changes in domestic, spatial planning systems and policies in the member states (Faludi, 2002; Rivolin & Faludi, 2005). However, these transformations are not an outcome of a top-down imperative European policy but a product of complex restructuring processes of adaptation and resistance to change diverging nationally (Giannakourou, 2005).

A growing awareness of the usefulness of strategic spatial planning can be observed both in the academic community and in the policy-making arena. Strategic planning as a process refers to the institutional design and to the new forms of governance, where multiple actors at different levels participate in arenas of decision-making and action. Strategic planning is directed at integrated policies and outcomes that combine legitimacy and effectiveness (Albrechts, 2001; Healey, 1998).

8.4 Transforming Planning: The Perspective of Strategic, Collaborative, Sustainable Planning and Urban Governance

8.4.1 Planning Policies Under Transformation

Spatial planning system traditions and policies differ across the European territory. This is obvious in the EU Compendium of Spatial Planning Systems and Policies (CEC, 1997b), reflecting the differences between the member states. Comparative work on this topic (Newman & Thornley, 1996), focusing on institutional competencies and tools of the planning systems, has shown that these differences reflect the differences among the countries in terms of constitutional and institutional setups, central-local relations and functional and contextual relations of the actors involved in the planning process (Faludi, 2004, p. 155). This implies that there is no homogeneous 'planning paradigm' across Europe. However, we can identify specific common features concerning the planning policies developed in Europe over the last 50 years as well as the transforming processes of the contemporary planning policies.

8.4.1.1 'Conventional' Planning Policies of the Past

Traditionally, planning has been defined as a state competency and was managed and conducted through a central, regional or local government as a state process. The greatest change can be observed in strongly hierarchical planning systems such as those of the Napoleonic family (Newman & Thornley, 1996), but also in more heterarchical networked planning systems, such as that of the Netherlands.

Participation and vertical coordination were not absent from the process, but were confined to predefined actors whose competencies were explicitly stated (often legally). Participation may have taken a discretionary form (e.g., United Kingdom), or an institutional (e.g., Germany), and the degree of centralisation varied markedly.

Yet despite these differences and those of regulatory and legalistic traditions, most European planning systems treated planning as a more or less tightly defined statutory process of the regulation of land uses and development. Thus planning served as a control function to organise and regulate development.

8.4.1.2 Contemporary Planning Policies

Since the 1990s, planning has undergone a transformation as a result of the transition from government to governance and the Europeanisation pressures presented in the previous sections. From the state's point of view its unique 'planning competency' has waned and been replaced by planning as a 'strategic enabling function' (Tewdwr-Jones, 2002) of the state, which must facilitate and coordinate the now much broader policy-making process involving central and local government as well as many public organisations and private sector actors.

Horizontal and vertical coordination and cooperation, including with the EU, which until now took place in a discretionary or legally explicit fashion, are now incorporated as necessary characteristics of the policy-making process. Planning at the commencement of the twenty-first century should be viewed as a much broader all-encompassing activity, since it exists to coordinate policy, cement partnerships and facilitate much-needed change (Healey, 1998). "The key issue for policy-makers from now on will be how to reconcile the apparent irreconcilable tensions inherent within the new governance of planning and how to meet the perceived high expectations from a range of government tiers, agencies, organisations, businesses and the public on why planning exists and what planning, and indeed the new political processes more generally, is expected to deliver" (Tewdwr-Jones, 2002, p. 279).

The encouragement of participation from the grassroots up is taken as an explicit characteristic of contemporary planning policies enhancing the sense of policy ownership and inclusiveness and thus legitimacy. The multitude of actors and institutions involved in the participatory process, in combination with the parallel process of devolution of power from the central state, can lead to policy fragmentation. Partnership principles are being applied between the public and private actors involved, in order to reduce this fragmentation of policy arenas. To counter this fragmentation tendency, there is also pressure for a recentralisation of strategic planning to the state in order to improve coordination. The pressures for the recentralisation of strategic planning have also been a result of the entry of the environmental sustainability theme onto the world scene and the perceived need for state influence for the legitimate inclusion of this perspective into policy-making in all fields and at all scales. This inclusion of sustainability principles in all policy fields and at all scales is made all the more challenging by the self-sustaining rise of market-led principles. Market-led principles are now being systematically brought into the policy-making arena throughout Europe at all spatial scales.

Contemporary strategic planning also seeks to be more flexible, accommodating, dynamic, taking local specificities into account in order to develop customised spatial development agendas and new forms of governance to take these agendas forward. This flexibility is ideally complemented by a capacity to learn. Learning processes are increasingly proving vital for effective coordination within these new governance forms.

Finally, it should be noted that contemporary planning policies as described above have not overwritten those of the past but have added to them and modified them. For example, we may still have strongly hierarchical, Napoleonic-style planning systems in parts of Europe, but they encourage participation and engage market principles more directly than in the past.

8.4.2 Opportunities and Risks of the New Strategic Planning Policies

Picking out some of the inherent opportunities and risks of contemporary planning policy, as described above, we can see that they are closely related to the opportunities and risks of the transition from government to governance. Broadening participation patterns from the inhabitants of an area to public institutions from all tiers and private sector actors presents an opportunity for those policies which enjoy greater support and ownership. If operated transparently, this participatory process may also improve throughput legitimacy and accountability. The learning potential in the new systems of governance, when properly harnessed, promises more effective policy-making processes.

The risks associated with contemporary strategic planning relate largely to the fragmentation and compartmentalisation of policy fields, a result of the inclusion of a multitude of institutions and actors in the policy-making process, and the challenge of coordinating a focused spatial agenda in this context. There is also a risk associated with the shift and sharing of power from the state to the participants in this broader process and with the 're-scaling of governance arenas and networks' (Healey, 2006). State leadership is emphasised as a necessity for guiding and coordination if effectiveness and legitimacy are to be maintained. There is a risk that the new governance systems supported by contemporary strategic planning processes will be perceived as a panacea, overlooking the aforementioned risks and leading to a 'dual tension' between high politics (national agenda setting) and low politics (policy implementation) (Tewdwr-Jones, 2002). This problem may be acknowledged and addressed through the raising of awareness among participants of the purpose of the new governance systems in the broader policy-making apparatus.

8.4.3 Strategic, Collaborative, Sustainable Planning

In the ongoing efforts to establish strategic, collaborative, sustainable planning systems which facilitate learning through discursive processes and/or argument, the dual aim of effectiveness and legitimacy must be emphasised. This aim can only be reached with the aid of state leadership. Planners should be aware of the opportunities and risks associated with contemporary strategic spatial planning policies and the new forms of governance. Albrechts (2001) concisely summarises the characteristics of successful strategic spatial planning systems as defined by Healey (1997); Granados-Cabezas (1995); and Faludi (2000).

8.5 Strategic Planning and the Diversity of Small- and Medium-Sized Cities in Europe: Dilemmas and Perspectives

8.5.1 Different Definitions of Small- and Medium-Sized Cities in European Countries

There are three main approaches to defining urban areas in Europe. Definitions within these approaches vary widely resulting in disparities in the criteria defining Small and MEdium Sized TOwns (SMESTOs).

Using administrative boundaries, such as those of municipalities, to define the extent of urban areas, relates not only to the organisation of the country by the state but also to the scale at which local actors interact in governance systems. Often the administrative unit is defined as urban or rural depending on its population and, immediately here, we can see how the definition of an urban municipality may differ from country to country. In Switzerland a threshold population of 10,000 is required, while in Luxembourg or the Czech Republic only 2,000 is required, and in Austria a threshold of 20,000 is accepted (Table 8.2).

Table 8.2 Population thresholds for defining urban municipalities (ESPON 1.4.1, 2006, p. 42)

Country	Name	Definition of the agglomeration
Switzerland	Commune Urbaine Städtische Gemeinde	More than 10,000 inhabitants
Austria	Statutarstadt	More than 20,000 inhabitants
Czech Republic	–	More than 2,000 inhabitants
Spain	–	More than 10,000 inhabitants
Italy	–	More than 10,000 inhabitants
Slovakia	–	More than 5,000 inhabitants (combined with function as a centre)
Luxembourg	–	Population of communes with an administrative centre of more than 2,000 inhabitants

Table 8.3 Synthesis continuous built-up area (ESPON 1.4.1, 2006, p. 46)

Country	Distance threshold (m)	Population threshold
Finland, Sweden and Denmark	200	200 inhabitants
Norway	50	200 inhabitants
Wales and England	50	1,000 inhabitants
Scotland (urban settlements)	50	3,000 inhabitants
Greece	200	10,000 inhabitants
Ireland	200	50 occupied dwellings
Belgium	250	150 inhabitants (in the statistical sector) Population density > 500 inhabitants/km^2

Morphological characteristics such as the extent of built-up areas or population density are also sometimes used to define urban areas. This definition treats the settlement as a physical or architectural object. The extent of built-up areas is defined by the distance between buildings. A maximum distance of 50 m is permitted in the United Kingdom before a building is considered outside an urban area, while the in Belgium 250 m is taken as the maximum.

Again we can see that the definition of an urban area varies widely. Land uses accepted within morphological urban areas also vary across the EU. For example, public, commercial and industrial uses are excluded in France, while they are included in other EU countries, which could give the impression that urban areas in France are more fragmented. Where the aggregate population is used to define an urban area, differences are even more striking. While Belgium and the Nordic countries set a threshold of 200 inhabitants, Austria and Greece require 10,000 to consider an area 'urban' (Table 8.3).

Functional approaches define urban areas in terms of interactions between the urban core and the hinterland around it. These approaches are often related to commuting flows which define the spatial extent of a labour market. Otherwise, a variety of criteria are used to define SMESTOs such as the provision of goods, services and housing and the ability to retain particular levels of economic activity. These often relate to mobility and/or accessibility. Even symbolisation functions: using symbolic, cultural and image definitions of a settlement are employed to define a SMESTO.

8.5.2 Grasping the Diversities: ESPON Project 1.4.1

Having considered the above disparities in the definition of SMESTOs in Europe, and a variety of related factors such as agglomeration economies, competitiveness, human capital and exogenous and endogenous development theories, ESPON

Project 1.4.1 (ESPON 1.4.1, 2006) weighted and combined indicators from all three approaches (administrative, morphological and functional), to develop a typology. Four types of SMESTO emerged:

1. *dynamic and growing SMESTOs*, where most of the proposed quantitative indicators are positively related;
2. *declining SMESTOs*, where most of the proposed indicators are negatively related;
3. *restructuring SMESTOs*, where several indicators show deterioration of functions but a process of upgrading of the functions is ongoing;
4. *potential developing SMESTOs*, where new trends are emerging for different endowment resources (geo-physical, historical, location related, quality factors).

8.5.3 Highlighting Three Important Issues

8.5.3.1 The Principles of 'Strategic Planning': Awareness

Strategic planning, in essence, is a long-term vision or perspective of development in a defined area. The key characteristic of strategic planning is inter-sectoral cooperation at all spatial scales, enabling partnerships and discursive processes among actors and stakeholders, stimulating common action based on negotiation and bargaining and aiming for sustainable outcomes with a good balance of effectiveness and legitimacy. These principles of collaborative, sustainable, strategic planning correspond with the new participatory, sustainable governance arrangements and are common for effective and legitimate outcomes at all spatial scales. Therefore, planners at all scales and in settlements of all different sizes including SMESTOs and large metropolitan areas should be aware of these principles.

8.5.3.2 Local Embeddedness: Urban Dynamism and Crisis (Vicious Circles)

Local circumstances must be part of the core considerations of strategic planning. Not only should the size, population and economic drivers (labour force, particular industries, etc.) of a SMESTO be taken into consideration, but the governance and participatory traditions should also be closely observed in the promotion and coordination of new urban governance systems.

By employing local characteristics in the strategic plans for an area and in the new urban governance systems for particular SMESTOs, local dynamism and endogenous potentials may be coaxed and developed into stimulating, self-sustaining, socio-economic development. If this is not the case, and particularly, when a SMESTO lies in a region stagnating as a whole, then vicious circles can become established, where economic decline fuels unemployment and innovative stagnation, leading to depopulation and increasing poverty, loss of know-how and the shrinkage of markets, resulting in further economic decline (Fig. 8.1).

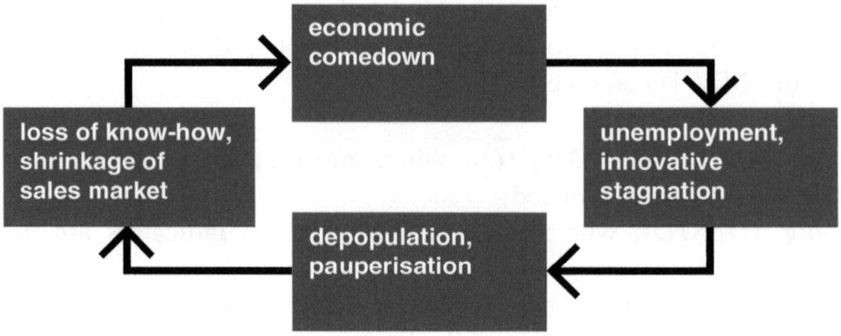

Fig. 8.1 Vicious circle of SMESTO decline (ESPON 1.4.1, 2006, p. 113)

8.5.4 Dilemmas and Perspectives

Networking versus individual solutions. Networking among stakeholders, and the encouragement to cluster enterprises and public and private partnerships for the stimulation of local production systems, is the most effective exit strategy from such vicious circles. The alternative of individual solutions to stagnation can work successfully, where a single industry or even an individual firm becomes the driver for an urban area's economic dynamism. However, where broader networking is sidelined and the narrow network of the industry or firm involved is the only networking activity, there can be the risk of a heavy dependence on this firm or industry and specialisation rather than diversification as discussed below. On the contrary, broader networking of stakeholders can lead to a more evenly distributed regional development.

Specialisation versus diversity. Comparative advantages in concrete sectors such as tourism can lead to specialisation. Diversity, on the other hand, is a more likely outcome of cooperative networking. To maximise and disperse the sustainable development of an area in the case of specialisation, this economically advantageous specialisation must be promoted in a way that integrates supporting industries (e.g., local products). In the case of tourism mentioned above, an example of a method for integrating other industries and dispersing the benefits in an area with a comparative advantage in tourism would be to pursue agro-tourism or eco-tourism, which integrate and stimulate the primary sector.

Urban governance rescaling. Institutions matter and new governance arrangements promoting participation for effectiveness and legitimacy are very important for local development. The new forms of governance can be a means for escaping administrative fragmentation and the confines of jurisdictional or administrative boundaries. By considering urban centres in a functional sense (Functional Urban Areas) (ESPON 1.4.1, 2006), and observing the hinterland with which they interact, or simply by observing the extent of the spatial interactions involved in a particularly dynamic industry which acts as the economic driver for an urban area, we can start to discern planning modes which will harness this potential. One approach through a hierarchical top-down initiative might be to create new institutions to replace older ones. The new institutions could be responsible for larger areas, for example, the functional area of an SMESTO rather than the urban core alone. This, however, may generate adverse reactions from the existing fragmented institutional structure. Another approach relying on heterarchical structures would stimulate the cooperation of existing municipalities on strategic planning and development issues.

8.6 Conclusion

Having observed and analysed the strengths and weaknesses of the shift from 'government' to 'governance', we turned to the transformation of the planning process from a more or less institutionalised government competency to an adaptive, participatory, multi-level process, in which government can serve as a regulator. The chapter highlights certain important considerations, especially in SMESTOs, where new forms of urban governance and contemporary planning processes can have a decisive positive or negative impact.

We noted that certain opportunities present themselves, in the shift from government to governance, for example, legitimacy gains through widening representation, effectiveness gains through broader participation, knowledge gains (learning), consensus gains and innovative restructuring. However, these opportunities can be jeopardised and outcomes may be the reverse, if we are not aware of certain risks. These include ineffectiveness, resulting from a lack of commitment of the large number of involved actors to a common goal; legitimacy loss, resulting from a loss of transparency through the dilution of responsibilities; a loss of comprehensiveness resulting from the fragmentation of policy areas and a loss of perspective of the broader policy context; and the loss of legitimacy in the pursuit of technocratically assessed efficiency. The aforementioned risks are compounded by a positive bias towards participatory methods of governance. In order to avoid the failure of governance it has first been suggested that different coordination modes – government, market and governance – should be employed in combination with effective meta-governance, harnessing each one's advantages as effectively as possible. Second, reflexive learning can help side-step some of the above problems. Finally, the complementarities between political leadership and democratic participation should be explored.

Similarly, the transformation of the planning process through the pursuit, once again, of participation presents opportunities and risks. The opportunities include enhanced legitimacy as a result of broader participation, whereas the risks include the potential compartmentalisation of policy fields and lack of horizontal and cross-sectoral coordination. With regard to the Europeanisation of domestic planning processes, not taking into consideration territorial specificities and local embeddedness can entail risks for effectiveness and legitimacy. As in the case of shift from government to governance, participation may be viewed as a panacea overlooking the risks involved. Political leadership is likewise presented as a necessity for coordination of effectiveness and legitimacy are to be promoted.

Thus, from the discussion of the shift from government to governance we can learn something new and apply it to the transformation of the planning process. Firstly, just as new forms of local governance present opportunities and risks, albeit different ones to those evident in a hierarchical government system, the same is true of the new forms of participatory planning, when compared with conventional planning processes of the past. Second, in both cases these opportunities and risks must be explored and we must be aware of them.

Finally, governance and contemporary planning process have not replaced government and conventional planning. They are a welcome addition, and complementarities between the old and the new should be sought if we are not to replace old problems with new ones in the pursuit of greater participation, effectiveness and legitimacy.

Notes

1. The pursuit of sustainable development through partnerships and community involvement at the urban level of governance by EU has powerful impetus from the follow-up of the idea of Local Agenda 21 (Action Plan adopted at the Rio Earth Summit in 1992).
2. Third Interim Report/ESPON Project 3.1 'Integrated Tools for European Spatial Development'.

References

Albrechts, L. (2001). In pursuit of new approaches to strategic spatial planning. A European Perspective. *International Planning Studies*, *6*(3), 293–310.

Bache, I. (2000). *Europeanization and partnership: Exploring and explaining variations in policy transfer*. Queen's Papers on Europeanization, Queen's University Belfast, Belfast.

Benz, A. (2000). Two types of multi-level governance: Intergovernmental relations in German and EU regional policy. *Regional and Federal Studies*, *10*(3), 21–44.

Benz, A., & Eberlein, B. (1998). *Regions in European Governance: The logic of multi-level interaction*. Working Paper PSC NO 98/31, European University Institute, Florence.

Bollen, F. (2000) Preparing for the EU structural funds: Role and opportunities for sub-national authorities and non-governmental organisationsIn Open Society Institute (Ed.) *European Union enlargement and the open society agenda: Local government and public administration* (pp. 53–88). Budapest: Open Society Institute, Local Government and Public Service Reform Initiative.

Borzel, T. (1999). Towards convergence in Europe? Institutional adaptation to Europeanization in Germany and Spain. *Journal of Common Market Studies*, *37*(4), 573–596.

CEC (Commission of the European Communities). (1997a). *Agenda 21 – The first 5 years*. Bruxelles: European Commission.

CEC (Commission of the European Communities). (1997b). *The EU compendium of spatial planning systems and policies*. Luxembourg: European Commission.

CEC (Commission of the European Communities). (2001a). *Implementing the habitat agenda. The European Union experience*. Report of the Commission of the European Commission. Bruxelles: European Union Press.

CEC (Commission of the European Communities). (2001b). *European governance: A white paper*. Bruxelles: European Commission.

CEC (Commission of the European Communities). (2001c). *Unity, solidarity, diversity for Europe, its people and its territory*. Second Report on Economic and Social Cohesion. Luxembourg: European Commission.

CSD (Committee on Spatial Development). (1999). *European spatial development perspective: Towards balanced and sustainable development of the territory of the EU (ESDP)*. Luxembourg: European Communities.
ESPON 1.4.1. (2006). *The role of small and medium-sized towns (SMESTO)*. Final Report[http://www.espon.eu/mmp/online/website/content/projects/261/410/index_EN.html; accessed 07/02/2007].
Faludi, A. (2000). *Strategic planning in Europe: Institutional aspects*. In W. Salet & A. Faludi (Eds.), *The revival of strategic spatial planning* (pp. 243–258). Amsterdam: Royal Netherlands Academy of Arts and Sciences.
Faludi, A. (2002). Positioning European spatial planning. *European Planning Studies*, *10*(7), 897–909.
Faludi, A. (2004). Spatial planning traditions in Europe: Their role in the ESDP process. *International Planning Studies*, *9*(2–3), 155–172.
Getimis, P. (2003). Improving European Union regional policy by learning from the past in view of enlargement. *European Planning Studies*, *11*(1), 77–87.
Getimis, P. (2005). European enlargement, territorial cohesion and ESDP. In *International conference of ministry of infrastructure and transport, present and future of the European spatial development perspective* (pp. 39–55). Italy: Italian Ministry of Infrastructure and Transport.
Getimis, P., & Georgandas, E. (2001). *Too many stakes, too few holders, but no partners: Governance failures, institutional inertia and the implementation of an URBAN community initiative in Greece*. Proceedings of the 8th international conference of the Regional Studies Association, Gdansk, 15–18th September 2001.
Getimis, P., & Grigoriadou, D. (2004). The Europeanization of urban governance in Greece: A dynamic and contradictory process. *International Planning Studies*, *9*(1), 5–25.
Getimis, P., & Kafkalas, G. (2002). Empirical evidence and comparative analysis of policy-making in the pursuit of innovation and sustainability. In H. Heilnet, P. Getimis, G. Kafkalas, R. Smith, & E. Swyngedouw (Eds.), *Participatory governance in multi-level context: Concepts and experience* (pp. 155–171). Opladen: Leske Budrich.
Giannakourou, G. (2005). Transforming spatial planning policy in Mediterranean countries: Europeanization and domestic change. *European Planning Studies*, *13*(2), 319–331.
Granados-Cabezas, V. (1995). Another methodology for local development? Selling places with packaging techniques: A view from the Spanish experience of city strategic spatial planning. *European Planning Studies*, *3*(2), 173–187.
Haus, M., Heinelt, H., & Stewart M. (Eds.). (2005). *Urban governance and democracy: Leadership and community involvement*. London: Routledge.
Healey, P. (1997). *Collaborative planning, shaping places in fragmented societies*. London: Macmillan.
Healey, P. (1998). Building institutional capacity through collaborative approaches to urban planning. *Environment and Planning A*, *30*(9), 1531–1546.
Healey, P. (2006). Transforming governance: Challenges of institutional adaptation and a new politics of space. *European Planning Studies*, *14*(3), 299–320.
Heinelt, H. (1996). Multi-level governance in the European Union and the structural funds. In H. Heinelt & R. Smith (Eds.), *Policy networks and European structural funds* (pp. 9–25). Aldershot: Avebury.
Heinelt, H., Lang, J., Malek, T., & Reissert, B. (2001). *Die Entwicklung der Europäischen Strukturfonds als Kumulativer Politikprozeß. Institutionalisierung und Veränderung von Politikinhalten im Mehrebenensystem der EU.* Report on a research project funded by the German Research Council (DFG) under the Programme Governance in the European Union, Darmstadt/Berlin.
Heinelt, H. & Smith, R. (Eds.). (1996). *Policy networks and European structural funds*. Aldershot: Avebury.
Hooghe, L. (1996). Building a Europe with the regions: The changing role of the European Commission. In L. Hooghe (Ed.), *Cohesion policy and European integration: Building multi-level governance* (pp. 89–128). Oxford: Oxford University Press.

Jessop, B. (2002). Governance and meta-governance: On reflexivity, requisite variety, and requisite irony. In H. Heilnet, P. Getimis, G. Kafkalas, R. Smith, & E. Swyngedouw (Eds.), *Participatory governance in multi-level context: Concepts and experience* (pp. 33–58). Opladen: Leske Budrich.

Keating, M., & Jones, B. (1995). *The European Union and the regions*. Oxford: Clarendon Press.

Kelleher, J., Batterbury, S., & Stern, E. (1999). *The thematic evaluation of the partnership principle*. Tavistock Institute, Evaluation Development and Review Unit [http://ec.europa.eu/regional_policy/sources/docgener/evaluation/doc/rathe/cov-cont.pdf; accessed 07/02/2007].

Knodt, M. (2002). Regions in multi-level governance arrangements: Leadership versus partnership. In J. Grote & B. Gbikpi (Eds.), *Participatory governance* (pp. 177–194). Opladen: Leske Budrich.

Marks, G. (1993). Structural policy and multi-level governance in the EC. In A. W. Cafruny & G. G. Rosenthal (Eds.), *The state of the European community, 2: The Maastricht debate and beyond* (pp. 391–410). Boulder: Lynne Rienner.

Moravcsik, A. (1995). Liberal intergovernmentalism and integration: A rejoinder. *Journal of Common Market Studies*, *4*(33), 611–628.

Newman, P., & Thornley, A. (1996). *Urban planning in Europe: International competition, national systems and planning projects*. London: Routledge.

Paraskevopoulos, C. (2001). *Interpreting convergence in the European Union*. London: Palgrave.

Paraskevopoulos, C., Getimis P., & Rees, N. (Eds.). (2006). *Adapting to EU multi-level governance: Regional and environmental policies in cohesion and CEE countries*. Aldershot: Ashgate.

Peter, J. (2000). The Europeanization of sub-national governance. *Urban Studies*, *37*(5–6), pp. 877–894.

Radaelli, C. (2000). *Whither Europeanization? Concept stretching and substantive change*. Paper presented at the Political Studies Association's annual conference, London.

Rivolin, U. J., & Faludi, A. (2005). The hidden face of European spatial planning: Innovations in governance. *European Planning Studies*, *13*(2), 195–215.

Staeck, N. (1996). The European structural funds. Their history and impact. In H. Heinelt & R. Smith (Eds.), *Policy networks and European structural funds* (pp. 46–73). Aldershot: Avebury.

Tewdwr-Jones, M. (2002). *The planning polity: Planning, government and the policy process*. London: Routledge.

Chapter 9
Socio-Economic Regeneration Initiatives and Strategic Governance in Old Industrial Towns Outside of Agglomerations

Thilo Lang

9.1 Introduction

In recent years, the concept of governance entered the debate about urban development. In particular, urban governance is often seen as a key to regeneration. The understanding of governance adopted here focuses on the processes of decision-making. Thereby the idea of governance is centered on the inclusion of other actors, besides traditional government, in the management of urban development. The applied understanding of governance places focus on 'strategic' forms of decision-making, that is, long-term visions and short-term actions linked to socio-economic regeneration. This understanding – which will be referred to as strategic governance – implies forms of coordination and partnership between public and private actors in a purposeful and collaborative way.

This chapter looks for forms of strategic governance in old industrial towns, drawing on a recently finalised empirical study conducted by the author. It is not so much the visioning process which is of interest in the context of this chapter but the relation between action (in the form of local initiatives), and forms of strategic governance.

In the first part, I discuss the processes of transformation and decline and how they relate to the overall global trends. The chapter also introduces a debate about urban governance. Particular emphasis is given to the links between processes of decline and local responses in terms of strategic governance. In the second part, I present a recent study about socio-economic regeneration in old industrial towns in England and Germany. In the final part, I discuss the findings of this study focussing on the role of local initiatives in urban governance and the role of strategic forms of governance for local regeneration.

T. Lang (✉)
Leibniz Institute for Regional Geography, 04329 Leipzig, Germany
e-mail: mail@thilolang.de

M. Cerreta et al. (eds.), *Making Strategies in Spatial Planning*,
Urban and Landscape Perspectives 9, DOI 10.1007/978-90-481-3106-8_9,

9.2 Issues of Local Development

9.2.1 *Responding to Processes of Transformation and Decline*

In Western Europe, there is a strong connection between writings about urban decline and those about economic restructuring in old-industrialised areas. In the globalised economy, only a few global cities and metropolitan regions are said to be the "control points of the global economic system" (Dicken, 2003, p. 240). Within the international system 'networked' cities and agglomerations (especially capital cities) are said to be the number one location for headquarters of multi-national companies and big national enterprises or sub-contractors, in particular in the financial sector. Consequently, current processes of internationalisation and global inter- and intra-firm relations tend to concentrate much of the world's most important trading activities in a relatively limited number of sub-national regions or agglomerations (Beaverstock, Smith, & Taylor, 1999; Scott & Storper, 2003).

Current tendencies of globalisation are likely to promote concentrated economic and demographic development in some metropolitan regions, which are thereby dominating national urban systems. There seems to be a selective concentration of growth potentials in a smaller number of regions (Krätke, 1990, p. 7). Amin and Thrift suggest that capital cities and core metropolitan regions can derive competitive advantage from the presence of many organisations in economic, political and cultural life (Amin & Thrift, 1995, p. 105). Developing local response to persisting socio-economic problems by providing local job opportunities is an important aspect of the quality of life in a region. However, forms of response are probably different in old industrial towns outside of agglomerations, which cannot profit from the development advantages of metropolitan regions (Dicken, 2003, p. 240; Cheshire, 1998, p. 106).

In terms of the management of uneven development, Painter and Goodwin point out that local governance can only be effective if it is part of a multi-scale system of regulation. As the causes of uneven development at least partly lie outside the local sphere of influence, local governance at best "can influence only the local half of the (unequal) relationship between global flows and local conditions" (Painter & Goodwin, 2000, p. 43). Local governance might have some possibilities to mitigate the social consequences of uneven development. In general, however, the contribution of local governance although seen as vital is limited in its stabilising capacity in a multi-scale mode of regulation.

9.2.2 *Urban Governance as a New Phenomenon?*

Also the ways in which towns and cities are governed have changed. It is said that instead of hierarchical government the focus is now on governance with more actors involved. Purely public modes of government are said to be more and more unable to respond to processes of decline (Denters & Rose, 2005a; Elander, 2002). Instead, new forms of governance are said to provide better ways to combat decline because they are constructed on the principle of partnership with its potential synergies. Dealing with urban problems in partnership with the public and the private sector theoretically joins up the strengths of different actors involved in urban regeneration. Governance is often seen as something better than government and is used to describe modern and effective ways of steering urban development. However, it is very unclear how such new modes of governance can help to cope with decline and in what way they can be better in terms of social inclusion and fighting unemployment.

Although governance mechanisms have always existed in the sense of controlling and directing, scholars generally use the term 'urban governance' to discuss the changes within the urban political scene of steering a city towards a broader involvement of private actors. There is a baseline agreement about these directly observable trends (Denters & Rose, 2005b). But there is an ongoing debate as to how to explain or understand these changes (Bevir & Rhodes, 2003). In recent decades, city governments have been said to be less and less able to manage urban development. Governing cities has become more difficult and those involved in urban government have been increasingly under pressure to adjust the urban political scene to more and more challenging development preconditions (Kearns & Paddison, 2000, p. 845). In this sense, most scholars see governance as a direct response to economic and social change and their consequences for policy agendas and policy networks (Healey, Cars, Madanipour, & de Magalhaes, 2002, p. 6; Denters & Rose, 2005a). There are at least three main points which are brought forward in this context (Healey et al., 2002; Kearns & Paddison, 2000; Newman & Verpraet, 1999):

1. processes of economic globalisation, structural economic and technological changes have lead to a loss of urban governments' control over urban economies. At the same time inter-urban competition has tightened (Krätke, 1990, p. 8), and attempts to develop a city's distinctive culture have been done to improve city's competitiveness in global markets;
2. national governments have become less able to assist cities in development problems. Neoliberal reforms have led to erosion of the welfare state in many European nations, despite rising social problems caused by enforced processes of structural change;
3. growing complexity and diversity of social life has revealed socio-spatial polarisation and social exclusion, often in the form of marginalisation of excluded people in particular neighbourhoods.

Another widespread notion is about a shift in the practice and mechanisms of regulating urban development, from the traditional understanding of authoritarian government towards governance as a process-based and cooperative government practice. The reasons for these shifts in the urban political scene might be seen not only in the rationality of local actors and organisations but also in the promotion of by national policy – in particular in Great Britain – or in the demand of other stakeholders such as private enterprises or NGOs to play a more active role in the planning process (Healey et al., 2002). These changes are often seen as a direct reaction to altered socio-economic conditions. The connected problems, such as social exclusion or economic decline, are unlikely to be managed through state led planning, traditional welfare policies or pure market means. In many countries, activating the voluntary sector to replace or complement social services became an issue. So too did more managerial and cost-effective practices of municipal services provision, leading to new forms of cooperation and privatisation (Denters & Rose, 2005a; Elander, 2002, p. 191).

9.2.3 Challenging the Governance Concept

The term governance is commonly understood as comprising strategic elements and the coordination of a multitude of actors and activities in a wider whole. However, many descriptions of governance are rather idealistic than empirically grounded. In various contexts, urban governance has been used as a normative model to promote the inclusion of civic actors as resource mobilisation strategy or to de-emphasise the influence of local governments (Pierre, 2005, p. 453). The policy network-based conception of urban governance as heterarchy, as explained by Jessop and Rhodes, appears to be too limited because in reality it cannot account for most of the forms of governance (Lowndes, 2001, p. 1962). The dominant patterns in partnership relations in the British new governance are said to be hierarchical, not coordinating and competitive (Davies, 2004, p. 582). "With a few exceptions, partnerships are bureaucratic, hierarchical and non-productive" (Davies, 2001, p. 14). It is just a logical consequence that most of the writings describing governance as a normative model also include the notion of 'governance failure' (Jessop, 1998, p. 43; Healey et al., 2002, p. 20; Coaffee & Healey, 2003). This discussion supplies arguments for an increasing role of local government. The governance debate, however, turns away attention from local government despite rising demands. Communal reforms in Germany (Gabriel & Eisenmann, 2005), for example, indicate increasing power of local governments. Maybe within the more complex world of governance the role of local governments has been changing towards enabling, coordinating, sub-contracting, controlling and legitimating. Maybe cooperation and innovation within local government are more important than outside of local government (Lang, 2005). There is surprisingly little research about these issues.

Urban governance cannot be understood as referring to one simple overall and integrative whole as dominant coordination mechanism. It is more like a collective name or container for a broad range of different approaches to deal with urban development – with the multi-actor, multi-sector approach as a uniting element. In reality, systems or arrangements of local governance are fragmented into different themes (e.g., culture and economic development), individual and collective interests as well as organisational and formal settings. Even single theme, such as socio-economic regeneration (which is the central policy field tackled in this chapter), is unlikely to be affiliated with one single coordination mechanism. Thus it is very unlikely that government will lose its role in the new forms of governance; it is just the relation between different actors which has been the object of change (Pierre, 2005,

p. 453). The reasons for these changing relations cannot automatically be seen in the demands of interest groups to play a bigger role in urban development; neither are changing forms of governance a purely local response to socio-economic challenges. New forms of local governance also reflect the given structure of national contexts and higher level policy interventions.

9.3 Case Study: Socio-Economic Regeneration in Eastern Germany and Northern England

9.3.1 A Comparative Study of Local Regeneration Initiatives in Old Industrial Towns

The study is about local actors' response to problems of inequality and unemployment. Policies to combat these problems are usually connected to concepts of socio-economic regeneration in England and economic and employment promotion in Germany. The research identified eight local socio-economic initiatives aiming at job creation and investigated their relation to local governance arrangements. In each of four old industrial towns, one social and one formal economy initiative was selected for further analysis (Fig. 9.1). Socio-economic initiatives were defined as non-routine local projects or schemes with a clear objective towards job creation. Such initiatives often focus specific local assets for the formal or the social economy. Examples of socio-economic initiatives range from the promotion of local economic networks to business incubators and community enterprises.

In Northern England and Eastern Germany, four old industrial towns have been selected for case studies. These towns with between 10,000 and 70,000 inhabitants

Towns	Town character	Social economy	Formal economy
Barrow-in-Furness (Northwest)	Isolated industrial town at the fringe of the Lake District national park	Community Action Furness (CAF)	Encouraging Entrepreneurship (E2)
Blyth (Northeast)	Former coal mining and harbour town ouside of the Newcastle travel to work area	Briardale Community Resource and Training Centre	Community Enterprise Centre (CEC): business incubator
Schwedt (Brandenburg)	GDR new town based on petrochemical industry	Local Initiatives for Local Activities (LILA)	Biofuels Initiative: industrial development programme
Wolgast (Mecklenburg-Vorpommern)	Remote one-industry-town based on a privatised GDR naval shipyard	Production-school: integration of young unemployed	Centrepoint: start-up support office

Fig. 9.1 Towns and initiatives under study

are located outside of the main agglomerations and are characterised by similar socio-economic challenges. The economic base of all towns is rooted in their industrial past. To date, the local economy does not provide employment on a satisfactory level. Unemployment and social deprivation have risen to serious problems and call for regeneration activities. Both regions, England's North and Germany's East have been characterised by continuing population loss and long-term socio-economic deprivation as well as processes of physical decline in the last decades.

Despite differing reasons for processes of industrial decline and socio-economic impacts, the towns are comparable in the need to respond to these processes.

Comparative study approach was applied to examine common themes rather than produce independent in depth narratives. Most urban governance studies have examined the impacts of particular forms of governance. Gissendanner (2003) looked at particular initiatives and posed the question to what extent their emergence can be explained with the existence of specific forms of governance or with other factors. Governance is seen as a way of managing urban development including other actors besides traditional government. The empirical part of the research has been mainly based on semi-structured interviews with 29 key actors and executives of the selected initiatives. Reputational analysis combined with a snowball survey served as a basis to identify and rank key actors in local socio-economic regeneration.

9.3.2 Classifying Relations from the Governance Perspective

Inspired by the utilised theoretical perspectives and the categories which emerged during the interpretation of the interviews with key decision-makers and the initiatives' executives, the selected initiatives were classified along the following dimensions: 'structures', 'origin' and 'norms'. Following the theoretical debate on governance and partnership, differences between the initiatives were expected because of their structural relation to individuals and organisations, who were playing an important role in local governance arrangements. This relation can be seen as particularly intense, if the initiative's origin is directly linked to such arrangements. Following the new institutional research perspective, the initiatives were finally analysed in terms of their normative acceptance among key decision-makers. Hence, mainly from the governance's point of view, the selected initiatives can be classified along three dimensions:

1. *structural dimension*, initiatives may be integrated in local governance arrangements via their management level or via their responsible bodies (+), or they may run parallel to these structures without major links (–);
2. *origin dimension*, initiatives' origins can be closely linked to (+) or be completely outside of local governance arrangements (–);
3. *normative dimension*, in terms of contents, objectives and ideological background, initiatives may be accepted by key decision-makers and supported as an explicit part of, or closely related to local strategies (+), just tolerated or even disapproved (–).

Dimensions	Type A	Type B	Type C	Type D
Structural dimension	+	–	–	–
Origin dimension	+	+	–	–
Normative dimension	+	+	+	–
	E2 Barrow	Briardale Blyth	CAF Barrow	LILA Schwedt
	Biofuels Schwedt	CEC Blyth	Centrepoint Wolgast	Prod.schule Wolgast

Fig. 9.2 Types of socio-economic initiatives

When we group the eight analysed initiatives, following the above dimensions, we obtain four types of relations between local initiatives and urban governance (Fig. 9.2):

1. *type A* – structurally integrated initiative with overlaps in terms of key actors and organisations, having unambiguous support of central decision-makers and originating within the local governance arrangement, that is, Encouraging Entrepreneurship (E2) initiative in Barrow and Biofuels Initiative in Schwedt;
2. *type B* – initiative originating in the inner circle of local governance, having broad support but only indirect links to key actors and organisations, that is, initiatives are performed outside of the central decision-taking structure, that is, Briardale Community Centre and Community Enterprise Centre (CEC) in Blyth;
3. *type C* – initiative is supported content-wise in general, but originates outside of the local governance arrangements without any overlaps with key actors or organisations, that is, Community Action Furness (CAF) in Barrow-in-Furness and Centrepoint in Wolgast;
4. *type D* – initiative's origin has nothing to do with local governance arrangement, content-wise support is ambiguous and there is no structural integration, that is, Local Initiatives for Local Activities (LILA) in Schwedt and Produktionsschule in Wolgast.

In particular type D and, to a lesser degree, type C initiatives have difficult standing in local governance and are thus hindered in their regeneration activities. Projects and initiatives which do not originate in an environment closely linked to the local governance arrangements need to fight for recognition among local actors. Such recognition is easier to gain, when there is content-wise support. Among the eight initiatives, there is no example where initiatives originating outside of local governance arrangements got structurally integrated. None of the executives in type C and D initiatives or the organisations they belong to play an important role in local governance arrangements.

9.3.3 *Specific Forms of Strategic Governance*

In England, Local Strategic Partnerships (LSPs) are expected to take the strategic lead over urban development and also socio-economic regeneration (ODPM, 2005). Thus, they are expected to play a key role in local governance. But both, Furness Partnership and Blyth Valley Strategic Partnership are rather unimportant when it comes to major decisions.

In Blyth, the LSP makes decisions on "some of the softer sort of community side or things like that. (. . .) It's not the real decision-maker (. . .) on economic stuff in Blyth Valley" (local community representative). But it would certainly be the arena where community activists bring in their ideas to get support, "I took my project to the partnership meeting and said this is what we plan to do." In Barrow-in-Furness, the opinion about 'Furness Partnership' is exactly the same.

In Barrow, the main actors of the regional Northwest Development Agency (NWDA), the urban regeneration company West Lakes Renaissance and the local business development agency Furness Enterprise are all involved in Barrow Task Force which "coordinates the overall strategy (. . .) of all the agencies – national and regional and local – for Barrow". The group decides about projects, sets priorities for overall regeneration and puts the case for funding. The task force was put in place to address problems within economic development and "the action that needs to be taken to stabilise the Furness economy".[1] In the task force, traditional actors (such as councillors, the leader and executive officers) are just members among others. The group has been established and chaired by the NWDA, which reports to central government. In Blyth, all these strategic functions can still be seen as a core element of the local council's work.

In Germany, there is nothing similar to these partnerships and task forces. In both German cases, decision-making is dominated by public policy with the mayors as central figures. In Schwedt, in terms of infrastructure and industrial projects, the collaboration with Industrie und Handelskammer Nord (IHK Nord) might be seen as a positive example of collaboration with private actors. As an informal coordination group of the sub-regional chamber of commerce, IHK Nord unites important local business actors.[2]

The group counts nine members and meets every couple of weeks to discuss major problems of urban and regional development. The aim of this informal cooperation is to strengthen Schwedt's role as industrial location and to coordinate the communication with the Land Brandenburg. Gradually, the group became very important for local governance and is nowadays probably the most important forum for decision-making in Schwedt. "Of course it is important. Well, somebody must generate the ideas. Somebody must stand the initiatives and must confirm their importance. And this cannot be the politician" (key decision-maker in local administration). The aim to install a similar forum in Wolgast in form of a round table has not been realised so far. A new attempt has been initiated by the local agency of social affairs, Sozialagentur, in form of a strategic task force uniting all actors in the field of economic development. Generally, however, decision-making in Wolgast is very traditional. Cooperative elements are restricted to coordination with the county council and within the public policy coordination unit (Stabstelle).

9.4 The Results: Strategic Governance and Local Initiatives

9.4.1 The Role of Initiatives in Local Governance

In Encouraging Entrepreneurship (E2) in Barrow and Briardale Community Centre in Blyth, the main partners are the key players in local decision-making. The Community Enterprise Centre in Blyth and the Biofuels-initiative in Schwedt also have their origins in local governance arrangements, but the main cooperation structures include further organisations and actors. From the beginning, in Blyth Valley Borough Council, CEC has been seen as an important initiative to achieve strategic regeneration objectives, "we need the Community Enterprise Centre to be part, a 'major' sort of 'part' on delivering a step change, a transformational sort of change" (regeneration officer). In practical terms, however, cooperation with regional and national support agencies such as Business Link, Princess Trust and Social Enterprise Northumberland have been crucial for the success of the idea.

A key difference between these 'integrated' initiatives and the other four initiatives is that the 'non-integrated' initiatives first had to fight for recognition, "In the very early days, it was very hard to even get into the town hall, talk to them. (...) Suddenly 4 years along the line, 1997, we were still there. So, then they realised what we were setting up and what we were doing (...). So, we gradually started to get invited" (staff of social initiative). LILA (Schwedt) is well networked on different spatial levels in women related working groups. The main contact person for LILA to discuss general development questions is the council's equal rights representative, who, however, plays a marginal role in local governance. Community Action Furness in Barrow is included in its own networks of support and works and is, to certain degree, independent of local actors. The most important cooperation partners of CAF are outside of Barrow. CAF received most of its support from big national charitable trust organisations. Also the initiatives in Wolgast are linked to partners outside of the local governance context. Centrepoint is integrated in a regional network of similar initiatives with its headquarters in Rostock. Main ideas originate in this network, which also maintains international contacts. Via their sub-regional section, Produktionsschule is well connected with a national youth charity organisation. Initially, there were links to organisations in Denmark which developed similar initiatives. There are further plans to set up a sub-regional business advisory council to extend their production activities.

Not all executives of the selected initiatives see their work in a wider urban regeneration context. The managers of the social economy initiatives perceive their work via social impacts but do not follow general objectives of urban regeneration. These initiatives are less integrated in the system of urban governance, whereas there is a clear recognition of the formal economy initiatives. Three out of the eight executives have been recognised as key decision-makers. Having links to (and between) key decision-makers might help successful implementation of initiatives but cannot be seen as precondition for their establishment. There is a number of successful socio-economic initiatives which at the outset were not linked to key decision-makers or organisations involved in local governance (such as Centrepoint in Wolgast or CAF in Barrow). Having these links by integrating key actors in the initiative's organisation structure is no guarantee for an integration in local decision-making and better support on the local level (as with LILA in Schwedt).

9.4.2 Regeneration Agencies, Strategic Governance and Local Initiatives

In each of the four towns studied, there was space for non-state-led local initiatives to develop. This could be seen as a sign of a non-authoritarian mode of state regulation which is prevalent in Germany and in the United Kingdom. In some cases, these initiatives emerged with the clear support of or in cooperation with key individuals involved in local socio-economic governance. In this respect, local initiatives might also be regarded as an output of complex interactions of a multitude of state and non-state actors on different vertical levels (local, regional, national, European). This becomes visible when we look at the involved actors (and funding streams) of the eight initiatives studied.

In some cases, particular forms of governance may encourage or support the formation of new projects or initiatives. In many more cases, however, local initiatives are the outcome of the work of particular agencies (such as Furness Enterprise in Barrow or Sozialagentur in Wolgast) or community organisations (such as the women's association in Schwedt, Creutzfeldt-Jakob Disease (CJD) association in Wolgast and Churches Together in Barrow).

The relevant activities of these organisations led to LILA, Produktions-schule and CAF, but were not linked to particular forms of governance as the agencies are not necessarily well networked in local governance arrangements as the example of Sozialagentur in Wolgast shows. In other cases, the traditional council was instrumental for the emergence of particular initiatives (most noticeable in Blyth Valley).

A widespread view is that cooperative forms of governance (such as strategic partnerships) make it possible to access the innovative potential of a multitude of actors in urban governance. Is there any proof for such notions? Compared to the power of particular agencies to initiate and set up new initiatives, local forms of governance seem to play a marginal role in supporting the emergence of local socio-economic initiatives. In all towns there are cooperative elements of decision-making for local socio-economic regeneration.

The identified forms of strategic governance (LSPs, task forces, regeneration initiatives and round tables) are usually dominated by the public sector, in Germany clearly by the local councils. In none of the identified forms of strategic governance, a direct link has been identified to any of the studied initiatives. This also applies to the local strategic partnerships, which did not really lead to policy outcomes in terms of local initiatives.

All towns show cooperative elements of local governance, however, there is clear dominance of public actor-based decision-making. Blyth shows the highest degree of interaction with non-council actors in terms of community involvement, Schwedt in terms of local industry, Barrow in terms of public agencies and partnerships and Wolgast in terms of the relation to the county council. These different orientations are, to a degree, in line with favouring different kinds of local initiatives. If local governance arrangements are mainly based on a close relation between public actors and business actors, there might be a local preference for business initiatives. If they are based on the integration of communities, there might be a preference for community initiatives. The main supporting criteria for the emergence of socio-economic initiatives would then be the basic orientation of key actors in local decision-making or the paradigms they are following (e.g., community led regeneration in Blyth or industrial development in Schwedt).

9.5 Conclusions

The understanding of governance as it is referred to in this chapter excludes normative implications and is built upon empirical findings about a widened field of public and private actors involved in urban development. This understanding of governance denies innovation, creativity and experimentation as an integral part of governance. Urban governance can be seen as a form of managing urban development including other actors besides traditional government. Then, decision-making must be seen as multi-actor, multi-sector and not as purely based on state authoritarian mechanism. Such understanding implies a view of organisations as institutions relying on processes, formal rules, informal practices and on influences of individual actors as well as on the overall structures and specific local cultures.

It should be an objective of urban regeneration to leverage response to local problems in all possible ways. Hence, it is important for all levels of policy-making to acknowledge the regeneration potentials of non-state initiatives. However, the development of local initiatives aiming at a reduction of socio-economic deprivation and social inclusion cannot be seen simply as installation of some forms of strategic governance by pulling together a number of different actors (e.g., local strategic partnerships). As part of more complex local governance arrangements, these forms of strategic governance may help horizontal communication and collaboration between a number of involved actors and vertically with the regional and the national authorities. Nevertheless, none of the analysed initiatives in the mentioned study can be regarded as the output of such forms of strategic governance. However, it must be considered as potentially helpful for the successful implementation and operation of local initiatives. It may also help receive support from individuals or organisations linked to local governance arrangements.

In contrast to Germany, there is a wealth of strategic processes and strategic partnerships in the United Kingdom, although with very little impact. The numerous guidelines and directives imposed by the central state probably must be regarded as counterproductive and hampering local regeneration rather than supporting it. In particular, the idea of compulsory partnerships must be criticised as ineffective. It is not sufficient to create new forms of strategic governance when it comes to supporting local socio-economic initiatives. Factors of success must be related to other issues – mainly in the institutional sphere (Lang, 2008). Strategies for local regeneration, putting the main emphasis on creating new forms of governance, are likely to fall short in terms of producing policy outcomes. There are some arguments for a positive relation between specialised agencies and socio-economic initiatives as well as between general (normative), orientations and projects fitting the local agenda. Hence, the debate about specific forms of strategic governance seems to be over-emphasised in the context of local regeneration. Instead, it might be helpful to investigate the role of local mobilisation strategies and specialised agencies with a clear task towards concrete local regeneration activities.

Notes

1. Source: nwda-cms.amaze.co.uk/DocumentUploads/012003BarrowED.doc (NWDA Press Release 21 January 2003).
2. Originally, the group was installed by the former owner and manager of the local paper mill. When he came to Schwedt in 1993, he complained about the poor infrastructure. In 1995, the coordination group was installed to move things forward and fight jointly for better infrastructure.

References

Amin, A., & Thrift, N. (1995). Globalisation, institutional 'thickness' and the local economy. In P. Healey, S. Cameron, S. Davoudi, S. Graham, & A. Madanipour (Eds.), *Managing cities. The new urban context* (pp. 91–108). Chichester: Wiley.

Beaverstock, J. V., Smith, R. G., & Taylor, P. J. (1999). A roster of world cities, globalization and world cities study group and network. *Cities*, *16*(6), 445–458.

Bevir, M., & Rhodes, R. A. W. (2003). *Interpreting British Governance*. London: Routledge.

Cheshire, P. (1998). New processes of convergence and divergence in the European urban system. In J. Egeln & H. Seitz (Eds.), *Städte vor Neuen Herausforderungen* (pp. 103–130). Baden-Baden/Mannheim: Nomos-Verlag/ZEW.

Coaffee, J., & Healey, P. (2003). 'My voice: My place': Tracking transformations in urban governance. *Urban Studies*, *40*(10), 1979–1999.

Davies, J. S. (2001). *Partnerships and regimes. The politics of urban regeneration in the UK*. Aldershot: Ashgate.

Davies, J. S. (2004). Conjuncture or disjuncture? An institutionalist analysis of local regeneration partnerships in the UK. *International Journal of Urban and Regional Research, 28*(3), 570–585.

Denters, B., & Rose, L. E. (2005a). Towards local governance? In B. Denters & L. E. Rose (Eds.), *Comparing local governance. Trends and developments* (pp. 246–262). Basingstoke: Palgrave Macmillan.

Denters, B., & Rose, L. E. (Eds.). (2005b). *Comparing local governance. Trends and developments.* Basingstoke: Palgrave Macmillan.

Dicken, P. (2003). *Global shift: Reshaping the global economic map in the 21st century.* New York: Guilford Press.

Elander, I. (2002). Partnerships and urban governance. *International Social Science Journal, 54*(172), 191–204.

Gabriel, O. W., & Eisenmann, S. (2005). Germany: A new type of local government? In B. Denters & L. E. Rose (Eds.), *Comparing local governance. Trends and developments* (pp. 119–138). Basingstoke: Palgrave Macmillan.

Gissendanner, S. (2003). Methodology problems in urban governance studies. *Environment and Planning C, 21*(5), 663–685.

Healey, P., Cars, G., Madanipour, A., & de Magalhaes, C. (2002). Transforming governance, institutionalist analysis and institutional capacity. In G. Cars, P. Healey, A. Madanipour, & C. de Magalhaes (Eds.), *Urban governance, institutional capacity and social milieux* (pp. 6–28). Aldershot: Ashgate.

Jessop, B. (1998). The rise of governance and the risks of failure: The case of economic development. *International Social Science Journal, 50*(155), 29–45.

Kearns, A., & Paddison, R. (2000). New challenges for urban governance. *Urban Studies, 37*(5–6), 845–850.

Krätke, S. (1990). Städte im Umbruch. Städtische Hierarchien und Raumgefüge im Prozess Gesellschaftlicher Restrukturierung. In R. Borst, S. Krätke, M. Mayer, R. Roth, & F. Schmoll (Eds.), *Das Neue Gesicht der Städte. Theoretische Ansätze und Empirische Befunde aus der Internationalen Debatte* (pp. 7–38). Basel: Birkhäuser.

Lang, T. (2005). Socio-economic regeneration outside of agglomerations. Local economic development in the Baltic Sea region. In T. Lang, M. Sonntag, & E. Tenz (Eds.), *Small and medium-sizes cities in the Baltic Sea region: Socio-economic and cultural approaches to urban development* (pp. 13–94). Leipzig: Kirchhof und Franke.

Lang, T. (2008). *Institutional perspectives on local development in Germany and England. A comparative study about regeneration in old industrial towns experiencing decline.* Doctoral dissertation, University of Potsdam, Potsdam.

Lowndes, V. (2001). Rescuing aunt sally: Taking institutional theory seriously in urban politics. *Urban Studies, 38*(11), 1953–1971.

Newman, P., & Verpraet, G. (1999). The impacts of partnership on urban governance: Conclusions from recent European research. *Regional Studies, 33*(5), 487–491.

ODPM (Office of the Deputy Prime Minister). (2005). *Local strategic partnerships: Shaping their future. A consultation paper.* London: Office of the Deputy Prime Minister.

Painter, J., & Goodwin, M. (2000). Local governance after fordism: A regulationist perspective. In G. Stoker (Ed.), *The new politics of British local governance* (pp. 33–53). London: Palgrave Macmillan.

Pierre, J. (2005). Comparative urban governance. Uncovering complex causalities. *Urban Affairs Review, 40*(4), 446–462.

Scott, A., & Storper, M. (2003). Regions, globalization, development. *Regional Studies, 37*(6–7), 579–593.

Chapter 10
When Strategy Meets Democracy: Exploring the Limits of the 'Possible' and the Value of the 'Impossible'

Valeria Monno

10.1 Delving into the Imaginative Gap Affecting Strategic Planning

This chapter is about the diffuse perception of an increasing malaise in the ability to imagine radically different urban and regional developments which currently affects collaborative/relational strategic planning processes (Healey, 2006). Such inability is particularly visible in declining urban areas characterised by a profound socio-economic and environmental crisis. Advocates of the relational approach describe such failures as a result of the inability of local institutional contexts to learn and change thus shifting the model away from the good norms and rules.

Within this narrative on the strategic planning failure, cities in which the replication of successful models of strategic planning do not seem to work properly and the hoped-for success is hard to achieve are often reported as belonging to 'another age', or as guilty bodies, with their bodies of local unchangeable stratified knowledge, practices and routines that keep them in a locked-in status. These cities, to paraphrase Bauman (2004), can be thought of as 'waste places' unable to align themselves to globalising imaginaries and narratives of a good city and the successful politics they deserves. As such they are relegated to occupy the space of exception to good norms and rules. Consequently, these cities are forced to change not by following their own desires, but by embracing the current feelings of what a good city is and the right way to manage it.

But is this the case? Or as Friedmann (1998) often warns us the narrative of cultural and institutional barriers as a source of the failure of imagination could be considered only one way of looking for what we would like to see happen in cities? Following this doubt, in this chapter, I contrast the narrative of cultural and institutional barriers as a source of the imaginative gap characterising strategic planning episodes with the story of one of many cities which might be judged as 'unable' to

---

V. Monno (✉)
Department of Architecture and Town Planning, Polytechnic of Bari, 70125 Bari, Italy
e-mail: valeria.monno@libero.it

M. Cerreta et al. (eds.), *Making Strategies in Spatial Planning*,
Urban and Landscape Perspectives 9, DOI 10.1007/978-90-481-3106-8_10,

learn and change and argue that the relational strategic planning approach can also function as a governing paralysing meta-cultural frame.[1]

The first part of the chapter discusses the meaning of imagination within the relational strategic planning approach and adopts a Deleuzean cartography to visit the complexity of the everyday urban life. The second part of the chapter concerns the city of Taranto, its stories and suffering, its beauty and irreversible cancer and its plans and desires. This city's everyday life is described either as a complex cartography of engaging, fighting, cooperating, ignoring trajectories of evolution and change (Amin & Thrift, 2002) or as a set of lines of thinking and acting, each of them characterised by its own movement and inhabited by actants (human and non-human), forces and relations (Deleuze & Parnet, 1977). These enclose imaginations of urban futures and models of governance transformation. In their interplay, the trajectories constitute/create the tissues or the relational complexity on which the urban life is articulated. The third part of the chapter argues that strategic planning can offer a comfort zone delimitated by the space of possibilities within which socio-economic and environmental crises can be anesthetised and treated as a set of problems and solutions more or less known.

The chapter concludes by arguing that the difficulties experienced in imagining radically different urban futures in the field of strategic planning figure in its conceptualisation of the imagination as the construction of executable possibilities, which ignores the imagination of the 'impossible'. This prevents the differences/tensions between what is considered possible and impossible emancipatory urban imaginations from emerging as legitimate sources of change.

10.2 Strategy-Making and the Narrative of Imagining New Possibilities

Images in urban planning have always played a relevant role as tangible representations of desired perfect future states. At its conception they took the form of a utopic urban form (Sandercock, 1998) embedding the values and techniques to be used to transform the unjust geographies of cities and their everyday lives. These were conceived as transgressive and subversive repositories of hope challenging the status quo and taken-for-granted conceptions of urban life. However, utopian physical images were intensively instrumentally used as a powerful framework for modernising cities seen as fixed and centred (Bridge, 2006) rather than producing in them an emancipatory social change (Pinder, 2002).

Post-modernist critiques have shown the illusiveness of utopian technocratic images since they unavoidably impose and fix a specific space and social order (Harvey, 1996; Hayden, 1995; Rodwin, 1981) thus reproducing existing injustices. These criticisms also reflected the need of changing the idea of a city as a 'fix' into one embracing flows and differences (Castells, 1996; Young, 1990). By the 1990s, urban imagination had taken the form of a process with weak links to spatial constructions. It could be termed the 'utopia in becoming' (Sandercock, 1998) that substitutes the modernist physical image with a political progressive project which realises itself in the making. "From this perspective any emancipatory politics calls for a living utopianism of process as opposed to the dead utopianism of spatialised urban form" (Harvey, 1996, p. 436). Against the technocratic-utopian planning, radical planning is not concerned about imagining the future. It works in the present to grasp prospective futures of the cities (Friedman, 2002) and fight against injustice by empowering local communities through an unconstrained encounter between expert and experiential knowledge (Sandercock, 1998).

In Europe, the relational strategic planning approach (Albrechts, 2009) seemed to provide a robust theory to a more pluralist and pragmatic conception of utopia in the making. Led by a conceptualisation of place as fragmented (Healey, 1997) and inspired by theories of the deliberative democracy and a relational conception of places (Massey, 1994, 2005), this pluralist-democratic version of strategic planning abandons utopianism and its emphasis on conflict to adopt a conception of planning as consensual practice (Feinstein, 1999). In contrast to the *hinc et nunc* ideal of radical planning (Friedman, 1994), the relational perspective conceives planning as a practice not only aimed at managing existing relations but at imagining and opening up future possibilities for improving the conditions of daily life existence (Healey, 2009). New urban possibilities have to be searched through placed-focused argumentative/consensual/persuasive strategy-making processes (Healey, 2006) aimed at the social construction of a shared vision.

Planning in a fragmented society implies capturing the dynamic and relational 'nature' of places (Massey, 1994, 2005) by means of development of a politics which encounters the actors who populate the multiple networks that the place is embroiled in and provide them with inclusive and collaborative/deliberative egalitarian arenas. Within them discursive struggles can take place in order to change actors' convictions and make them converge in a shared vision for an improved quality of places. A relational strategy-making activity does not only aim at capturing the place's relational complexity, but it also tries to change established and dynamic relations among actors and the relationships between the actors and places. Improving the quality of place "is more than just producing collective decisions. It is about shifting and re-shaping convictions" (Healey, 1997, p. 244). Strategy-making and changes are indeed inescapably linked (Davoudi & Strange, 2008).

The relational strategic planning "invents, or creates, futures – in relation to the context, the social and cultural values to which a particular place/society is historically committed – as something new rather than as a solution arrived at as a result of existing trends" (see Chapter 1, this volume). "A willed future is a clear reaction against the future as a mere extension of the here and now. On the other hand, the future cannot be so open that anything is possible, as though we could achieve anything we want to achieve. Conditions and constraints on 'what is' and 'what is not' possible are placed by the past and the present. These conditions and constraints have to be questioned and challenged in the process, given the specific context of place and time. So, in order to imagine differently the conditions and constraints for the future, we need to deal with history and to overcome history. Therefore, we also need an exploratory approach. The interrelation between the normative and the exploratory approach, defines the boundaries of a fairly large space between openness and fixity" (see Chapter 1, this book).

Strategy-making is a crucial activity for mobilising and filtering concepts of spatial organisation (strategic frames) having sufficient allocative, authoritative and imaginative force to shape both the materialities and identities of particular places and the networks which transect and give value to them (Healey, 2006, p. 527). The consensual feature prevents the vision from remaining a mere utopia. In this way, 'permanences' are created in the dynamic relational dialectics of urban life. From such a perspective imagining a different place quality is about combining an appreciation of the open, dynamic, multiple and emergent nature of social relations with some degree of a stabilising force (Healey, 2006).

Despite the enormous success which relational strategic planning has obtained in recent years its results in terms of improvement of the quality of places have been modest (Gaffikin & Sterrett, 2006; Healey, 2006). As advocates of the strategic planning approach now admit, current endeavours in strategic spatial planning are experiencing difficulties in creating emancipatory spatial imaginations. However, for them, this is firstly due to contextual, institutional and cultural barriers. These are signalled by the persistence of traditional physicalist concepts about spatial order which fail to capture the dynamics and tensions of relations coexisting in particular places and a rhetorical commitment to inclusivity which limits perceptions of diversity and causes deliberate exclusions (Healey, 2006).

Critics have instead argued that the relational approach to spatial planning is in effect unable to generate alternative ways of confronting the consequences of uneven development because of substantive rather than contextual constraints (Allmendinger & Haughton, 2009; Flyvbjerg, 1998, 2002; Huxley, 2000; Purcell, 2009; Yiftachel, 1998). The narrative of change embedded in the strategic planning approach is one articulated around the missing link between knowledge and power; the missing link between the planning and the dynamics of urbanisation (Beauregard, 1990) – or as Bridge (2006) has called this phenomenon it is a matter of an 'urbanism without cities' – the violence of the consensual logic (Mouffe, 2000, 2005; Swyngedouw, 2009); and a methodological individualism which is unable to face issues and dilemmas concerning the collective action (Monno forthcoming; Young, 2001). I would like to add another criticism to these which, perhaps, is less practised. It stresses the limits of an idea of strategic planning as a practice which, by conceiving the future as something that cannot be so open that anything is possible, delimits the concept of imagination as a possible, executable future. At this idea of future, 'utopian' critics oppose the utopia to show how strategic planning traces the boundaries between the possible and the impossible (Pinder, 2002, 2005; Baeten, 2002).

This chapter draws on the utopia criticism but with some differences. It is not so much concerned about the definition of what could be today a utopia. It is, instead, interested in dealing with the relationships between the relational strategic planning and what is considered an 'impossible' change. In some ways, it is similar to an exploration of the function of utopia in history (Jameson, 2005), even if it is not concerned with utopia as the 'Other' but with imaginations as a framework to see the difference between what is gained and lost through the planning process. The understanding of the differences/tensions between what is retained as possible (the shared vision) and what is considered an 'impossible' radical change (the imagination which appears impossible to be carried out) might enable citizens to understand the material and immaterial collective 'costs' that the relational strategic planning produces and, thus, help them to make a more conscious choice about their own future.

This perspective draws on a conception of the 'impossible' as an emancipatory imagination which is based on a presupposition of equality: the equality of people qua speaking people (Rancière, 1998). This "is an an-archic equality in the sense that it exists through the inability of any political order to count the communal parts and to distribute the shares of the common between them under the harmonious geometrical governance of some *arkhe* (the principle of Justice, of the Good) without there being a fundamental wrong [le tort] done; a miscount, which is then where the politics begins" (Arsenjuk, 2007 p. 1).

As such the 'impossible' imagination is political. As Rancière explains (1998, pp. 32–33): "nothing is political in itself. But anything may become political if it gives rise to a meeting of these two logics [police logic, which is opposed to egalitarian/political logic]. The same thing – an election, a strike, a demonstration – can give rise to politics or not give rise to politics. A strike is not political when it calls for reforms rather than a better deal or when it attacks the relationships of authority rather than the inadequacy of wages. It is political when it reconfigures the relationships that determine the workplace in its relation to the community. The domestic household has been turned into a political space not through the simple fact that power relationships are at work in it but because it was the subject of an argument in a dispute over the capacity of women in the community."

10.3 A Deleuzian Interpretation of Cities: Seeing the Urban as a Cartography of Trajectories of Thinking and Action

> What distinguishes the map from the tracing is that it is entirely oriented toward an experimentation in contact with the real. The map does not reproduce an unconscious closed in upon itself; it constructs the unconscious. (. . .) The map is open and connectable in all of its dimensions; it is detachable, reversible, susceptible to constant modification (Deleuze & Guattari, 1987, p. 12).

How to inquire into the relationships between the strategic planning and what is considered an 'impossible' change? What is the conjunction point between the two? I see this conjunction as being the city itself, its everyday life rhythms and flows of nomadic and no-nomadic lives, its 'nature' metabolic (Heynen, Kaika, & Swyngedouw, 2006) and perpetual being embroiled in imagining/representing beyond what already is (Castoriadis, 1987, 1997) within powerful disciplining boundaries (Foucault, 1991). This is a conceptualisation of the city as a heterogeneous place, the troubled coming-together of a multitude of human and non-human agents and their struggle between what it is and their dreams of something different, which trace the city as a set of trajectories of evolution and change (Amin & Thrift, 2002). This is also a conceptualisation of the urban complexity whose exploration needs to break usual ways of interpretation of urban life. Deleuze and Guattari's philosophy, despite its obscure and esoteric language articulated around key concepts which defy any systematised knowledge, can represent a theory enabling such exploration.

Reading Deleuze's and Deleuze and Guattari's writings for a planner is always fascinating since it deals with a question which is at the base of any planning endeavour about 'how might we live differently in a city or in a neighbourhood'? In a world that holds banality to be a virtue and originality a disease, Deleuze's engagement with the question 'how might one live?' continually challenges us to think about what other possibilities life holds open for us and how we might think about things in ways that would open up new regions of life (May, 2005, p. 3). For him this implied thinking the unthinkable (Deleuze & Parnet, 1977): thinking outside the codes to create a constellation of concepts functioning as a toolbox enabling the encounter and experimentation of new possibilities of living together.

At the same time Deleuze's theory of creation rather than of discovery (May, 2005) forces a reader to rethink the cartography of the urban space as oriented not so much to explain the true nature of things, but to know reality through its dynamics. As such, it opens the possibility to an unusual journey into the urban complexity which pushes us to abandon the dualism one/many individual/collective resistance/dominion and rethink the urban as "multitude so that we can cease treating the multiple as a numerical fragment of a lost Unity or Totality or as the organic element of a Unity or Totality yet to come, and instead distinguish between different types of multiplicity" (Deleuze & Guattari, 1987, p. 32). "A multiplicity has neither subject nor object, only determinations, magnitudes, and dimensions that cannot increase in number without the multiplicity changing in nature (the laws of combination therefore increase in number as the multiplicity grows)" (Deleuze & Guattari, 1987, p. 8).

They call the different types of multiplicity 'assemblage': "An assemblage is precisely this increase in the dimensions of a multiplicity that necessarily changes in nature as it expands its connections" (Deleuze & Guattari, 1987, p. 8). Specifically, an assemblage is a multiplicity which is made up of many heterogeneous terms and which establishes *liasions*, relations between them. These *liasions* link an assemblage's material content (passions, actions, bodies) and enunciations (laws, plans, statements) in a non-linear relation (Deleuze & Guattari, 1987). Thus an assemblage "it is never filiations which are important, but alliances, alloys; these are not successions, lines of descent, but contagions, epidemics, the wind" (Deleuze & Parnet, 1977, p. 69). The reciprocal movement among forces (*liasions*) which operates inside and between different assemblage determines their acting as disciplining or transgressive (out of the order and opening up to a new order).

The concept of assemblage does more than focus on the urban complexity a set of distributed actors and their relationships. Groups, collectives and agents within the assemblage constitute a topography of changing field of forces (*liasions*) that cross or engage with each other to different extents over time rather than act as a set of static, predefined positions and interests. Within it, there is no single central governing power, nor an equally distributed power, but rather there is power acting as a field of forces or as plurality in transformation (McFarlane, 2009). The assemblage evolves and changes through the practice of reassembling and disassembling which modifies their field of forces (*liasions*).

The Deleuzean and Guttarian cartography of social space is a map oriented at capturing the movement of "the lines that we are" (Deleuze & Parnet, 1977, p. 124). The line (*liasion* or field of forces) as opposed to the 'point' is a dynamic element. It is a true becoming, a turning point between the past and future, always silently working. As such, a line belongs to geography since it is orientations, directions, entries and exits. Thus "the map is open and connectable in all of its dimensions; it is detachable, reversible, susceptible to constant modification. (...) A map has multiple entryways, as opposed to the tracing, which always comes back 'to the same' " (Deleuze & Guattari, 1987, p. 12).

Though surely trivialising their complex thought, in this chapter I take the Deleuzean and Guttarian cartography as a conceptual tool to inquire into the relationships between the strategic planning and what is considered an 'impossible' change. Through this map it is possible to identify the trajectories whose realisation is considered possible or impossible in relation to a specific strategic planning process. Thus I describe Taranto through a map representing the 'lines' which city of Taranto is. On the map, Taranto appears as an ensemble of assemblages which, individually, can be represented by means of the line or field of forces characterising each of them. I call the line 'trajectory of thinking and action'. However, each trajectory is seen in this chapter from a particular angle. It considers only the force of imagination. This is characterised as the experience of future in relation to the memory and the experience of the present. Actors are hybrid. They move inside and through the different trajectories which in some moments represent their feelings and desires. Consequently, each trajectory can be shared or rejected by different actors in different moments. Strategic planning in the map is an event in which the theory meets a tool and a context and materialises itself into a practice. The strategic plan encounters the different trajectories which constitute the tissue of urban complexity because of its necessity of re-defining the borders between the possible and impossible. The map represents the complexity of Taranto through five trajectories: fear, identity, movements, experimentation and planning.

This way of describing imaginations is oriented to understand how and with what trajectories a strategic plan interacts and is intended to evaluate how such interaction shapes the direction of change. In fact, if citizens were able to understand the differences/tensions between what is retained as possible (the plan) and what is considered a radical change (the imagination which seems impossible to be carried out), then they might be able to understand the material and immaterial collective 'costs' that the direction of change traced by the plan produces and make more conscious choices about their own future as part of a collectivity.

10.4 Taranto: Is this a City Unable to Change?

Taranto, a southern Italian medium-sized city, is well known in Italy as the 'steel town'. Since the 1960s, the fate of this city has been associated with a large steel plant (ILVA), which in the 1970s employed more than 20,000 workers. Once owned by the state, in 1996 the steel plant was privatised. Even after the huge job cuts during the 1980s and 1990s and following its privatisation, this steel plant is still one of the largest in Europe. Before the financial crisis it employed around 13,000 workers. With the rise of global economy, Taranto has experienced a continuous social and economic decay. The industrial development which would have made the city and its hinterland a wealthy and modern territory has not only failed to promote local development, but it has also reduced both the Taranto and the sub-region into a polluted and isolated area populated mainly by a working class (Barbanente & Monno, 2004).

In 1998 the Province of Taranto was declared at risk of environmental crisis by the National Government (D.P.R. 23.04.1998, GU number 196 30 November 1998) and thus in an urgent need of an environmental plan aimed at cleaning up the whole area. The plan has never been carried out, and subsequent national laws have reduced the area to be cleaned to few scattered polluted sites. Since then, the environmental crisis has worsened, up to the point that Taranto is currently one of the most polluted and polluting cities in Europe. In order to face the social–economic decline of this city and its hinterland, the port of Taranto was enlarged and modernised. And in fact, at the beginning of 2000 it became one of the most important transhipment ports in the Mediterranean. Despite this, it only employs around 500 workers.

The city is characterised by an underequipped public administration, sometimes transacted by corruption, and entrapped in an old logic of political power. Only recently have some local governments started trying to transform a top-down style of government into a more democratic governance-based one. NGOs have flourished around issues such as health and environmental protection and cultural renaissance.

10.4.1 Trajectories of Change: Telling Taranto Through the 'Lines that It Is'

The map which describes Taranto as trajectories of thinking and action is based on the results of an ongoing research focused on possibilities of a radical environmental regeneration of this city. Its purpose is moving beyond the *habitus*[2] as explanation of its decline to begin to explore the role played out by existing/emerging urban imaginations transecting this city. During the last 30 years different kinds of descriptions have inquired into its decline. Some have explained it as a result of the loss of local identity determined by the industrial culture. Others have emphasised the distorted nature of modernisation in southern Italy (Rinella, 2002). In more recent accounts, Taranto appears as the city of clouds (Vulpio, 2009), the city of dioxin and cancer or of a port city blocked by local and national blind politics. However, these accounts thus focused on Taranto's urban pathologies tend to think of this city as a homogenous place and identify exogenous forces and a local inability to change its *habitus* as being causes of the decline.

In what follows I abandon this point of view and to describe Taranto as a map of trajectories of thinking and action which draws on the results of a research aimed at identifying the circulating imagination on the future of this city. The research has been grounded on walking and asking through the 'visible city' and its lived spaces. At the same time, it has been based on a series of strategic choice experiments involving students attending my classes and local stakeholders. Interviews have been carried out with the aim of identifying the imaginations flowing into the always in flux "palimpsest of overlapping, coinciding, colliding shifting meanings" (Friedmann, 1999, p. 7) and 'different possible worlds' (Rajchman, 1998, p. 117). The strategic choice experiments have been used as heuristic tools aimed at highlighting the relevant trajectories of change rather than means to construct a future image of the city. Obviously, the map is not intended to give a comprehensive and comprehensive account of the city. It just recounts the trajectories which the research has been able to recognise and meet.

In what follows Taranto is taken as a typical example of a city considered unable to change. As such, this city is seen as a useful case to evaluate whether the failure of strategic planning is only due to cultural barriers or whether it also depends on its concept of the future as a possible future.

10.4.1.1 Fear

Jacques Rancière (2004) argues that what is at stake in politics, just as it is in aesthetics, is the distribution of the sensible. Politics happens not only through the disruption of a certain aesthetic organisation of sense experience but also through the eruption of a distinct aesthetics. "It is thus that the task of politics becomes one of producing and forcing into everyday experience a distinct organization of the sensible, conditioned by a distinct aesthetics" (Wolfe, 2006 p. 1).

In Taranto the colour red of coal represents the local partition of the sensible. This is the colour of silence, of an ecology of fear of losing one's own job, the imagination of the big disaster (Davis, 1999) – the closure of ILVA – which shapes the social, economic and political rhythm of everyday urban life and the production of the nature of the city (Heynen et al., 2006). It keeps people silent even when ILVA threatens new dismissal without notice. At the same time it has the power to free the public landscape from the bad dreams: nothing has changed since the steel plant arrived in Taranto.

The colour red is also the colour of the local population's distrust of in public institutions and its retreat from the democratic life. It is the colour of the crisis of democracy which worries so many scholars. Citizens either delegate decisions, rather than participating, or do not react to the corruption of public administrators. Instead, they try to profit from this state of affairs. Despite the dead sea, the polluted air and the increasing rate of death by cancer, people only protest if someone proposes to close ILVA. Everything else (like dioxin, the opening of a new risky industrial activity, etc.) leaves the people indifferent so long as it does not threaten their job. This trajectory is a memory of survival and resignation, which thinks about the future through the 'benefit' of the present. Within it the experience of future cannot be disconnected by the constraints shaping current everyday life. The desire of change is looked for outside of Taranto, or in the many myths of consumerism or the social climbing from the working- to the middle-class.

10.4.1.2 Identity

Identity draws on nostalgia, a word which in politics usually hints at "a problem of the imperfect assimilation of the categories and practices of history, that is, the condition of those who did not have what in modernity gradually became the dominant relationship to the past". Such a definition is based on a conception of history as necessarily emancipatory, progressive and rationally comprehensible. In it a social conflict and injustice are concealed in "idealized representations of the past" (Natali, 2004, p. 10). But at the same time it can be a source of a critical thinking and as such can induce change (Natali, 2004). In Taranto nostalgia operates in both these two meanings. Nostalgia acts both as a sweet paralysing memory of a lost hope and success and as a weak critical thinking.

This trajectory sees the future as the overcoming of the current disillusions through the replication of the model of Taranto as a growth Pole. The present is lived as a suffering generated by exogenous causes which are beyond the local people's control. The port of Taranto (again a mega-project funded by the state) is seen as a symbol of the lost industrial grandeur of the city partially replacing it with a more attractive and post-modern skyline. It expresses the desire of a powerful economic political coalition to expand to the Mediterranean Sea the influence of Taranto as a city-region. It believes that the ecological modernisation can solve this city's decline. Sustainability, when it is evoked, is usually conceptualised a soft sustainability which subordinates issues of environmental and social justice to competitiveness.

In fact it has also supported the transformation of large Taranto areas into a free-market zone which will attract new capitals, while subtracting further public spaces from the city.

10.4.1.3 Movements

New 'origins' for a local politics already exist in the city. They are embedded in the everyday experience which contributes to and promotes forms of political action, which question prevalent ways of thinking about the city in the light of a radical future imagination: Taranto without ILVA. Within this trajectory there is no disconnection between the past, the present and the future. Experiencing the future means acting in the present with a critical memory. This has allowed the myth of Taranto's grandeur to be rewritten as a story of death rather than a new progressive life. This trajectory resists univocal integration into emerging hegemonic forms of political domination by practising a form of action which could be termed cooperative-autonomy. To paraphrase Gandy (2005, p. 33), this trajectory proposes a new kind of human agency to the intersection between technological change and the reformulation of the public sphere inherited from the industrial city.

Taranto Sociale, Traranto Viva, Peacelink, The Committee of Taranto's Blog are as virtual as material alliances which speak a new language based on human flourishing, environmental, economic and social care and justice, and solidarity. Their aim is to connect different people to constitute a new active public in the city and a new way of imagining the city as no longer being necessarily managed by a centralised power which acts as an collective actor. The city is imagined as led by a critical alliance between citizens and institutions. They do not intend to construct a good community but only connect the vital forces of the city to change the passivity, individualism and egoism currently shaping urban life.

These networks are attempting to reassemble the social through a patient practice of knowing and acting which reshapes the force field governing Taranto's decline. In fact their arguments/contestation are always based on a practice of knowing and not intended to be included in decisional arenas. They practice what has been recently described as the 'civic science' (Scott & Clive Barnett, 2009). For this trajectory "there is an inseparability between action and experience: (. . .) every act of knowing brings forth a world" (Maturana & Varela, 1987, p. 26). The networks have carried out many surveys on urban air quality and cancer disease and have denounced and demonstrated how and to what extent ILVA ignores humans and non-humans' right for health. Their civic science stopped a re-gasification plant being localised in Taranto on a site too close to the petrochemical plant thus avoiding the risk of a catastrophic explosion. For these reasons these networks' knowledge claims represent a direct challenge to well-established powers.

10.4.1.4 Experimentation

'Taranto does not sleep', 'Reawakening Taranto' are expressions currently circulating in the city. They are injected into the urban life by a group of people 'PUNTO E A CAPO' who are trying to make the younger generation take a creative approach to urban politics. These kinds of associations can be termed as performing the 'outside' politics or strategies of political engagement after representation, that is, strategies of dis-identification (Stephenson & Papadopoulos, 2006) which are performed by actors who circumvent the ascription of both the dominant trajectories of identity and fear. Perhaps, as Derrida (2001) maintains there is a link between the dis-identification and the forgiving of the unforgivable: dis-identification as forgiving the city for not having reacted to the decline and suicide of the city. After the failure, the city does not need a new vision which risks producing a new collapse of the ability to imagine a different Taranto. This trajectory thinks about the past in a critical way in order to change the city in the present. Instead, what is needed consists in changing its *habitus*. The city needs to bridge the cultural gap by turning silence into action, that is, into a practical coping with the wicked problems of the city through creative writing or other forms of art. The few bookshops existing in a city in which there is not even a theatre have been transformed into the meeting points in order to enact this creative endeavour. Art can play a relevant role in bringing about a change. These associations represent a shared idea of the city as a laboratory. This is seen as a way both to break the rhythm of routines and to manage stubborn conflicts avoiding direct mediations and negotiation. Facing stubborn conflicts through formal meetings aimed at reconciling adversary interests to solve conflicts is a slow process of everyday action. Forgiveness is eminently an act, an experience.

10.4.1.5 Planning

This trajectory is characterised by a static memory of planning. It has neither a perception nor systematised knowledge of the current needs of Taranto's population. It has no imagination concerning the future of the city and planning opportunities to act and change Taranto's decline. It only tries to use innovative planning tools and opportunities to occasionally obtain funds and manage Taranto's decay. The alteration of institutional memory which followed the industrialisation has even resulted in a refusal to imitate other cities' management practices, giving preference to imitation of one's own past, a phenomenon called *automorphism*, in contrast to the *isomorphism* prevalent in modern management. This leaves the organisation field empty. If it is easy to bring new solutions, it can be very difficult to make them operative (Czarniawska, 2002). This trajectory has been marked by the ineffectiveness, patronage and only a few episodes of an innovative practice. Among these the Rehabilitation Plan of the extremely decaying old city, the participative Poseidonia project which aimed at protecting the Mar Piccolo, the Urban II European Initiative funded by the EU–FESR and an integrated program 'Contratto di Quartiere Salinella'. The master plan was approved in the 1970s is still regulating Taranto's development.

10.4.2 Strategic Planning in Taranto

Strategic planning is not compulsory in Italy: it is a voluntary agreement. At the beginning of the new century, following the stories of success of strategic plans carried out in Europe (Barcellona, Bilbao, etc.), different Italian cities started to make their own strategic plans. In this period Taranto too produced its first urban strategic plan.

However, the strategic planning approach spread throughout Southern Italy after the ICEP (Interministerial Committee for Economic Programming) resolution number 20 was issued in 2004. The resolution redistributed Funds for Underexploited Areas (FUA) among the Southern Italian underexploited urban areas in order to contain their decline and peripherisation process in an enlarged European Space. It was specifically aimed at improving Southern Italian medium-sized cities and the infrastructure systems of metropolitan areas to transform them into good European cities competing as nodes or poles of development for larger sub-regional areas in the global economy.

Against an old and inflexible national planning normative apparatus, which still forces cities to plan their development through master plans, the resolution aimed at exclusively funding projects and programmes included in a strategic plan. Due to the absence of any national laws concerning strategic spatial planning, guidelines were drawn up to define a possible methodological approach to the strategic planning. The guidelines define a strategic plan as a voluntary and collaborative effort of strategy-making aimed at balancing different and competing interests in a flexible and open shared vision, which is the result of a governance process among local/relevant actors.

The process is seen as articulated in phases. It starts with the analysis of the context. Next it focuses on formulation of a preliminary vision which represents the main issues to be discussed and renegotiated within governance processes and ends with the approval of a shared vision and a set of projects allowing its implementation. The leaders of strategy-making processes can be only those cities able to play the role of nodes or poles of development.

As far as the Apulia Region is concerned, the Regional Government promoted a competition in 2005 to allocate the FUA funds among urban areas which intended to plan their development by means of a strategic plan. The competition was won by nine strategic plans. Two years later, in 2007, the new Regional Government decided to update this list in order to integrate the FUA funds not yet allocated with the European Structural Funds (ESF) 2007–2013 and promote a more efficient and sustainable local territorialised development. Basically, this updating process added a new strategic plan to the list, changed the urban areas into sub-regional areas and gave a strong economic imprinting to the strategic planning process. Thus, it drew up compulsory guidelines resembling the national ones to help the sub-regions to develop their own strategic plan.

The Taranto sub-regional Strategic Plan is one of the ten sub-regional strategic plans funded by the Regional Government. It includes the Taranto Provincial Government, 28 out of the 29 municipalities of the Taranto Province and is led by the Taranto Municipality, the capital of the Province. In this rural sub-region Taranto is the only medium-sized city with more than 200,000 inhabitants having the economic and institutional qualities required to lead the sub-regional strategy-making process. Most of municipalities are small cities gravitating around the Taranto urban area or rural towns currently trying to profit from tourism. Among them, only four have a population of between 20,000 and 30,000; the others having populations ranging from 1,500 to 20,000 inhabitants. Around 500,000 inhabitants live in the whole sub-region. In practice, the sub-regional strategic plan is a strategic plan mainly aimed at sustaining and promoting the city of Taranto's competitiveness.

The first strategic plan was intended to face the social-economic decline of this city and its hinterland. It was commissioned by the Taranto Municipality to a group of experts. The plan, which was made following the technocratic tradition, associated the industrial image of the city with a lively, post-industrial one: that of Taranto as a port city playing the role of a crucial node in the relational space of the global economy. The enlargement and modernisation of the port of Taranto, which had made it one of the most important in the Mediterranean, was seen as a feasible alternative to the industrial 'monoculture'. It would have to favour the emergence of a new regional development path more suitable to the post-industrial and knowledge-based economy on which so many European cities have flourished (Barbanente & Monno, 2004).

Despite the risk that the steel plant could compromise the post-industrial global image of Taranto, none of the relevant stakeholders questioned its permanence in the area. The plan intercepted and strengthened the identity of Taranto as a pole of growth and set up a framework aimed at enhancing this city's competitiveness. Citizens knew about the plan only when it was officially shown to the city. However, they did not react. In fact, the strategic plan neither changed the industrial monoculture nor did it directly influence the land use since in Italy strategic planning is not compulsory. At that time, the impulse towards experimentation and radical change was weak or almost invisible. Even though the vision promoted a new, more modern image of the city, the strategic plan continued a regional and urban development based on the ecology of fear – the fear Taranto's inhabitants have of losing a stable job. The steel plant was (and still is) seen by the local population as the only opportunity for a stable job. The plan has never been implemented but its vision, which implicitly accepted the environmental decay in Taranto produced by the steel plant, came to catalyse the new dominant urban imagination.

The second strategic planning process was started in Taranto in 2007. It began after the setting up of the Planning Board (PB) as prescribed by the Apulia Regional Guidelines. The strategy was constructed through a governance process including all the public institutions concerned, a socio-economic partnership representing local needs and interests, and an enlarged public to be informed and listened to through public assemblies.

The governance process was neither opposed by citizens nor by local NGOs. On the one hand, an enlarged citizens' participation was not organised due to the local underequipped public administration and also in order to limit the possibility of irresolvable conflicts arising. On the other hand, citizens, even citizens who understood the relevance of a strategic planning process, were not interested in participating in it: "this was business as usual". As far as NGOs were concerned, most of them were struggling to obtain an environmental monitoring system and a regional law which could force the steel plant to contain the pollution within the thresholds imposed by the National and European Union resolutions and laws. As one of the most well-known environmental activists said to me to justify such an absence: "Although I have been invited I cannot participate in any process." Citizens and relevant actors belonging to the trajectories of experimentation and radical change were not very interested in negotiating their imaginations.

Analogously no professional organisations contested the strategy-making process. In fact, out of the formal governance process a Technical-Scientific Observatory (TSO) having the role of monitoring the quality of the strategic planning process was set up. It was managed by the Provincial Architects Corporation whose head is the public official responsible for the Taranto sub-regional strategic plan. The observatory included local professionals, academicians and socio-economic actors. Although the setting up of the TSO as a certificatory institution was a decision taken autonomously by architects and the public official responsible for the PB, neither the policy actors nor the socio-economic partnership ever questioned such a role. The fact that it existed as the node of networks of expertise not directly involved in the strategy-making process which act as supervisor of the goodness of the planning process was a sufficient explanation to legitimise it.

As a result, both the governance process and the vision were proclaimed a success by the PB and policy actors. It had been the result of a consensual innovative governance process which had changed the unfair local decision-making practices and routines. Its consensual character signalled a change in the actors' knowledge frames and convictions. At the same time, it was considered a new fresh imagination of the city. The vision basically proposed the ideal of Taranto as a local and transnational node in the global relational space which well complied with the image of a competitive sustainable and creative good European city fed by National and European funds and the old myth of a regional development based on Taranto as a port city. The role of Taranto had also been rethought in a multi-level (global and local) and multi-functional perspective which exploited the opportunity of present (the steel plant included). This was despite the fact that the vision of the Taranto sub-region as a translocal and local node had not substantially changed the previously one which had been constructed by adopting a stronger technocratic approach.

Once agreement on the vision had almost been reached, thematic forums were organised by the Taranto Board (TB) to mobilise both relevant institutional and socio-economic stakeholders' local creativity and translate the vision into projects. In most cases no innovative ideas came out of the forum, only old projects awaiting funding, such as the regeneration of old Taranto's Navy Yard, the construction of a logistic platform close to the Taranto port and the dredging of the port seabed. The first represented the symbol of a new tourist renaissance, the second, the necessary link between the port and local economy and the third, the necessity to adapt the port to the dimensions of the huge new ships. The only new proposal was presented by a group of academics who came up with the idea of constructing a new scientific-technological pole in Taranto. Since this proposal filled the knowledge gap underlying the vision it was accepted by all the relevant stakeholders and taken as one of the symbols of the Taranto renaissance.

Under the pressure of funding mechanisms, a first call for projects to be funded was announced and almost 400 were submitted by the municipalities and NGOs, plus about 700 hundred by private actors. In order to avoid the choice of projects being compromised by the interference of powerful coalitions it was decided to set up a Scientific-Technical Committee (STC) composed of academics and technicians working in key local institutions. The Committee was to evaluate the projects and select a first group which could be funded in the short-term. It is at this point that the strategic planning process intersects the trajectories of movements and experimentation. Some of the STC technicians and academics had been active in promoting the cleaning of Taranto or the rehabilitation of its beautiful historical and cultural heritage. Some of them as doctors know the rate of death in the city very well. Others had contrasted the possible localisation in the city of a regasification plant, which due to its proximity to the petrochemical plant would be a source of another risk for the city.

What policy actors and the PB expected from this group of experts was a choice based on the result of the governance process. In some way, the STC would be the legitimating-actor for the vision and for some symbolic or 'necessary' projects which emerged in the forums. Contrary to all expectations, the STC decided that the evaluation should be considered as advice rather than a final decision. Thus, after a quick look at the list of projects the STC decided to adopt a simplified multi-criteria evaluation method which could easily be understood by all the relevant stakeholders and an enlarged public. The projects were to be evaluated on the basis of different criteria such as sustainability, integration among projects, coherence with the vision and others. The judgement to be assigned to any project in relation to a specific criteria would be the output of a dialogical confrontation among the experts. This would also favour an intriguing mixture of scientific and professional knowledge. The STC decided to begin by evaluating the institutional stakeholder's projects.

At first, the evaluation process appeared to be characterised by good agreement between the STC and the TB. Yet, as the process proceeded, the agreement slowly faded away, despite the fact that during the many meetings organised to explain the evaluation method nobody had objected to it. The set of projects which ranked high in evaluation did not correspond with those considered necessary by local stakeholders. Projects such as the regeneration of the old Navy Yard, the dredging of the port seabed and many others showed on deeper examination problems of environmental compatibility or poor integration with the overall strategy or lack of coherence with the vision.

The evaluation appeared as a direct challenge to the vision emerging from the governance process. The set of projects selected favoured a diffuse environmental and social requalification based on the development of multiple cognitive capacities engaged in redefining the industrial vocation. It also challenged the idea of a sustainable development of the sub-region as a result of the use of eco-technologies and a practice based on an efficient funding of mega-projects. For example, if the exclusion of the dredging project unveiled the environmental impacts produced by the enlargement of the port, the exclusion of the Navy Yard regeneration project showed the weakness of the vision in challenging the industrial monoculture. Similarly, the inclusion of a deprived neighbourhood regeneration project signalled the social weakness of the vision.

To sum up, by showing that the vision left the sub-regional economic, social and environmental metabolism untouched, the evaluation potentially reopened the consensus achieved for contestation. New actors who up to that moment had not been fully involved in the governance process started to make themselves heard. Despite the fact that the evaluation was to be considered only as advice, its reopening some crucial development options to a more careful examination, determined a new but deeper knowledge controversy among experts and policy-makers. The PB and most of the policy actors rejected the evaluation by arguing that it had no validity since it had not been discussed with them previously. The vision was not changed since it was considered the only possible, executable shared vision. Consequently, almost every project was funded.

10.5 Relationships Between Strategic Planning and the Trajectories

Taranto is currently traced by multiple trajectories of thinking and action. Each of them connects the past, present and future in a specific way which gives rise to different imaginations. Among them the trajectories of fear, identity and planning adopt a linear conception of time in which memory, the present and future are stocks related in a sequential movement. Even from different perspectives, these trajectories see the future as a prosecution of the past and the change as a disconnection to be represented as something real and to be controlled by means of the parameters dictated by the present. Difference between the past, present and future is a matter of difference between absolute qualities. Obviously, the qualities of future can be better than the current ones only when they derive from a reasonable mediation between the accumulated experience, the immediate necessities and the future uncertainty. The imagination is constrained by limits and possibilities enclosed in the present experience since the future is beyond control. In this way imagination neither questions the memory nor does it challenge Taranto's urban metabolism and pathologies as necessary costs to be paid to get ahead. Concepts such as sustainability, tools such as strategic planning or occasional funds offer the looked-for/desired solution to this city's problems, a safe path showing how things should be done in order to go from here to there.

In contrast to these trajectories, the movements and experimental lines are based on a conception of time as not linear. Past, present and future are always interconnected and reciprocally shaping each other. The future is conceived as an experience in the present based on a critical memory. The nebulous future permeating the experimental trajectory induces people to question their urban story. This one, together with the impossible imagination of 'Taranto without the ILVA', has rewritten the memory and slowly de-constrained the future by freeing it from the limits of a present so strongly compromised by the fear of losing one's own job and acceptation of whatever health and environmental damage ILVA produces. These trajectories ask: how can we live differently in this city? How can we reconceptualise the city in which we live? What kind of urban development can avoid injustices and violence? What kind of city are we going to live in together? What does socio-economic and environmental justice mean, and how are they related? What are the forces shaping the city? In asking these questions, inequalities and power relations, far from being something 'out there' or 'out of control', became visible forces and relations that can be discussed and acted upon. At the same time, imaginations such as 'Taranto without the ILVA' challenges citizens to imagining a life lived and not yet lived. This creates a stream of connections that allows citizens to feel for a moment not what a better world might look like, but what it might feel like, and how that hopeful utopic sentiment might become a motivation for social change (Dolan, 2005). By mixing a critical thinking with the imagination of a life lived and not yet lived, these trajectories come to represent a horizon of action which allows future changes to be given a meaning, evaluated and oriented.

The two strategic plans are the result of the tendency of strategic planning to dialogue only with trajectories which consider the future as being not so open that everything is possible. For this reason strategic planning has to pre-filter the multiplicity of imaginations circulating in a city and ignore those which resemble a utopia or appear too disordering. Under these premises, in Taranto the two strategic planning processes easily intersect and incorporate the trajectories of fear, planning and identity; but in doing so, they exclude any appreciation of the values embedded in the movements and experimental trajectories. They offers a comfort zone to fear and identity by delimiting the space of possibilities as that within which socio-economic and environmental crises can be anesthetised and treated as a set of problems and solutions more or less known. The first strategic plan deliberately excludes everything which is beyond the local economic and political imagination and constructs a future which re/dis-orients the public interest on a new already existing place of work: the port. This offers an escape from the nightmare of unemployment as a negotiated dream among fears, opportunities and existing solutions to urban decay.

The second strategic planning process is more akin to what that Weiner (2009) has defined a form of hegemonic imagination. This is one which has the power to condition our waking dreams by claiming that it allows an infinite variety of thoughts, ideas, dreams and visions to be produced, while in practice it limits the freedom of considering the future as radically open. The first step of this strategic planning which consists of a technical diagnosis of pathologies affecting Taranto function as the tracing of borders between what is possible and impossible: the absence of alternative sources of jobs to ILVA excludes both the trajectory of movements and experimentation. The former is considered unrealistic because of its being concerned with reimaging Taranto without ILVA, while the latter is seen as lacking of any ability to design the future. Thus, analogously to the first strategic plan it intercepts the trajectories of fear, identity and planning. Yet, in contrast to the first strategic planning process it is the result of a mediation rather than a mere negotiation. The planning process tries to change the *habitus*, but it does this by adopting the burden of the present and the limits of what can be executable as the horizon to construct the future. Again it changes the aesthetic image of the city – the current partition of sensible – without changing the logic and power underlying it.

The controversy which emerged between the PB and STC shows the borders between the possible and impossible traced by planning as grounded in social and individual established categories of the real and thus their synchronism with the present. In this strategic planning process the "change is possible at the level of representation, but transformative discourses that operate at diachronic levels are dismissed as vulgarly utopian. Fantasising rather than imagining is the muscular technology of the imagination and, as such, escape and/or adaptation becomes the only sanctioned response to repression" (Weiner, 2009, p. 149). At the same time it shows how, despite its not being considered realisable, just like another utopia, the impossible imagination has slowly intruded inside the social context and has started orienting the small and big changes which have occurred in the city. By selecting the projects which are coherent with the idea of Taranto without ILVA, the evaluation produced by the STC gives a framework of action to the impossible imagination against which the official executable/possible vision can be evaluated and thus contested as nothing more than the reproduction of the present. The Taranto experince shows how strategic planning can also act as a paralysing meta-frame.

10.6 Conclusions

This chapter has tried to see if the relational strategic planning can also act as a paralysing meta-cultural frame which, instead of liberating energies, is constrained by its conception of the role of the imagination as a social construction of a possible future. It states that the future cannot be so open that anything is possible. In this way strategic planning traces the boundaries between the possible and the impossible. Specifically, this chapter has analysed the interactions between the relational strategic planning and what is considered the 'impossible' change in order to understand to what extent the definition of such boundaries limits the real change. What has emerged is the necessity not to pre-filter the concept of impossibility but to grasp its embedded value.

Following Deleuze and Guattari's theory (1987), this chapter has described the city as a heterogeneous place, a coming-together of different trajectories of actions and thinking each of them characterised by its own imagination of the future. This has allowed an analysis of how and with which trajectories the strategic plan interacts in order to evaluate the direction of change should take. This way of description is not only concerned with the inclusion of hope within strategic planning or the necessity of an agonistic mediation (Gunder & Hillier, 2007; Gunder, 2003), which will return to the anesthetisation of imagination. It is thought to take the concept of difference as a variation of tension rather than an absolute quality. At the same time, the impossible imagination is not only conceived to explore desires and possibilities with the potential of disrupting political horizons (Pinder, 2002). This way of description is thought to allow an understanding of the differences/tensions between what is retained as possible (the shared vision/strategic frame) and what is considered a radical change to arise. This could reposition planning into the political field and smooth its obsession with executable politics.

The value of the impossible lies in its representing and embedding the instances of change and its strength in its ability to highlight and stress the differences and the tensions between what has been gained and what has been lost. These differences and tensions can be the motivation for a collective action, especially when environmental issues are at stake. In such situations, it becomes crucial to define what kind of model of development we are going to carry out. Moreover, this could become much more crucial in contexts in which the resilience to change is too easily considered to be caused only by cultural and institutional barriers, rather than the strategic planning itself. More specifically this could highlight when the resilience is caused by the inability of strategic planning to give the right value to the 'impossible' imagination. In fact, if we as citizens, so embroiled in the lines which trace a city, were able to understand the differences/tensions between what is retained as possible and what is considered a radical change, then we might be able to understand the material and immaterial collective 'costs' that the direction of change of a strategic frame produces and make a more conscious choice of our own future. " [D]emocracy consists in this, that society does not halt before a conception, given once and for all, of what is just, equal, or free, but rather institutes itself in such a way that the question of freedom, of justice, of equity, and of equality might always be posed anew within the framework of the 'normal' functioning of society" (Castoriadis, 1990, p. 87).

The Taranto case might be an example of how the idea that strategic planning can also act as a paralysing meta-frame. The impossible imagination of 'Taranto without the ILVA' and that of dis-identification have rewritten the memory of city in a critical way and slowly de-constrained and freed the future from the limits of a present so strongly compromised by the fear and the obsessive search for a future linked to mega-projects. This impossible imagination has slowly intruded inside the social context thus orienting the small and big changes which have occurred in the city, despite the fact that it was considered an utopia not realisable.

If the imagination of an 'impossible' change is ignored, then there could be the risk that relational strategic planning will always act as a paralysing meta-cultural frame. Furthermore, the relations between the demand of a change and planning could tend to become exclusively mediated by power. A possible consequence could be an increasing retreat of citizens from participating in planning and make planning something superfluous to the most effective power games.

Notes

1. A meta-cultural frame mediates people's sense-making (Schön & Rein, 1994).
2. *Habitus* is the system "of durable, transposable dispositions, structured structures predisposed to function as structuring structures, that is, as principles which generate and organize practices and representations" (Bourdieu, 1990, p. 53).

Acknowledgements I want to thank my brother Emilio for spending with me sleepless nights discussing doubts and fissures in my thought. He has the power to remind me of the difficulty and joy of doing research.

References

Albrechts, L. (2009). From strategic spatial plans to spatial strategies. *Planning Theory and Practice*, *10*(1), 133–149.

Allmendinger, P., & Haughton, G. (2009). Critical reflections on spatial planning. *Environment and Planning A*, *41*(11), 2544–2549.

Amin, A., & Thrift, N. (2002). *Cities: Reimagining the urban*. Cambridge: Polity Press.

Arsenjuk, L. (2007). *On Jacques Rancière* [http://www.eurozine.com/articles/2007-03-01-arsenjuk-en.html; accessed October 2009].

Baeten, G. (2002). The spaces of utopia and dystopia: Introduction. *Geografiska Annaler B, 84* (3–4), 14–142.

Barbanente, A., & Monno, V. (2004). Changing images and practices in a declining 'growth pole' in Southern Italy: The 'steel town' of Taranto. *Städte im Umbruch, 2*, 36–44.

Bauman, Z. (2004). *Wasted lives*. Cambridge: Polity Press.

Beauregard, R. A. (1990). Bringing the city back. *Journal of the American Planning Association, 56*(2), 210–215.

Bourdieu, P. (1990). *The logic of practice*. Stanford: Stanford University Press.

Bridge, G. (2006). *Urbanism without cities* [http://www.dur.ac.uk/j.m.painter/ CSCR/Bridge.pdf, accessed October 2009].

Castells, M. (1996). *The rise of the network society*. Oxford: Blackwell.

Castoriadis, C. (1987). *The imaginary institution of society*. Oxford: Polity Press.

Castoriadis, C. (1990). The Greek and the modern political imaginary. In C. Castoriadis & D. C. Ames (Eds.), (D. C. Ames, Trans.), [1997], *World in fragments: Writings on politics, society, psychoanalysis, and the imagination*. Stanford, CA: Stanford University Press.

Castoriadis, C. (1997). Imaginary and imagination at the crossroad. In C. Castoriadis (2005) *Figures of the thinkable* (pp. 123–152) (translated and edited anonymously as public service) [http://www.costis.org/x/castoriadis/Castoriadis-Figures_of_the_Thinkable.pdf, accessed October 2009].

Czarniawska, B. (2002). Remembering while forgetting: The role of automorphism in city management in Warsaw. *Public Administration Review, 62*(2), 163–173.

Davis, M. (1999). *Ecology of fear: Los Angeles and the imagination of disaster*. New York: Vintage Books.

Davoudi, S., & Strange, I. (Eds.). (2008). *Conceptions of space and place in strategic spatial planning*. London: Routledge.

Deleuze, G., & Guattari, F. (1987). *A thousand plateaus: Capitalism and schizophrenia*. Minneapolis, MN: University of Minnesota Press.

Deleuze, G., & Parnet, C. (1977). *Dialogues*. Paris: Flammarion.

Derrida, J. (2001). *On cosmopolitanism and forgiveness*. New York: Routledge.

Dolan, J. (2005). *Utopia in performance. Finding hope at the Theater*. Ann Arbor, MI: University of Michigan Press.

Feinstein, S. (1999). Can we make the cities we want? In S. Body-Gendrot & R. Beauregard (Eds.), *The urban moment* (pp. 249–272). Thousand Oaks, CA: Sage.

Flyvbjerg, B. (1998). *Rationality and power: Democracy in practice*. Chicago: University of Chicago Press.

Flyvbjerg, B. (2002). Bringing power to planning research. One researcher's praxis story. *Journal of Planning Education and Research, 21*(4), 357–366.

Foucault, M. (1991). Governmentality. In G. Burchell, C. Gordon, & P. Miller (Eds.), *The Foucault effect: Studies in governmentality* (pp. 87–104). London: Harvester Wheatsheaf.

Friedmann, J. (1994). *Counterpoint: The utility of non-Euclidean planning. Journal of the American Planning Association, 60*(3), 377–379.

Friedmann, J. (1998). Planning theory revisited. *European Planning Studies, 6*(3), 245–253.

Friedmann, J. (1999). The city of everyday life. Knowledge/power and the problem of representation. *DISP, 136/137*(1–2), 4–11.

Friedmann, J. (2002). *The prospect of cities*. Minneapolis, MN: Minnesota University Press.

Gaffikin, F., & Sterrett, K. (2006). New visions for old cities: The role of visioning in planning. *Planning Theory and Practice, 7*(2), 159–178.

Gandy, M. (2005). Cyborg urbanization: Complexity and monstrosity in the contemporary city. *International Journal of Urban and Regional Research, 29*(1), 26–49.

Gunder, M. (2003). Passionate planning for the others' desire: An agonistic response to the dark side of planning. *Progress in Planning, 60*(3), 235–319.

Gunder, M., & Hillier, J. (2007). Planning as urban therapeutic. *Environment and Planning A, 39*(2), 467–486.

Harvey, D. (1996). *Justice, nature and the geography of difference*. Cambridge: Oxford and Blackwell.
Hayden, D. (1995). *The power of place. Urban landscape as public history*. Cambridge, MA: MIT Press.
Healey, P. (1997). *Collaborative planning, shaping places in fragmented societies*. London: Macmillan.
Healey, P. (2006). Relational complexity and the imaginative power of strategic spatial planning. *European Planning Studies*, *14*(4), 525–546.
Healey, P. (2009). The pragmatic tradition in planning thought. *Journal of Planning Education and Research*, *28*(3), 277–292.
Heynen, N., Kaika, M., & Swyngedouw, E. (2006). Urban political ecology: Politicizing the production of urban nature. In N. Heynen, M. Kaika, & E. Swyngedouw (Eds), *In the nature of the city* (pp. 1–20). Abingdon: Routdledge.
Huxley, M. (2000). The limits of communicative planning. *Journal of Planning Education and Research*, *19*(4), 369–377.
Jameson, F. (2005). *Archaeologies of the future*. London: Verso.
Massey, D. (1994). *Space, place and gender*. Cambridge: Polity Press.
Massey, D. (2005). *For space*. London: Sage.
Maturana, H. R., & Varela, F. J. (1987). *The tree of knowledge: The biological roots of human understanding*. Boston, MA: Shambala.
May, T. (2005). *Gilles Deluze: An introduction*. Cambridge: Cambridge University Press.
McFarlane, C. (2009). Translocal assemblages: Space, power and social movements. *Geoforum*, *40*(4), 561–567.
Monno, V. (forthcoming). Tracing relational complexity. In G. De Roo, J. Hillier, & J. van Wezemael (Eds.), *Planning and complexity. In depth analysis*. Farnham Surrey: Ashgate.
Mouffe, C. (2000). *The democratic paradox*. London & New York: Verso.
Mouffe, C. (2005). *On the political, thinking in action*. London: Routledge.
Natali, M. P. (2004). History and the politics of Nostalgia. *Iowa Journal of Cultural Studies*, *5*(5), 10–25.
Pinder, D. (2002). In defence of Utopian urbanism: Imagining cities after the end of Utopia. *Geografiska Annaler B*, *84*(3–4), 229–241.
Pinder, D. (2005). *Visions of the city*. New York: Routledge.
Purcell, M. (2009). Resisting neoliberalization: Communicative planning or counter-hegemonic movements? *Planning Theory*, *8*(2), 140–165.
Rajchman, J. (1998). *Constructions*. Cambridge: MIT Press.
Rancière, J. (1998). *Disagreement: Politics and philosophy*. Minneapolis, MN: University of Minnesota Press.
Rancière, J. (2004). *The politics of aesthetics: The distribution of the sensible*. London and New York: Continuum.
Rinella, A. (2002). *Oltre l'Acciaio: Taranto Problemi e Progetti*. Bari: Progedit.
Rodwin, F. L. (1981). *Cities and cities planning*. New York: Plenum Press.
Sandercock, L. (1998). *Towards cosmopolis. Planning for multi-cultural cities*. Chichester: Wiley.
Schön, D., & Rein, M. (1994). *Frame reflection*. New York: Basic Books.
Scott, D., & Clive Barnett, C. (2009). Something in the air: Civic science and contentious environmental politics in post-apartheid South Africa. *Geoforum*, *40*(3), 373–382.
Stephenson, N., & Papadopoulos, D. (2006). Outside politics/continuous experience. *Ephemera*, *6*(4), 433–453.
Swyngedouw, E. (2009). The antinomies of the post-political city: In search of a democratic politics of environmental production. *International Journal of Urban and Regional Research*, *33*(3), 601–620.
Vulpio, C. (2009). *La Città delle Nuvole*. Milano: Edizioni Ambiente.
Weiner, E. J. (2009). Time is on our side: Rewriting the space of imagination. *Situations: Project of the Radical Imagination*, *30*(1), 127–150.

Wolfe, K. (2006). From aesthetics to politics: Rancière, Kant and Deleuze. *Journal of Aesthetics, 4* [http://www.contempaesthetics.org/newvolume/pages/article.php?articleID=382, accesed Octobe009].
Yiftachel, O. (1998). Planning and social control: Exploring the dark side. *Journal of Planning Literature, 12*(4), 395–406.
Young, I. M. (1990). *Justice and the politics of difference*. Princeton: Princeton University Press.
Young, I. M. (2001). Activist challenges to deliberative democracy. *Political Theory, 29*(5), 670–690.

Chapter 11
Impossible Sustainability and the Post-political Condition

Erik Swyngedouw

Well, my dear Adeimantus, what is the nature of tyranny? It's obvious, I suppose, that it arises out of democracy
Plato

Barbarism or socialism
Karl Marx

Kyoto or the Apocalypse
Green saying

11.1 The Question of 'Natures'

"Nature does not exist"... or..."When vegetarians will eat meat!"

The Guardian International reported (13 August 2005) how a University of Maryland scientist had succeeded in producing 'cultured meat'. Soon, he said, "it will be possible to substitute reared beef or chicken with artificially grown meat tissue. It will not be any longer necessary to kill an animal in order to get access to its meat. We can just rear it in industrialised labs". A magical solution, so it seems, that might tempt vegetarians to return to the flock of animal protein devotees, while promising yet again (after the failed earlier promises made by the pundits of pesticides, the green revolution and now genetic engineering and Genetically Modified – GM – products) the final solution for world hunger and a more sustainable life for the millions of people who go hungry now. Meanwhile, NASA is spending about US $40 million a year on how to recycle wastewater and return it to potable conditions, something that would of course be necessary to permit space missions of long duration, but which would be of significant importance on earth as well. At the same time, sophisticated new technologies are developed for sustainable water harvesting,

E. Swyngedouw (✉)
School of Environment and Development, The University of Manchester,
Manchester M60 1QD, UK
e-mail: erik.swyngedouw@manchester.ac.uk

M. Cerreta et al. (eds.), *Making Strategies in Spatial Planning*,
Urban and Landscape Perspectives 9, DOI 10.1007/978-90-481-3106-8_11,

for a more rational use of water or for a better recycling of residual waters, efforts defended on the basis of the need to reach the Millennium Development Goals that promise, among others, a reduction by half of the 2.5 billion people that do not have adequate access to safe water and sanitation.

In the mean time, other 'Natures' keep wrecking havoc around the world. The Tsunami disaster comes readily to mind, as do the endless forest fires that blazed through Spain in the summer of 2005 during the country's driest summer since records started, killing dozens of people and scorching the land; HIV continues its genocidal march through Sub-Saharan Africa, summer heat waves killed thousands of people prematurely in 2004 in France.

In 2006, Europeans watched anxiously the nomadic wanderings of the avian flue virus and waits, almost stoically, for the moment it will pass more easily from birds to humans. While all this is going on, South Korea's leading bio-tech scientist, Hwang Woo Suk proudly presented, in August 2005, the Seoul National University Puppy (SNUPPY) to the global press as the first cloned dog (a Labrador) while a few months later, in December 2005, this science hero was forced to withdraw a paper on human stem cells from *Science* after accusations of intellectual fraud (later confirmed, prompting his resignation and wounding South-Korea's great biotech dream). In the United Kingdom, male life expectancy between the 'best' and 'worst' areas is now more than 11 years and the gap is widening with life expectancy actually falling (for the first time since the Second World War) in some areas.[1] Tuberculosis is endemic again in East London, obesity is rapidly becoming the most seriously lethal socio-ecological condition in our fat cities (Marvin & Medd, 2006), and, as the ultimate cynical gesture, nuclear energy is again celebrated and iconised by many elites, among whom Tony Blair, as the world's saviour, the ultimate response to the climatic calamities promised by continuing carbon accumulation in our atmosphere while satisfying our insatiable taste for energy.

This great variety of examples all testify to the blurring of boundaries between the human and the artificial, the technological and the natural, the non-human and the cyborg-human; they certainly also suggest that there are all manner of Natures out there. While some of the above examples promise 'sustainable' forms of development, others seem to stray further away from what might be labelled as sustainable. At first glance, Frankenstein meat, cyborg-waters and stem cell research are exemplary cases of possibly 'sustainable' ways of dealing with apparently important socio-environmental problems while solving significant social problems (animal ethics and food supply on the one hand, dwindling freshwater resources or unsustainable body metabolisms on the other). Sustainable processes are sought for around the world and solutions for our precarious environmental condition are feverishly developed. Sustainability, so it seems, is in the making, even for vegetarians.

Meanwhile, as some of the other examples attest, socio-environmental processes keep on wrecking havoc in many places around the world. 'Responsible' scientists, environmentalists of a variety of ideological stripes and colours, together with a growing number of world leaders and politicians, keep on spreading apocalyptic and dystopian messages about the clear and present danger of pending environmental catastrophes that will be unleashed if we refrain from immediate and determined action. Particularly the threat of global warming is framed in apocalyptic terms if the

Global warming and ozone loss: Apocalypse soon
(The Independent, 17.02.2006)

Sea levels likely to rise much faster than was predicted
(The Independent, 17.02.2006)

Global warming is causing the Greenland ice cap to disintegrate far faster than anyone predicted
(The Independent, 17.02.2006)

Global warming 30 times quicker than it used to be
(The Independent, 17.02.2006)

Climate change: on the edge
(The Independent, 17.02.2006)

Water wars
(The Independent, 28.02.2006)

The four horsmen of industrial society: war, over-population, climate change and peak oil
(Energy Bulletin 12.01.2006)

Pentagon warns Bush of apocalyptic climate change by 2020

Fig. 11.1 The desire of the apocalypse: some media headlines

atmospheric accumulation of CO_2 (which is of course the classic 'side effect' of the accumulation of capital in the troposphere) continues unheeded. Figure 11.1 collects a sample of some of the most graphic doomsday media headlines on the theme. The world as we know it will come to a premature end (or be seriously mangled) unless we urgently reverse, stop or at least slow down global warming and return the climate to its status quo ante. Political and regulatory technologies (such as the Kyoto Protocol) and CO_2 reducing techno-machinery (like hybrid cars) are developed that would, so the hope goes, stop the threatening evolution and return the earth's temperature to its benevolent earlier condition. From this perspective, sustainability is predicated upon a return, if we can, to a perceived global climatologic equilibrium situation that would permit a sustainable continuation of the present world's way of life.

So, while one sort of sustainability seems to be predicated upon feverishly developing new Natures (like artificial meat, cloned stem cells or manufactured clean water), forcing Nature to act in a way we deem sustainable or socially necessary, the other type is predicated upon limiting or redressing our intervention in Nature, returning it to a presumably more benign condition, so that human and non-human sustainability in the medium- and long-term can be assured. Despite the apparent contradictions of these two ways of 'becoming sustainable' (one predicated upon preserving Nature's status quo, the other predicated upon producing new Natures), they share the same basic vision that techno-natural and socio-metabolic interventions are urgently needed if we wish to secure the survival of the planet and much of what it contains. But these examples also show that Nature is not always what it seems to be. Frankenstein meat, dirty water, bird flue virus symbiosis, stem cells, fat bodies, heat waves, tsunamis, hurricanes, genetic diversity, CO_2, to name just a few, are radically different things, expressing radically different Natures, pushing in radically different directions, with radically different consequences and outcomes, and with radically different human/non-human connectivities. If anything, before we can even begin to unpack 'sustainability', the above examples certainly suggest that we urgently need to interpolate our understandings of Nature, revisit what we mean by Nature, and, what we assume Nature to be.

11.1.1 Surrendering Nature, Indeterminate Natures

Slavoj Žižek suggests in *Looking Awry* that the current ecological crisis is indeed a radical condition that not only constitutes a real and present danger, but, equally importantly, "questions our most unquestionable presuppositions, the very horizon of our meaning, our everyday understanding of Nature as a regular, rhythmic process" (Žižek, 1992, p. 34).[2] It raises serious questions about what were long considered self-evident certainties. He argues that this fundamental threat to our deepest convictions of what we always thought we knew for certain about Nature is co-constitutive of our general unwillingness to take the ecological crisis completely serious. It is this destabilising effect that explains the fact that the typical, predominant reaction to it still consists in a variation of the famous disavowal, "I know very well – that things are deadly serious, that what is at stake is our very survival – but just the same I don't really believe, (. . .) and that is why I continue to act as if ecology is of no lasting consequence for my everyday life" (Žižek, 1992, p. 35). The same unwillingness to question our very assumptions about what Nature is (and even more so what Natures might 'become') also leads to the typical obsessive reactions of those who *do* take the ecological crisis seriously. Žižek considers both the case of the environmental activist, who in his or her relentless and obsessive activism to achieve a transformation of society in more ecologically sustainable ways expresses a fear that to stop acting would lead to catastrophic consequences. In his words, obsessive acting becomes a tactic to stave off the ultimate catastrophe, that is *if I stop doing what I am doing, the world will come to an end in an ecological Armageddon.* Others, of course, see all manner of transcendental signs in the 'revenge of Nature', read it as a message that signals our destructive intervention in Nature and urge us to change our relationship with Nature. In other words, we have to listen to Nature's call, as expressed by the pending environmental catastrophe, and respond to its message that pleas for a more benign, associational relation with Nature, a post-human affective connectivity, as a cosmo-political 'partner in dialogue'. While the first attitude radically ignores the reality of possible ecological disaster, the other two, which are usually associated with actors defending 'sustainable' solutions for our current predicament, are equally problematic in that they both ignore, or are blind to the inseparable gap between our symbolic representation (our understanding) of Nature and the actual acting of a wide range of radically different, and often contingent, Natures. In other words, there is – of necessity – an unbridgeable gap, a void, between our dominant view of Nature (as a predictable and determined set of processes that tends towards a (dynamic) equilibrium – but one that is disturbed by our human actions and can be 'rectified' with proper sustainable practices) and the acting of Natures as an (often) unpredictable, differentiated, incoherent, open-ended, complex, chaotic (although by no means unordered or un-patterned) set of processes. The latter implies the existence not only of many Natures, but, more importantly, it also assumes the possibility of all sorts of possible future Natures, all manner of imaginable different human–non-human assemblages and articulations, and all kinds of different possible socio-environmental becomings.

The inability to take Natures seriously is dramatically illustrated by the controversy over the degree to which disturbing environmental change is actually taking place and the risks or dangers associated with it. Lomborg's *The Sceptical Environmentalist* captures one side of this controversy in all its phantasmagorical perversity (Lomborg, 1998), while climate change doomsday pundits represent the other. Both sides of the debate argue from an imaginary position of the presumed existence of a dynamic balance and equilibrium, the point of 'good' Nature, but one side claims that the world is veering off the correct path, while the other side (Lomborg and other sceptics) argues that we are still pretty much on Nature's course. With our gaze firmly fixed on capturing an imaginary 'idealised' Nature, the controversy further solidifies our conviction of the possibility of a harmonious, balanced, and fundamentally benign *one* Nature if we would just get our interaction with it right, an argument blindly (and stubbornly) fixed on the question of where Nature's rightful point of benign existence resides. This futile debate, circling around an assumedly centred, known and singular Nature, certainly permits – in fact invites – imagining ecological catastrophe at some distant point (global burning or freezing through climate change, resource depletion, death by overpopulation). Indeed, imagining catastrophe and fantasising about the final ecological Armageddon seems considerably easier for most environmentalists than envisaging relatively small changes in the socio-political and cultural-economic organisation of local and global life here and now. Or put differently, the world's premature ending in a climatic Armageddon seems easier to imagine (and sell to the public) than a transformation of (or end to) the neo-liberal capitalist order that keeps on practicing expanding energy use and widening and deepening its ecological footprint.

It is this sort of considerations that led Slavoj Žižek controversially to state that "Nature does not exist" (1992, p. 34). Of course, he does not imply that there are no such 'things' as quarks or other subatomic particles, black holes, tsunamis, sunshine, trees or Human Immunodeficiency (HI) viruses. Even less would he decry the radical effects of CO_2 and other greenhouse gases on the climate or the lethal consequences of water contamination for the world's poor. On the contrary, they are very real, many posing serious environmental problems, occasionally threatening entire populations (e.g., Acquired Immune Deficiency Syndrome, AIDS), but he insists that the Nature we see and work with is necessarily radically imagined, scripted, symbolically charged; and is radically distant from the Natures that are there, which are complex, chaotic, often unpredictable, often radically contingent, risky, patterned in endlessly complex ways, ordered along 'strange' attractors. In other words, there is no balanced, dynamic equilibrium-based Nature out there that needs or requires salvation in name of either Nature itself or of an equally imagined universal human survival. Nature simply does not exist. There is nothing foundational in Nature that needs, demands or requires sustaining. The debate and controversies over Nature and what do with it, in contrast, signals rather our political inability to engage in directly political and social argument and strategies about re-arranging the social co-ordinates of everyday life and the arrangements of socio-metabolic organisation (something usually called capitalism) that we inhabit. Imagining a benign and 'sustainable' Nature avoids asking the politically sensitive, but vital, question as to what kind of socio-environmental arrangements do we wish to produce, how can this be achieved and what sort of Natures do we wish to inhabit.

11.2 What Is Environment: Impossible Sustainability? Undesirable Sustainability?

11.2.1 Desiring (Un)sustainability

So, if Nature does not exist, what, then, to say about sustainability, a concept (and associated set of fuzzy practices) that is deeply indebted and intensely connected to the particular imaginations of Nature suggested above. Since its early definition in the *Brundtland Report* (1987), the concept (but not much of the practice) of sustainability has really taken off. A cursory glance at both popular and academic publications will quickly assemble a whole array of 'sustainabilities': sustainable environments, sustainable development, sustainable growth, sustainable wetlands, sustainable bodies, sustainable companies, sustainable processes, sustainable incomes, sustainable cities, sustainable technologies, sustainable water provision, even sustainable poverty, sustainable accumulation, sustainable markets and sustainable loss. I have not been able to find a single source that is against sustainability. Greenpeace is in favour, George Bush Jr. and Sr. are, the World Bank and its chairman (a prime war monger in Iraq) are, the pope is, my son Arno is, the rubber tappers in the Brazilian Amazon are, Bill Gates is, the labour unions are. All are presumably concerned about the long-term socio-environmental survival of (parts of) humanity; most just keep on doing business as usual. The clear and present danger posed by the environmental question is obviously not dramatic enough to be taken seriously in terms of embarking on a different socio-environmental trajectory. That is left to do some other time and certainly not before the day after tomorrow. Of course, this cacophony of voices and imaginations also points to the inability to agree on the meaning or, better, to the lack of a singular Nature. There are obviously multiple imaginations that mobilise or appropriate sustainability as radically and truthfully theirs, based on equally imaginative variations of what constitutes Nature.

Environmentalists (whether activists or scientists) invariably invoke the global physical processes that threaten our existence and insist on the need to re-engineer Nature, so that it can return to a 'sustainable' path. Armed with their charts, formulas, models, numbers and grant applications, to which activists usually add the inevitable pictures of scorched land, factories or cars emitting carbon fumes, dying animals and plants, suffering humans, apocalyptic rhetoric, and calls for subsidies and financial support, scientists, activists, and all manner of assorted other human and non-human actants enter the domain of the social, the public and, most importantly, the political. Thus Natures enter the political.

A particular and symbolically enshrined nature enters the parliament of politics, but does so in a duplicitous manner. It is a treacherously deceitful Nature that enters politics, one that is packaged, numbered, calculated, coded, modelled, represented by those who claim to possess, know, understand, speak for the 'real Nature'. In other words, what enters the domain of politics is the coded and symbolised versions of nature mobilised by scientists, activists, industrialists and the like. This is particularly evident in examples such as the debate over GM organisms, global climate change, Bovine Spongiform Encephalopathy (BSE), biodiversity loss and other equally pressing issues. Invariably, the acting of Nature – as scripted by the bearers of Nature's knowledge – enters the political machinery as coded language that also already posits its political and social solution and does not tolerate, in the name of Nature, dissent other than that framed by its own formulations. It is in this sense of course that the argument about climate change is exclusively formulated in terms of believers and non-believers, as a quasi-religious faith, but the weapons of the struggle in this case are matters of fact like data, models and physico-chemical analysis. And the solutions to the question of sustainability are already pre-figured by the way in which Nature is made to speak. Creeping increases in long-term global temperatures, which will cause untold suffering and damage, are caused by CO_2 output. Hence, the solution to future climate ills resides in cutting back on CO_2 emissions. Notwithstanding the validity of the role of CO_2 in co-constituting the process of climate change, the problems of the future calamities the world faces are posited primarily in terms of the physical acting of one of Nature's components CO_2 as is its solution found in bringing CO_2 within our symbolic (socio-economic) order, futilely attempted with the Kyoto agreement or other neo-liberal market-based mechanisms. Questioning the politics of climate change in itself is already seen as an act of treachery, as an unlawful activity, banned by Nature itself.

11.2.2 Undesirable Sustainability: Environmental Politics as Post-politics

Although there may be no Nature, there certainly is a politics of Nature or a politics of the environment. The collages of apparently contradictory and overlapping vignettes of the environmental conditions outlined above share one common threat that many of us, Bush and Blair, my son and Greenpeace, Oxfam and the World Bank, agree on. The world is in environmental trouble. And we need to act politically now.

Both the 2004 tsunami and New Orleans's Katrina brought the politicisation of Nature home with a vengeance. Although the tsunami had everything to do with the earth's geodetic acting out and with the powerless of South East Asia drowning in its consequences and absolutely nothing with climate change or other environmentally degrading practices, the tsunami calamity was and continues to be staged as a socio-environmental catastrophe, another assertion of the urgent need to revert to more 'sustainable' socio-environmental practices. New Orleans' socio-environmental disaster was of a different kind. While there may be a connection between the number and intensity of hurricanes and climate change, that of course does account neither for the dramatic destructions of poor people's lives in the city nor for the plainly blatant racist spectacles that were fed into the media on a daily basis in the aftermath of the hurricane's rampage through the city. The imaginary staged in the aftermath of the socio-environmental catastrophe of New Orleans singled out disempowered African Americans twice, first as victims, then as criminals. Even the *New York Times* conceded that 80% of the reported 'crimes' taking place in unruly and disintegrating New Orleans in the aftermath of the hurricane's devastations were based on rumour and innuendo – a perverse example of how liberal humanitarian concern is saturated with racialised coding and moral disgust with the poorest and most excluded parts of society. Of course, after the poor were hurricaned out of New Orleans, the wrecked city is rapidly turning into a fairy-tale playground for urban developers and city boosters who will make sure, this time around, that New Orleans will be rebuilt in their image of a sustainable capitalist city: green, white, rich, conservative and neo-liberal (Davis, 2006).

The popular response to Katrina, the barrage of apocalyptic warnings of the pending catastrophes wrecked by climate change and the need to take urgent remedial action to engineer a retro-fitted balanced climate are perfect examples of the tactics and configurations associated with the present post-political condition, primarily in the United States and Europe. Indeed, a politics of sustainability, predicated upon a radically conservative and reactionary view of a singular – and ontologically stable and harmonious – Nature is necessarily one that eradicates or evacuates the 'political' from debates over what to do with Natures. The key political question is one that centres on the question of what kind of Natures we wish to inhabit, what kinds of Natures we wish to preserve, to make or, if need be, to wipe off the surface of the planet (e.g., like the HI virus) and on how to get there. The fantasy of 'sustainability' imagines the possibility of an originally fundamentally harmonious Nature, one that is now out-of-synch but, which, if 'properly' managed, we can and have to return to by means of a series of technological, managerial and organisational fixes. As suggested above, many, from different social, cultural and philosophical positionalities, agree with this dictum. Disagreement is allowed, but only with respect to the choice of technologies, the mix of organisational fixes, the detail of the managerial adjustments and the urgency of their timing and implementation. Nature's apocalyptic future, if unheeded, symbolises and nurtures the solidification of the post-political condition. And the excavation and critical assessment of this post-political condition nurtured and embodied by most of current Western socio-environmental politics is what we shall turn to next.

11.3 The Post-political Post-democratic Condition: Evacuating Socio-Environmental Politics

11.3.1 Post-political!?

Slavoj Žižek and Chantal Mouffe, among others, define the post-political as a political formation that actually forecloses the political, that prevents the politicisation of particulars (Žižek, 1999a, p. 35, 2006; Mouffe, 2005). A situation or an event becomes "political when a particular demand (cutting greenhouse gases, stopping the exploitation of a particular resource and so on) starts to function as a metaphoric condensation of the global opposition against Them, those in power, so that the protest is no longer just about that demand, but about the universal dimension that resonates in that particular demand. (...) What post-politics tends to prevent is precisely this metaphoric universalisation of particular demands: post-politics mobilizes the vast apparatus of experts, social workers and so on, to reduce the overall demand (complaint) of a particular group to just this demand, with its particular content – no wonder that this suffocating closure gives birth to 'irrational' outbursts of violence as the only way to give expression to the dimension beyond particularity" (Žižek, 1999b, p. 204). In Europe and the United States, in particular, such post-political arrangements are largely in place. Post-politics reject ideological divisions and the explicit universalisation of particular political demands. Instead, the post-political condition is one in which a consensus has been built around the inevitability of neo-liberal capitalism as an economic system, parliamentary democracy as the political ideal, humanitarianism and inclusive cosmopolitanism as a moral foundation. As Žižek (1999b, p. 198) puts it: "[i]n post-politics, the conflict of global ideological visions embodied in different parties which compete for power is replaced by the collaboration of enlightened technocrats (economists, public opinion specialists, etc.) and liberal multiculturalists; via the process of negotiation of interests, a compromise is reached in the guise of a more or less universal consensus. Post-politics thus emphasizes the need to leave old ideological visions behind and confront new issues, armed with the necessary expert knowledge and free deliberation that takes people's concrete needs and demands into account."

Post-politics is thus about the administration of social or ecological matters, and they remain of course fully within the realm of the possible, of existing socio-ecological relations. "The ultimate sign of post-politics in all Western countries", Žižek (2002, p. 303) argues, "is the growth of a managerial approach to government: government is reconceived as a managerial function, deprived of its proper political dimension". Post-politics refuses politicisation in the classical Greek sense, that is, as the metaphorical universalisation of particular demands, which aims at 'more' than negotiation of interests: "[t]he political act (intervention) proper is not simply something that works well within the framework of existing relations, but something that *changes the very framework that determines how things work.* (...) [A]uthentic politics (...) is the art of the *impossible* – it changes the very parameters of what is considered 'possible' in the existing constellation (emphasis in original)" (Žižek, 1999b, p. 199).

A genuine politics, therefore, is "the moment in which a particular demand is not simply part of the negotiation of interests but aims at something more, and starts to function as the metaphoric condensation of the global restructuring of the entire socials space" (Žižek, 1999b, p. 208). It is about the recognition of conflict as constitutive of the social condition, and the naming of the socio-ecological spaces that can become.

The political becomes for Žižek and Rancière the space of litigation (Žižek, 1998), the space for those who are *not-all*, who are uncounted and unnamed, not part of the 'police' (symbolic or state) order. A true political space is always a space of contestation for those who have no name or no place. As Diken and Laustsen (2004, p. 9) put it: "[p]olitics in this sense is the ability to debate, question and renew the fundament on which political struggle unfolds, the ability to radically criticise a given order and to fight for a new and better one. In a nutshell, then, politics necessitates accepting conflict". A radical-progressive position "should insist on the unconditional primacy of the inherent antagonism as constitutive of the political" (Žižek, 1999a, p. 29).

Post-political parliamentary rule, in contrast, permits the politicisation of everything and anything, but only in a non-committal way and as non-conflict. Absolute and irreversible choices are kept away; politics becomes something one can do without making decisions that divide and separate (Thomson, 2003). A consensual post-politics arises thus, one that either eliminates fundamental conflict (i.e., we all agree that climate change is a real problem that requires urgent attention) or elevates it to antithetical ultra-politics. Those who deny the realities of a dangerous climate change are blinded radicals that put themselves outside the legitimate social (symbolic) order. The same 'fundamentalist' label is of course also put on those who argue that dealing with climate change requires a fundamental reorganisation of the hegemonic neo-liberal-capitalist order. The consensual times we are currently living in have thus eliminated a genuine political space of disagreement. However, consensus does not equal peace or absence of fundamental conflict (Rancière, 2005a, p. 8). Under a post-political condition, "[e]verything is politicised, can be discussed, but only in a non-committal way and as a non-conflict. Absolute and irreversible choices are kept away; politics becomes something one can do without making decisions that divide and separate. When pluralism becomes an end in itself, real politics is pushed to other arenas" (Diken & Laustsen, 2004, p. 7), in the present case to street rebellion and protest, and terrorist tactics (e.g., animal liberation movement in the UK).

Difficulties and problems, such as environmental concerns that are generally staged and accepted as problematic need to be dealt with through compromise, managerial and technical arrangement and the production of consensus. "Consensus means that whatever your personal commitments, interests and values may be, you perceive the same things, you give them the same name. But there is no contest on what appears, on what is given in a situation and as a situation" (Rancière, 2003, § 4). The key feature of consensus is "the annulment of dissensus (...) the 'end of politics'" (Rancière, 2001, § 32). The most utopian alternative to capitalism left to our disposal is to develop post-political alternatives to creating a more just and sustainable society, since it would not make any economic sense not to do so. Of course, this post-political world eludes choice and freedom (other than those tolerated by the consensus). And in the absence of real politicisation of particulars, the only position of real dissent is that of either the traditionalist (those stuck in the past who refuse to accept the inevitability of the new global neo-liberal order) or the fundamentalist. The only way to deal with them is by sheer violence, by suspending their 'humanitarian' and 'democratic' rights. The post-political relies on either including all in a consensual pluralist order and on excluding radically those who posit themselves outside the consensus. For them, as Agamben (2005) argues, the law is suspended; they are literally put outside the law and treated as extremists and terrorists.

The environment and debates over the environment and Nature are not only perfect expressions of such a post-political order, but in fact, the mobilisation of environmental issues is one of the key arenas through which this post-political consensus becomes constructed, when "politics proper is progressively replaced by expert social administration" (Žižek, 2005a, p. 117, 2005b). The fact that Bush does not want to play ball on the climate change theme is indeed seen by both the political elites in Europe and the environmentalists as a serious threat to the post-political consensus. That is why both political elites and opposition groups label him as a radical conservative. Bill Clinton, of course, embodied the post-political consensus in a much more sophisticated and articulated manner, not to speak of his unfortunate successor, Al Gore, who, in May 2006, resurfaced as a newborn climate change warrior.[3]

The post-political environmental consensus, therefore, is one that is radically reactionary, one that forestalls the articulation of divergent, conflicting and alternative trajectories of future socio-environmental possibilities and of human–human and human–nature articulations and assemblages. It holds on to a harmonious view of Nature that can be recaptured while re-producing if not solidifying a liberal-capitalist order for which there seems to be no alternative. Much of the sustainability argument has evacuated the politics of the possible, the radical contestation of alternative future socio-environmental possibilities and socio-natural arrangements, and silences the radical antagonisms and conflicts that are constitutive of our socio-natural orders by externalising conflict. In climate change, for example, the conflict is posed as one of society versus CO_2. In fact, the sustainable future desired by 'sustainability' pundits has no name. While alternative futures in the past were named and counted (e.g., communism, socialism, anarchism, libertarianism and liberalism), the desired sustainable environmental future has no name and no process, only a state or condition. This is as exemplified by the following apocalyptic warning in which the celebrated quote from Marx's Communist Manifesto and its invocation of the 'the spectre of communism that is haunting the world' (once the celebrated name of hope for liberation) is replaced by the spectre of Armageddon: "[a] spectre is haunting the entire world: but it is not that of communism. (. . .) Climate change – no more, no less than Nature's payback for what we are doing to our precious planet – is day by day now revealing itself. Not only in a welter of devastating scientific data and analysis but in the repeated extreme weather conditions to which we are all, directly or indirectly, regular observers, and, increasingly, victims" (Levene, 2005 p. 1).

Climate change is of course not a politics, let only a political programme or socio-environmental project; it is pure negation, the negativity of the political; one we can all concur with, around which a consensus can be built, but which eludes conflict, evacuates the very political moment. By doing so, it does not translate Marx's dictum for the contemporary period, but turns it into its radical travesty.

11.3.2 From the Post-political to Post-democracy

There is of course a close relationship between the post-political condition and the functioning of the political system. In particular, the post-political threatens the very foundation upon which a democratic polity rests.

Indeed, the French philosopher Jacques Rancière defines this kind of consensual post-politics as harbouring a 'post-democracy', rather than seeing it as forming a deepened democracy: "[p]ost-democracy is the government practice and conceptual legitimation of a democracy *after the demos*, a democracy that has eliminated the appearance, miscount, and dispute of the people and is thereby reducible to the sole interplay of state mechanisms and combinations (. . .) It is the practice and theory of what is appropriate with no gap left between the forms of the State and the state of social relations" (Rancière, 1995, pp. 142–153; also in Mouffe, 2005, p. 29).

In this post-democratic post-political era, in contrast, adversarial politics (of the left/right variety or of radically divergent struggles over imagining socio-environmental futures) are considered hopelessly out of date. Although disagreement and debate are of course still possible, they operate within an overall model of consensus and agreement (Crouch, 2004). There is indeed, in the domain of environmental policies and politics, a widespread consensus that Nature and the Environment need to be taken seriously, and that appropriate managerial-technological apparatuses can and should be negotiated to avoid imminent environmental catastrophe. At the same time, of course, there is hegemonic consensus that no alternative to liberal-global hegemony is possible.

This post-political frame is of course politically correlative to the theoretical argument, advanced most coherently by sociologists like Ulrich Beck (1994) or Anthony Giddens (1991). They argue that adversarial politics organised around collective identities that were shaped by the internal relations of class-based capitalism are replaced by an increasingly individualised, fragmented, 'reflexive' series of social conditions.

For Beck, for example, "simple modernization ultimately situates the motor of social change in categories of instrumental rationality (reflection), 'reflexive' modernization conceptualizes the motive power of social change in categories of the side-effect (reflexivity). Things at first unseen and unreflected, but externalized, add up to structural rupture that separates industrial from 'new modernities' in the present and the future" (Beck, 1997, p. 38). From this perspective, "the distinction between danger (characteristic of pre-modern and modern societies) and risk (the central aspect of late modern risk society) refers to technological change. However, the transition from danger to risk can be related to the (...) process of the weakening of the state. In risk society what is missing is an authority that can symbolise what goes wrong. Risk is, in other words, the danger that cannot be symbolised" (Diken & Laustsen, 2004, p. 11; see also Žižek, 1999b, pp. 322–347); that what has no name. Politicisation, then, is to make things enter the parliament of politics (Latour, 2004), but the post-democratic condition does so in a consensual conversation in tune with the post-political evacuation of real antagonism. The environmental apocalypse in the making puts the state on the spot (cfr. BSE, avian flu, climate change), yet exposes the impotence of the state to 'solve' or 'divert' the risk and undermines the citizens' sense of security guaranteed by the state.

It is these 'side-effects' identified by Ulrich Beck (e.g., the accumulation of CO_2) that are becoming the key arenas around which political configuration and action crystallise, and of course, (global) environmental problems are the classic example of such effects, unwittingly produced by modernisation itself, but now requiring second 'reflexive' modernisation to deal with. The old left/right collective politics that were allegedly generated from within the social relations that constituted modernity are no longer, if they ever were, valid or performative. This, of course, also means that the traditional theatres of politics (state, parliament, parties, etc.) are not any longer the exclusive terrain of the political: "the political constellation of industrial society is becoming unpolitical, while what was unpolitical in industrialism is becoming politicals" (Beck, 1994, p. 18). It is exactly the side effects (the risks) of modernising globalisation that need management, that require politicisation. A new form of politics (what Rancière, Žižek and Mouffe exactly define as post-politics) thus arises, what Beck calls sub-politics: "'[s]ub-politics' is distinguished from 'politics' in that (a) agents outside the political or corporatist system are allowed also to appear on the stage of social design (this group includes professional and occupational groups, the technical intelligentsia in companies, research institutions and management, skilled workers, citizens' initiatives, the public sphere, etc.) and (b) not only social and collective agents but individuals as well compete with the latter and each other for the emerging power to shape politics" (Beck, 1994, p. 22).

Chantal Mouffe (2005, pp. 40–41) summarises Beck's prophetic vision of a new democracy as follows: "[i]n a risk society, which has become aware of the possibility of an ecological crisis, a series of issues which were previously considered of a private character, such as those concerning the lifestyle and diet, have left the realm of the intimate and the private and have become politicized. The relation of the individual to Nature is typical of this transformation since it is now inescapably interconnected with a multiplicity of global forces from which it is impossible to escape. Moreover, technological progress and scientific development in the field of medicine and genetic engineering are now forcing people to make decisions in the field of 'body politics' hitherto unimaginable. (...) What is needed is the creation of forums where a consensus could be built between the experts, the politicians, the industrialists and citizens on ways of establishing possible forms of co-operation among them. This would require the transformation of expert systems into democratic public spheres."

This post-political constitution, which we have elsewhere defined as new forms of autocratic governance-beyond-the-state (Swyngedouw, 2005), reconfigures the act of governing to a stakeholder-based arrangement of governance in which the traditional state forms (national, regional or local government) partakes together with experts, NGOs and other 'responsible' partners (Crouch, 2004). Not only is the political arena evacuated from radical dissent, critique and fundamental conflict, but the parameters of democratic governing itself are being shifted, announcing new forms of governmentality, in which traditional disciplinary society is transfigured into a society of control through disembedded networks (like *The Kyoto Protocol*; *The Dublin Statement*, *The Rio Summit*, etc.). These new global forms of 'governance' are expressive of the post-political configuration (Mouffe, 2005, p. 103): "[g]overnance entails an explicit reference to 'mechanisms' or 'organized' and coordinated activities' appropriate to the solution of some specific problems. Unlike government, governance refers to 'policies' rather than 'politics' because it is not a binding decision-making structure. Its recipients are not 'the people' as collective political subject, but 'the population' that can be affected by global issues such as the environment, migration, or the use of natural resources" (Urbinati, 2003, p. 80).

Anthony Giddens (1991, 1994, 1998) has also been a key intellectual interlocutor of this post-political consensus. He argues that globalised modernity has brought in its wake all manner of uncertainties as a result of human's proliferating interventions in Nature and in social life, resulting in an explosive growth of all sorts of environmental and life-related issues. The ensuing "life politics is about the challenges that face collective humanity" (Giddens, 1994, p. 10). What is required now, in a context of greater uncertainty but also with enhanced individual autonomy to make choices, is to generate active 'trust' achieved through a 'dialogic democracy'. Such 'dialogic' mode is exactly the consensual politics Jacques defines as post-democratic (Rancière, 1995, 2005b). As Chantal Mouffe (2005, p. 45) maintains, "[a]ctive trust implies a reflexive engagement of lay people with expert systems instead of their reliance on expert authority". Bruno Latour, in his politics of Nature, of course equally calls for such new truly democratic cosmo-political constitution through which both human and non-human actants enter in a new public sphere, where matters of fact are turned into matters of concern, articulated and brought together through heterogeneous and flat networks of related and relationally constituted human/non-human assemblages (Latour, 2004, 2005). Nothing is fixed, sure or given, everything continuously in doubt, negotiated, brought into the political field. Political space is not a contingent space where that what has no name is brought into the discussion, is give a name and is counted, but rather things and people are 'hailed' to become part of the consensual dialogue, of the dialogic community. The question remains of course of 'who does what sort of hailing'. Thinking about true and false, doubt and certainty, right or wrong, friend or foe, would no longer be possible, the advent of a truly cosmopolitan order in a truly cosmo-political (Stengers, 2003) constitution looms around the corner as the genuine possibility in the new modernity.

In the domain of the environment, climate change, biodiversity preservation, sustainable socio-technical environmental entanglements and the like exemplify the emergence of this new post-political configuration: they are an unexpected and unplanned by-product of modernisation, they affect the way we do things and, in turn, a new politics emerges to deal with them. This liberal cosmo-political 'inclusive' politics suggested by Beck and his fellow-travellers as a radical answer to unbridled and unchecked neo-liberal capitalist globalisation, of course, is predicated upon three assumptions.

1. The social and ecological problems caused by modernity/capitalism are external side effects; they are not an inherent and integral part of the de-territorialised and re-territorialised relations of global neo-liberal capitalism. That is why we speak of the excluded or the poor, and not about social power relations that produce wealth and poverty, or empowerment and disempowerment. A strictly populist politics emerges here; one that elevates the interest of the people, Nature or the environment to the level of the universal rather than aspiring to universalise the claims of particular natures, environments or social groups or classes.
2. These side effects are posited as global, universal and threatening: they are a total threat, of apocalyptic nightmarish proportions.
3. The 'enemy' or the target of concern is thereby of course continuously externalised. The 'enemy' is always vague, ambiguous and ultimately vacant, empty and unnamed (CO_2, gene pools, desertification, etc.). They can be managed through a consensual dialogical politics. Demands become depoliticised or rather radical politics is not about demands but about things.

11.4 Environmental Populism Versus a Politics of the Environment

The post-political condition articulates, therefore, with a populist political tactic as the conduit to instigate 'desirable' change. Environmental politics and debates over 'sustainable' futures in the face of pending environmental catastrophe are a prime expression of the populist ploy of the post-political post-democratic condition.

In this part, we shall chart the characteristics of populism (see, among others, Canovan, 1999; Laclau, 2005; Mouffe, 2005; Žižek, 2005a, 2005b) and how this is reflected in mainstream environmental concerns.

First, populism invokes *the* Environment and *the* people if not Humanity as whole in a material and philosophical manner. All people are affected by environmental problems and the whole of humanity (as well as large parts of the non-human) is under threat from environmental catastrophes. At the same time, the environment is running wild, veering off the path of (sustainable) control. As such, populism cuts across the idiosyncrasies of different human and non-human Natures and their specific 'acting outs', silences ideological and other constitutive social differences and papers over conflicts of interests by distilling a common threat or challenge to both Nature and Humanity.

Second, populism is based on a politics of 'the people know best' (although the latter category remains often empty, unnamed), supported by an assumedly neutral scientific technocracy and advocates a direct relationship between people and political participation. It is assumed that this will lead to a good, if not optimal, solution, a view strangely at odds with the presumed radical openness, uncertainty and undecidability of the excessive risks associated with Beck's or Giddens' second modernity.

The architecture of populist governing takes the form of stakeholder participation or forms of participatory governance that operates beyond-the-state and permits a form of self-management, self-organisation and controlled self-disciplining (Crouch, 2004; Dean, 1999; Lemke, 1999; Swyngedouw, 2005), under the aegis of a non-disputed liberal-capitalist order.

Third, populism customarily invokes the spectre of annihilating apocalyptic futures if no direct and immediate action is taken. The classic racist invocation of Enoch Powell's notorious 1968 *Streams of Blood* speech to warn of the immanent dangers of unchecked immigration into the United Kingdom has of course become the emblematic populist statement as are many of the slogans assembled in Fig. 11.1. If we refrain from acting (in a technocratic-managerial manner now), our world's future is in grave danger.

Fourth, populist tactics do not identify a privileged subject of change (like the proletariat for Marx, women for feminists or the 'creative class' for competitive capitalism), but instead invoke a common condition or predicament, the need for common humanity-wide action, mutual collaboration and co-operation. There are no internal social tensions or internal generative conflicts. Instead the enemy is always externalised and objectified. Populism's fundamental fantasy is of an intruder, or more usually a group of intruders, who have *corrupted* the system. CO_2 stands here as the classic example of a fetishised and externalised foe that requires dealing with if sustainable climate futures are to be attained.

Problems therefore are not the result of the 'system', of unevenly distributed power relations, of the networks of control and influence, of rampant injustices or of a fatal flow inscribed in the system, but are blamed on an outsider. That is why the solution can be found in dealing with the 'pathological' phenomenon, the resolution for which resides in the system itself. It is not the system that is the problem, but its pathological syndrome (for which the cure is internal). While CO_2 is externalised as the socio-climatic enemy, a potential cure in the guise of the Kyoto principles is generated from within the market functioning of the system itself. The 'enemy' is, therefore, always vague, ambiguous, socially empty or vacuous, and homogenised (like CO_2); the 'enemy' is a mere thing, not socially embodied, named and counted.

Fifth, populist demands are always addressed to the elites. Populism as a project always addressed demands to the ruling elites; it is not about changing the elites, but calling the elites to undertake action. A non-populist politics is exactly about obliterating the elite, imagining the impossible nicely formulated in the following joke: an IRA man in a balaclava is at the gates of heaven when St Peter comes to him and says, "I'm afraid I can't let you in". "Who wants to get in?" the IRA man retorts. "You've got 20 min to get the fuck out".

Sixth, no proper names are assigned to a post-political populist politics (Badiou, 2005a). Post-political populism is associated with a politics of not naming in the sense of giving a definite or proper name to its domain or field of action. Only vague concepts like climate change policy, biodiversity policy or a vacuous sustainable policy replaces the proper names of politics. These proper names, according to Rancière (1995) (see also Badiou, 2005b), are what constitutes a genuine democracy, that is a space where the unnamed, the uncounted and, consequently, unsymbolised become named and counted. Consider, for example, how class struggle in the nineteenth and twentieth century was exactly about naming the proletariat, its counting, symbolisation and consequent entry into the techno-machinery of the state.

Seventh, populism becomes expressed in particular demands (get rid of immigrants, reduce CO_2) that remain particular and foreclose universalisation as a positive socio-environmental project. In other words, the environmental problem does not posit a positive and named socio-environmental situation, an embodied vision, a desire that awaits its realisation, a fiction to be realised. In that sense, populist tactics do not solve problems, they are moved around. Consider, for example, the current argument over how the nuclear option is again portrayed as a possible sustainable energy future and as an alternative to deal both with CO_2 emissions and peak oil. It hardly arouses the passions for what sort of better society might arise from this.

In sum, post-political post-democracy rests, in its environmental guise, on the following foundations. First, the social and ecological problems caused by modernity/capitalism are external side effects; they are not an inherent and integral part of the relations of global neo-liberal capitalism. Second, a strictly populist politics emerges here; one that elevates the interest of an imaginary 'the People', 'Nature' or 'the Environment' to the level of the universal rather than aspiring to universalise the claims of particular socio-natures, environments or social groups or classes. Third, these side effects are constituted as global, universal and threatening: they are a total threat. Fourth, the 'enemy' or the target of concern is thereby of course continuously externalised and disembodied. The 'enemy' is always vague, ambiguous, unnamed and uncounted and ultimately empty. Fifth, the target of concern can be managed through a consensual dialogical politics and, consequently, demands become depoliticised.

11.5 Producing New Environments: A Politics of Socio-Natures

A true politics for Jacques Rancière (but also for others like Badiou, Žižek or Mouffe) is a democratic political community, conceived as "[a] community of interruptions, fractures, irregular and local, through with egalitarian logic comes and divides the police community from itself. It is a community of worlds in community that are intervals of subjectification: intervals constructed between identities, between spaces and places. Political being-together is a being-between: between identities, between worlds (. . .) Between several names, several identities, several statuses" (Rancière, 1998, pp. 137–138).

Ranciere's notion of the political is characterised in terms of division, conflict and polemic (Valentine, 2005, p. 46). Therefore, "democracy always works against the pacification of social disruption, against the management of consensus and 'stability' (. . .). The concern of democracy is not with the formulation of agreement or the preservation of order but with the invention of new and hitherto unauthorised modes of disaggregation, disagreement and disorder" (Hallward, 2005, pp. 34–35).

The politics of sustainability and the environment, therefore, in their populist post-political guise are the antithesis of democracy and contribute to a further hollowing out of what for Rancière and others constitute the very horizon of democracy as a radically heterogeneous and conflicting one. For that reason, as Badiou (2005a) argues, a new radical politics must revolve around the construction of great new fictions that create real possibilities for constructing different socio-environmental futures. To the extent that the current post-political condition, which combines apocalyptic environmental visions with a hegemonic neoliberal view of social ordering, constitutes one particular fiction (one that in fact forecloses dissent, conflict and the possibility of a different future), there is an urgent need for different stories and fictions that can be mobilised for realisation. This requires foregrounding and naming different socio-environmental futures, making the new and impossible enter the realm of politics and of democracy and recognising conflict, difference and struggle over the naming and trajectories of these futures. Socio-environmental conflict, therefore, should not be subsumed under the homogenising mantle of a populist environmentalist-sustainability discourse, but should be legitimised as constitutive of a democratic order.

In the final paragraphs of this chapter, I shall outline what constitutes for me, the key drivers of conflict and where the possibilities for different fictions and consequently new environmental futures, might tentatively reside.

Processes of socio-physical entanglement (what Marx called 'metabolic circulation') transform both social and physical environments and produce specific, differentiated and unique social and physical milieus with new and distinct qualities (Swyngedouw, 2006). In other words, environments are combined socio-physical constructions that are actively (by both humans and non-humans) and historically produced, in terms of both social content and physical-environmental qualities. Whether we consider the making urban parks, natural reserves or skyscrapers, they each contain and express fused socio-physical processes that contain and embody particular metabolic and social relations (Heynen, Kaika, & Swyngedouw, 2006). There is, in this sense, no single Nature, no One-All, but rather a great variety of distinct and often radically different (if not antagonistic) Natures. There is nothing a priori unnatural or unsustainable, therefore, about produced environments like cities, genetically modified organisms, dammed rivers or irrigated fields. The world is a 'cyborg world', part natural part social, part technical part cultural, but with no clear boundaries, centres or margins. The type and character of physical and environmental change and the resulting socio-environmental flows, networks and practices are not independent from the specific historical social, cultural, political or economic conditions and the institutions that accompany them. All socio-spatial processes are invariably also predicated upon the circulation and metabolism of physical, chemical or biological components. Non-humans of course play an active role in mobilising socio-natural circulatory and metabolic processes. It is these circulatory conduits that link often distant places and ecosystems together and permit relating local processes with wider socio-metabolic flows, networks, configurations and dynamics. These socio-environmental metabolisms produce a series of both enabling and disabling socio-environmental conditions. Of course, such produced milieus often embody contradictory or conflicting tendencies. While environmental (both social and physical) qualities may be enhanced in some places and for some humans and non-humans, they often lead to a deterioration of social, physical and/or ecological conditions and qualities elsewhere. Processes of metabolic change are, therefore, never socially or ecologically neutral. This results in conditions under which particular trajectories of socio-environmental change undermine the stability or coherence of some social groups, places or ecologies, while their sustainability elsewhere might be enhanced. Social power relations (whether material or discursive, economic, political and/or cultural) through which metabolic circulatory processes take place are particularly important. It is these power geometries, the human and non-human actors and the socio-natural networks carrying them that ultimately decide who will have access to or control over, and who will be excluded from access to or control over, resources or other components of the environment and who or what will be positively or negatively enrolled in such metabolic imbroglios. These power geometries, in turn, shape the particular social and political configurations and the environments in which we live. Henri Lefebvre's 'Right to the City' also invariably implies a 'Right to Metabolism'.

Questions of socio-environmental sustainability are fundamentally political questions revolving around attempts to tease out who (or what) gains from and who pays for, who benefits from and who suffers (and in what ways) from particular processes of metabolic circulatory change. Such politicisation seeks answers to questions about what or who needs to be sustained and how this can be maintained or achieved. This includes naming socio-environmental trajectories, enrolling them in a political process that is radically differentiated and oppositional. Clearly, Bush's notion of and desire for sustainability is not that of a Chinese peasant, a *maquiladora* woman worker or a Greenpeace activist. It is important to unravel the Nature of the social relationships that unfold between individuals and social groups and how these, in turn, are mediated by and structured through processes of socio-ecological change. In other words, environmental transformation is not independent from class, gender, ethnic or other power struggles. Socio-ecological 'sustainability' can only be achieved by means of a democratically (in the sense of a genuine political space) organised process of socio-environmental (re)-construction. The political programme is to enhance the democratic content of socio-environmental construction by means of identifying the strategies through which a more equitable distribution of social power and a democratically more genuine mode of the production of Natures can be achieved.

A radical socio-environmental political programme, therefore, has to crystallise around imagining new ways to organise processes of socio-metabolic transformation. This requires first of all a radical re-politicisation of the 'economic' as it is exactly the latter that structures socio-metabolic processes. But this is predicated upon traversing the fantasy that the 'economic' is the determining instance of the political, recapturing the political mains foregrounding the political arena as the decisive material and symbolic space, as the space from which different socio-environmental futures can be imagined, fought over and constructed. This, of course, turns the question of sustainability radically to a question of democracy and the recuperation of the horizon of democracy as the terrain for the cultivation of conflict and the naming of different socio-environmental futures.

Notes

1. See www.statistics.gov.uk; accessed August 30, 2006.
2. Page referred to the 2002 edition of the book.
3. *The Independent*, 22 May 2006[0].

References

Agamben, G. (2005). *State of exception*. Chicago, IL: The University of Chicago Press.

Badiou, A. (2005a) *Politics: A non-expressive dialectics*. Conference *is the politics of truth still thinkable?* The Birkbeck Institute for the Humanities, Birkbeck College, University of London, London, November 25–26th, 2005 [http://w-w.lacan.com/zi-zpopulism.htm; accessed 12/09/2006].

Badiou, A. (2005b). *Meta-politics*. London: Verso.

Beck, U. (1994). The reinvention of politics: Towards a theory of reflexive modernization. In U. Beck, A. Giddens, & S. Lash (Eds.), *Reflexive modernization. Politics, tradition and aesthetics in the modern social order* (pp. 1–55). Cambridge: Polity Press.

Beck, U. (1997). *The reinvention of politics: Rethinking modernity in the global social order*. Cambridge: Polity Press.

Canovan, M. (1999). Trust the people! Populism and the two faces of democracy. *Political Studies, 47*(1), 2–16.

Crouch, C. (2004). *Post-democracy*. Cambridge: Polity Press.

Davis, M. (2006) *Who is killing New Orleans*. The Nation [http://www.common-dreams.org/views06/0324-34.htm; accessed 24/03/2006].

Dean, M. (1999). *Governmentality. Power and rule in modern society*. London: Sage.

Diken, B., & Laustsen, C. B. (2004) *7/11, 9/11, and post-politics*. Online paper, Department of Sociology, Lancaster University, Lancaster [http://www.a.aa-aarg.org/text/3169/711-911-and-post-politics; accessed 30/08/2006].

Giddens, A. (1991). *Modernity and self-identity*. Cambridge: Polity Press.

Giddens, A. (1994). *Beyond left and right*. Cambridge: Polity Press.

Giddens, A. (1998). *The third way*. Cambridge: Polity Press.

Hallward, P. (2005). Jacques Rancière and the subversion of mastery. *Paragraph, 28*(1), 26–45.

Heynen, N. C., Kaika, M., & Swyngedouw, E. (Eds.). (2006). *In the nature of Cities. The politics of urban metabolism*. London: Routledge.

Laclau, E. (2005). *On populist reason*. London: Verso.
Latour, B. (2004). *Politics of nature: How to bring the sciences into democracy*. Cambridge: Harvard University Press.
Latour, B. (2005). *Reassembling the social: An introduction to actor-network-theory*. Oxford: Oxford University Press.
Lemke, T. (1999). The birth of bio-politics. Michel Foucault's lectures at the college de France on neo-liberal governmentality. *Economy and Society*, *30*(2), 190–207.
Levene, M. (2005) *Rescue! history. A manifesto for the humanities in the age of climate change. An Appeal for Collaborators* [http://www.crisis-forum.or-g.uk/rescue_history.htm; accessed 12/09/2006].
Lomborg, B. (1998). *The skeptical environmentalist. Measuring the real state of the world*. Cambridge: Cambridge University Press.
Marvin, S., & Medd, W. (2006). Metabolisms of obecity: Flows of fat through bodies, cities, and sewers. *Environment and Planning A*, *38*(2), 313–324.
Mouffe, C. (2005). *On the political*. London: Routledge.
Rancière, J. (1995). *La Mésentente. Politique et Philosophie*. Paris: Editions Galilée.
Rancière, J. (1998). *Disagreement*. Minneapolis, MN: University of Minnesota Press.
Rancière, J. (2001). Ten theses of politics. *Theory and Event*, *5*, 3.
Rancière, J. (2003). Comment and responses. *Theory and Event*, *6*, 4.
Rancière, J. (2005a). *Chroniques des Temps Consensuels*. Paris: Seuil.
Rancière, J. (2005b). *La Haine de la Démocratie*. Paris: La Fabrique éditions.
Stengers, I. (2003). *Cosmopolitiques*. Paris: La Découverte.
Swyngedouw, E. (2005). Governance innovation and the citizen: The Janus face of governance-beyond-the-state. *Urban Studies*, *42*(11), 1991–2006.
Swyngedouw, E. (2006). Circulations and metabolisms: (hybrid) natures and (cyborg) cities. *Science as Culture*, *15*(2), 105–121.
Thomson, A. J. P. (2003) *Re-placing the opposition: Rancière and Derrida*. Conference *Fidelity to the Disagreement*, Goldsmith's College, University of London, London, September 16–17th, 2003.
Urbinati, N. (2003). Can cosmopolitan democracy be democratic? In D. Archibugi (Ed.), *Debating cosmopolitics* (pp. 67–85). London: Verso.
Valentine, J. (2005). Rancière and contemporary political problems. *Paragraph*, *28*(1), 46–60.
Žižek, S. (1992). *Looking awry: An introduction to Jacques Lacan through popular culture*. Cambridge: MIT Press.
Žižek, S. (1998). For a leftist appropriation of the European legacy. *Journal of Political Ideologies*, *3*(1), 63–78.
Žižek, S. (1999a). Carl Schmitt in the age of post-politics. In C. Mouffe (Ed.), *The challenge of Carl Schmitt* (pp. 18–37). London: Verso.
Žižek, S. (1999b). *The ticklish subject. The absent centre of political ontology*. London: Verso.
Žižek, S. (2002). *Revolution at the gate: Žižek on Lenin, the 1917 writings*. London: Verso.
Žižek, S. (2005a). Against human rights. *New Left Review*, *34*, 115–131.
Žižek, S. (2005b) *Against the populist temptation*. Conference *Is the Politics of Truth still Thinkable?* The Birkbeck Institute for the Humanities, Birkbeck College, University of London, London, November 25–26th, 2005 [http://www.la-can.com/zi-zpopulism.htm; accessed 12/09/2006].
Žižek, S. (2006). *The parallax view*. Cambridge: MIT Press.

Part III
Cognition Dynamics and Knowledge Management in Strategy-Making

All empirical knowledge – scientific and technical as well as personal – is validated, before an action is taken on it, by talking about the evidence. The construction of knowledge must therefore be regarded as an intensely social process, with its own interpersonal and group dynamics. Because human beings pursue ends, have desires, and want their wishes to be accepted by others, communication processes are structured both politically and theoretically. The knowledge that we have about the world is in part a reflection of our passions. When we say that we 'know about the world', we are talking of stories in which, by relating facts, experiences, beliefs, and visions in a narrative, we attempt to make sense of the world. We construct the world out of these stories in a process that is at once individual and social.

Friedmann, J. (1987). *Planning in the public domain: From knowledge to action* (p. 43). Princeton, NJ: Princeton University Press.

Chapter 12
Futures Studies and Strategic Planning

Abdul Khakee

12.1 Introduction

Futures studies vary all the way from artistic and philosophical descriptions of the future to quantified socio-economic analysis. They differ with regard to their relationship with planning and decision-making, from autonomous studies to integrated parts of a planning document.

In recent years, a great number of futures studies on global, national and regional development have been published. However, few of them have become integral components of planning for the future. Part of the reason has been that the envisioned future has often been depicted "in a vacuum rather than at the end of a path commencing in the present" (Huber, 1978, p. 180). This lack of relationship to planning and policy-making is regrettable especially because an increasing number of private and public organisations regard futures studies as an important and complementary activity to the planning of current operations. For example, an increasing number of urban governments in Sweden and elsewhere have become aware that amid the current political and economic uncertainties, the middle-term planning (normally for 4 or 5 years) has become more and more like crisis management and has to be supplemented by long-term structural studies. Similar examples can be found among the large number of American impact assessment studies and studies of regional and local development in North Western Europe (Holling, 1978; Jain, Urban, & Stacey, 1981).

There is no one single way of developing a model for futures-oriented planning. In fact, there is very little in the literature about how the results of a futures study can be used in operational plans and how futures studies need to be modified in order to be useful in a planning system. An important requirement in such a model, however, seems to be that futures studies should provide perspectives for policies or proposals in a plan.

A. Khakee (✉)
Department of Urban Planning and Environment, Royal Institute of Technology,
Stockholm 100 44, Sweden
e-mail: ablkhe@infra.kth.se

M. Cerreta et al. (eds.), *Making Strategies in Spatial Planning*,
Urban and Landscape Perspectives 9, DOI 10.1007/978-90-481-3106-8_12,

Linking futures studies to planning and decision-making processes is not only a matter of providing results to serve as inputs in the processes but also a question of organising futures studies in such a way that the experience of new ways of thinking can be transferred to planners in the course of the studies (Bell, 1997).

Since the first real attempts were made at institutionalising futures research in the early 1960s, a good deal has been written about constructing the images of the future, with emphasis on forecasting, modelling and scenario-generation (Amara, 1974, p. 290), but relatively little is known about how to link these images to the present and to successfully implement the delineated transition strategies. There is hardly any general knowledge or systematic experience about how futures studies are organised.

The aim of this chapter is to examine some important aspects of the relationship between futures studies and planning and to present some models where futures studies have been developed as an integral part of urban planning. The chapter is divided into four sections besides the introduction. The first section discusses differences and similarities between futures studies and planning. The second section presents some features of models that have been found to be useful for connecting futures studies to planning. This is followed by a section describing the experience of integrating futures studies and planning. The final section of the chapter presents some general conclusions about the requirements of imaginative and normative focus in urban planning and the improvement of the conceptual framework and operative features of urban planning.

12.2 Differences and Similarities Between Futures Studies and Planning

There are many variations in futures studies as well as in planning, with regard to aims (use of), methods, organisation and presentation of the results. The literature therefore is not unanimous about similarities and differences between these activities. Some authors (Huber, 1978; Schwarz, 1977) feel that the boundary between the two is often blurred, while others point out that there are differences between futures studies and planning (Cornish, 1969; McHale, 1970). But even where conceptual frameworks are presented to illustrate differences between the two activities, it is recognised that in reality the demarcation between them will be far less sharp and that the two approaches should be regarded as complementary (Shani, 1974, pp. 646–648).

In the planning literature various conceptual frameworks have been proposed to distinguish between long-term and strategic planning on the one hand and operational and management planning on the other. A comparison between the conceptual framework constructed by Anthony (1965, p. 67), distinguishing strategic planning and management planning, and the conceptual framework suggested by Shani (1974, p. 647), distinguishing planning and futures studies, show a considerable amount of closeness between strategic planning and future studies.

In this chapter, planning is defined to include a systematic making of decisions, preparation of programmes for their implementation and a measurement of performance against the programme. Futures studies, on the other hand, clarify the range of possible futures and create images of attainable and desirable futures. Defined in this fashion, there is a well-established tradition of middle-term planning in local government in many Western European countries.

This planning system has developed with the growth of the welfare state. As political and economic uncertainties have increased, so has the recognition of the inadequacies of these planning instruments to develop long-term strategies for development. Futures studies at the urban level are of a much more recent date and there is no well-developed approach to these activities. The tendency has been, however, to follow the tradition of futures research with the emphasis on studying a number of alternative scenarios to provide a basis for a public debate and/or for long-term political decisions (Gidlund, 1985, p. 29).

Shani's (1974) conceptual framework showing differences between planning and futures studies (Fig. 12.1) provides a useful starting point for analysing a proper

Characteristic	Planning	Future studies
1. Output	Set of decisions	Background and context for decisions
2. Extent of detail	Fairly detailed	Relatively undetailed
3. Organisational location	Within the policy-making organisational setting	Usually outside the policy-making organisational setting
4. Time element	Relatively limited	Relatively unlimited
5. Involvement in power struggle	High involvement	Low involvement
6. Time-span	Up to 5–10 years	Usually beyond 10 years
7. Techniques	Mainly data-based, rigorous, analytical and quantitative techniques	Mainly methods involving imagination, intuition and tacit knowledge
8. Mode of publication	Internal, occasionally public	Public, occasionally internal
9. Evaluation	Mainly based on performance	Mainly based on anticipation

Fig. 12.1 Schematic presentation of differences between planning and futures studies (Shani, 1974, p. 647)

relationship between futures studies and planning. In fact several of the characteristics have been useful in developing the Västerås and the Concerted Action model. The output of planning activities is a set of decisions to be implemented by the organisation, whereas a futures study results in a knowledge base on which the present policy alternatives can be evaluated. The output of futures studies, if properly handled, can be used as input in the planning process.

Since planning is action-oriented, it has to be fairly detailed, whereas futures studies provide a broad perspective of futures which need not to be detailed. In an organisation where futures studies are used as a complement to planning activities, the degree of detail for both can be adjusted to obtain a proper feedback between the two.

In Shani's (1974) framework the organisational location of futures studies is usually outside, whereas that of planning inside the policy-making organisation. This is one of the central issues for obtaining a suitable relationship between futures studies and planning. If the same persons are responsible for both futures studies and management planning, it is quite certain that routine will drive out analysis. On the other hand, an autonomous group, while being innovative, may lack a real understanding of development alternatives which are more relevant for the organisation. In order to assure the integration of futures studies and planning processes, one can, for example, supplement the internal recruitment of a futures studies group with special project groups, brainstorming sessions and external reference groups.

As regards the time element, planning has to be carried out within time constraints and with reference to the ongoing activities. Futures studies are relatively free from these constraints. Even if the differences in the time available for these two activities is of significance, the use of results from futures studies in the planning process makes it necessary that futures studies be carried out in some kind of temporal relationship to planning.

Planning activities are usually subject to political bargaining, compromise and reconciliation of conflicting interests, whereas the relative independence of futures studies from policy-making settings can remove them from power struggle. However, since normative aspects are inseparable parts of futures studies, it is desirable to involve policy-makers in these studies, provided proper arrangements are made, to take into consideration the subjective values and alternative scenarios are proto cover ideological differences (Khakee & Dahlgren, 1986). Planning is associated with a relatively short time-span, whereas futures studies with a long time-span. The difference in time horizon has not only quantitative but also qualitative implications since in the short-run, the future is viewed on the basis of quantitative and usually linear changes, whereas the long-run is defined by the relative lack of constraints and commitments so that consideration can be given to a qualitatively different society. While there are obvious reasons for the differences in the time-span between futures studies and planning, future scenarios can be so constructed that development is envisioned in suitable time intervals corresponding to the planning periods (Engellau & Ingelstam, 1978, pp. 72–73).

Shani (1974) contends that the most appropriate techniques for planning are data-based and rigorous, emphasising analytic and quantitative approaches, whereas techniques for futures studies are based on intuition and tacit knowledge. There are, however, many examples of futures studies which successfully combine the use of intuitive and rigorous techniques (Schwarz, Svedin, & Wittrock, 1982, pp. 7–11), whereas planning requires methods involving imagination and intuition as well as quantitative data and rigorous analysis. In fact, there is a considerable common ground with regard to the choice of techniques in policy analysis and futures studies.

In the public sector there is a need to stimulate public debate and increase public awareness of the future. These are among the most important objectives of futures studies. Not only should the results of futures studies be widely available; people should be induced to express opinions in order to enhance their commitment to work on a desirable future (Schwarz et al., 1982, pp. 55–61). In planning participation is regulated by legislation and it is only the affected parties that are encouraged to participate. A systematic citizen participation is, however, necessary, for both planning and futures studies. More recently, environmental concern has paved the way for a more extensive participation in planning for sustainable development.

One basis for evaluating a plan is its (successful) implementation. Plans are evaluated in terms of their costs and benefits. Futures studies are, however, more difficult to evaluate. Eventually, they can be appraised on the basis of their impact on planning and policy-making or in terms of desirable human values. In an organisation where planning and futures studies are regarded as complementary activities, two things have to be kept in mind: (1) futures studies should not be regarded as a direct prolongation of planning activities, or else many of the restrictions that surround planning activities will also be included in the futures studies; (2) futures studies should not become an autonomous activity as their interplay with the policy-making framework might (thus) be easily compromised, making their results unrealistic and of little use in planning and policy-making.

12.3 Models of Linking Futures Studies to Planning

Linking futures studies to the present by means of transition strategies has engaged a few futurologists. The models available for this purpose can be roughly classified as either general models, which present guidelines for such linkage, or models for developing specific transition strategies.

12.3.1 The French Prospective Model

According to the prospective school, the future is not a part of a predetermined temporal continuity but an entity quite separate from the past. It takes on meaning only insofar as it is related to present actions. The model emphasises the need to isolate 'future-bearing facts' (i.e., factors from which future realities emerge). The future-bearing facts can be used in the construction of alternative futures of, for example, the local government as well as in the discussion about the alternative futures of urban society (Cournand & Lévy, 1973).

The major features of the model are:

- isolate 'future-bearing facts', provocative ideas and desired goals;
- prepare a creative plan consisting of a series of possible futures based on future-bearing facts and then evaluated them in terms of desirable human and social values;
- derive a 'decision plan' consisting of a series of decisions based on the creative plan and evaluated in terms of present realities, probable and desirable futures.

12.3.2 The Futures-Creative Planning Approach

Ozbekhan's model introduces the concept of 'ideal ends' which are indicative of the most desirable outcomes. The alternative futures are evaluated in terms of these ends. This normative emphasis has been important in deriving a desirable image of the municipal government's future (Ozbekhan, 1973).

The major features of the future-creative model are:

- derivation of 'ideal ends' that are indicative of the most desirable outcomes;
- use of ideal ends as criteria for selecting among alternative objectives;
- use of objectives to design policies.

12.3.3 Futures-Oriented Urban Planning Model: The Västerås Model

The focus of the Västerås model (called after the municipality in Sweden for which the model was developed) has been threefold:

- generate knowledge about the future so as to evaluate possible consequences that action taken in the face of future uncertainties will have on the present;
- enhance the insight of planners and politicians through their direct participation in the futures studies, which would also increase the possibility of implementing the images of a more desirable future;
- establish futures studies as complementary activities to the existing urban planning system.

In order to reach these objectives, the Västerås model emphasises a successive choice of techniques as the issues under study are clarified and as planners and politicians involved in the study obtain better understanding of the approaches to solve various problems. The object of the futures study in Västerås has been to generate several possible futures of the urban services and the urban government (Khakee, 1985).

The Västerås model was constructed as six inter-related operations in order to ensure feedback from the futures studies in urban planning. These activities are:

- review of current planning and decision-making was undertaken by means of a questionnaire-survey among politicians in executive positions, heads of departments and other civil servants occupying important positions in the planning system. The review brought forward (some) major ideological, methodological and organisational shortcomings in the municipal planning process;
- methodological orientation involving brainstorming sessions with politicians and planners in order to elicit their reactions to various methodological issues (as shown in Fig. 12.1);
- back-view mirror analysis in order to: (1) increase interest among the employees in futures problems by letting some of them examine how their own involvement in the organisation's past activities has affected the development of the organisation; (2) provide an opportunity to examine the common ground between past, present and future beyond a few quantifiable variables; and (3) analyse the strengths and weaknesses of the existing organisation (Khakee, 1986);
- construction of alternative scenarios with the use of three techniques: futures autobiographies, scenario writing and trend extrapolation (Khakee, 1991);
- evaluation of alternative futures for the municipal development and urban amenities on the basis of back-view mirror analysis and the alternative scenarios.

12.3.4 The Concerted Action Model

The European Union–financed project, the Concerted Action, had the objective of studying sustainable use of natural resources in southern Mediterranean region. One part of the study was developing futures studies with the help of public and private stakeholders in order to have the widest possible appreciation of the issue in urban governance even after completion of the project. The entire project consisted of three case studies, Tunis, Izmir and Casablanca/Rabat, each with a set of specific objectives (Barbanente, Camarda, Grassini, & Khakee, 2007; Barbanente, Khakee, & Puglisi, 2002; Khakee, Barbanente, Camarda, & Puglisi, 2002). As stakeholder involvement was one of the major concerns in the Concerted Action, the methodological focus was therefore on participatory scenario building.

The method applied was 'future workshop'. Robert Jungk developed the framework in order to allow people to become involved in creating their own preferred future (Jungk & Mullert, 1996).

Future workshop is made up of three major phases, all involving interactive brainstorming sessions. The three phases have the following contents:

- *critique phase* – dissatisfactions and negative experiences with the current situation; organisation of problems in problem areas and selection of issues of greatest interest by a system of voting;

– *fantasy phase* – free generation of ideas relating to desires, dreams, fantasies and opinions about the future; selection of a number of ideas for further discussion in the implementation phase;
– *implementation phase* – positive, idealistic, innovative and often seemingly impractical ideas, confronted with the problems defined and elaborated in the critique phase; identification of obstacles and restrictions and of possible ways to overcome them in order to implement feasible ideas.

The Concerted Action model ensures interplay between futures studies and planning in the following manner:

– 'Futures Workshop' to derive alternative, desirable scenarios as well as current premises (possibilities and restrictions);
– use of the scenarios to assess various ways to overcome current restrictions and emulate possibilities;
– design of policies based on the scenario-based assessment.

12.4 Integrating Futures Studies and Planning

In this section we shall evaluate our experience in trying to integrate futures studies in planning, bearing in mind the differences and similarities between the two (see Fig. 12.1).

12.4.1 Output and Feedback in the Planning Process

The Västerås as well as the Concerted Action models resulted in a set of preferred future images from which transition strategies were derived. Instead of forecasting the likely trends of development from which goals and strategies are derived, as in conventional planning, strategies were derived from images that represent a rich source of knowledge with a strong input of desirable changes from participants who feel committed to take actions in order to achieve these changes. Backcasting is increasingly seen as an important approach for orientating in an uncertain future characterised by considerable amount of qualitative uncertainty. In the latter case it is difficult to make use of forecasting methods.

12.4.2 Techniques: Appreciation and Relationship

The use of Delphi and other brainstorming approaches in the Västerås and in the Concerted Action model requires that the participants have a good understanding of the applied techniques. This implies that the techniques have more lasting impact on the policy as they achieve consensus among the participants. Both models provide an interesting example of the education of the stakeholders in the science and art of futures studies. An important by-product of these models is the feeling of many stakeholders that the techniques used in futures studies are also useful in the middle-term planning. The techniques function as means of capturing stakeholders' imagination with respect to different issues and encouraging them to think systematically about how to implement the desired images.

12.4.3 Organisation

The fear that the participation of stakeholders, who at the same time have commitments in politics or administration or business or whatever, would result in predomination of the 'present' did not turn out to be the case in these two models. The organisation of various activities (brainstorming, work-shops, interview surveys, etc.), in conjunction with the construction of scenarios, has been an important factor to prevent such a development. Another factor was the use of external advisers with keen interest in future problems; they prevented the abandoning of even future images, even those that were highly improbable or undesirable.

12.4.4 Political Involvement

Political actors have their inner desires about what type of future they would like to have. But they have fears about how far ahead of the mandate period they can commit themselves. Their feelings are continuously re-shaped through interaction with outside realities but also within the futures-making processes. In our two approaches future studies helped in giving politicians new perspective on everyday life issues, far from the rhetoric of contemporary world and in that way, helped them to grasp new ideas for change in real-life processes. As a whole, futures studies seem to have been useful in the shaping of the cognitive and behavioural attitudes of the participating agents. Political involvement meant a higher adherence to real-life complexity and social expectations, inducing fewer uncontrolled results. This implied a relatively more systematic usability of the results from futures studies in conventional planning.

12.5 Concluding Remarks

Both the Västerås and Concerted Action studies indicate that in the long run, urban planning benefits from an imaginative and normative focus on the future. Futures studies make decision-makers aware of the great variety of possibilities lying ahead. This is, after all, what good planning is all about.

The Västerås as well as the Concerted Action model provide a self-feeding application of analysis and synthesis, whereby the present processes of the urban society and of the urban government can be constantly guided with reference to the future.

It provides thereby a new framework for decision-making apparatus and a basis for redirecting its institutions.

The two models with their various techniques provide flexible frameworks that enable participants to improvise and proceed step-by-step in an interactive process. In a way the models are examples of organisational or community learning. The participants are not only involved in generating knowledge about the future and its application to long-term planning but also contribute towards improving the operative features of the models. They certainly contribute towards improving the transparency of decision-making.

The Västerås and the Concerted Action model put considerable emphasis on defining what is desirable and on appreciating the inter-dependence between goals related to various policy areas. The application of a process approach to do this proved essential for devising effective links between the present and the distant ideal ends as well as in helping the coordination between sectors.

Both models strongly assert the potentials of future studies not as much as a discipline of forecast of ready-to-come trajectories of change, but as a discipline which is able to throw light on hidden dynamics of change which risk to be overwhelmed by macroscopic processes. In this respect, future studies are able to unveil many nuances between the polarised space of stereotyped future images, and thus display many unforeseen future possibilities through a recombination of identities and desires within non-hierarchical spaces of co-existence.

This chapter presented strong cases for the role of analysis in policy-making. The implementation of the Västerås as well as the Concerted Action models show that the role of analysis in futures studies of urban communities is tremendously important. It is difficult, however, to draw far-reaching conclusions about the relationship between futures studies and policy-making processes from only two models. Further research is required to analyse this relationship especially with reference to the organisational aspects of futures-oriented studies, the choice of approaches and the presentation and utilisation of the results.

References

Amara, R. (1974). The futures field: Functions, forms and critical issues. *Futures*, *6*(4), 289–301.

Anthony, R. N. (1965). *Planning and control systems: A framework for analysis*. Boston, MA: Harvard University Press.

Barbanente, A., Camarda, D., Grassini, L., & Khakee, A. (2007). Visioning the regional future: Globalization and regional transformation of Rabat/Casablanca. *Technological Forecasting and Social Change*, *74*(6), 763–778.

Barbanente, A., Khakee, A., & Puglisi, M. (2002). Scenario building for metropolitan Tunis. *Futures*, *34*(6), 583–596.

Bell, W. (1997). *Foundations of futures studies*. New Brunswick, NJ: Transaction Publishers.

Cornish, E. S. (1969). The professional futurist. In K. R. Jung &J. Galtung (Eds.), *Mankind 2000* (pp. 244–250). London: Allen and Unwin.

Cournand, A., Lévy, M. (Eds.). (1973). *Shaping the future: Gaston Berger and the concept of prospective*. New York: Gordon and Breach Science Publishers.

Engellau, P., & Ingelstam, L. (1978). *Prognoser och Politisk Framtidsplanering* (Prognoses and political futures planning). Stockholm: Secretariat for Futures Studies.

Gidlund, J. (Ed.). (1985). *Choosing local futures*. Stockholm: Swedish Council for Building Research.
Holling, C. S. (Ed.). (1978). *Adaptive environmental assessment and management*. New York: Wiley.
Huber, B. J. (1978). Images of the future. In J. Fowles (Ed.), *Handbook of futures research* (pp. 179–224). London: Greenwood Press.
Jain, R. K., Urban, L. V., & Stacey, G. C. (1981). *Environmental impact analysis: A new dimension in decision making*. New York: Van Nostrand-Reinhold.
Jungk, R., & Mullert, N. (1996). *Future workshops. How to create desirable futures*. London: Institute for Social Inventions.
Khakee, A. (1985). Futures-oriented municipal planning. *Technological Forecasting and Social Change*, *28*(1), 63–83.
Khakee, A. (1986). Backview mirror analysis in futures studies. *Omega, International Journal of Management Sciences*, *14*(5), 391–399.
Khakee, A. (1991). Scenario construction for urban planning. *Omega, International Journal of Management Sciences*, *19*(5), 459–469.
Khakee, A., Barbanente, A., Camarda, D., & Puglisi, M. (2002). With or without? Comparative study of preparing participatory scenarios using computer-aided and traditional brainstorming. *Journal of Future Research*, *6*(1), 45–64.
Khakee, A., & Dahlgren, L. (1986). Values in futures studies and long-term planning. *Futures*, *18*(1), 52–67.
McHale, J. (1970). *Typological survey of futures research in the US*. Binghampton: Center for Integrative Studies, State University of New York.
Ozbekhan, H. (1973). The emerging methodology of planning. *Fields Within Fields*, *10*(1), 63–80.
Schwarz, B. (1977). Long-range planning in the public sector. *Futures*, *9*(2), 115–127.
Schwarz, B., Svedin, U., & Wittrock, B. (1982). *Methods in futures studies: Problems and applications*. Boulder, CO: Westview Press.
Shani, M. (1974). Futures studies versus planning. *Omega, International Journal of Management Sciences*, *2*(6), 635–649.

Chapter 13
Managing Argumentative Discourses in Multi-Actor Environments

Nikos Karacapilidis

13.1 Introduction

Argumentative collaboration is critical for the creation, leveraging and utilisation of knowledge in various public administration issues. One of the most important advantages of modern organisations in today's complex political, economic, social and technological environment is their ability to leverage and utilise their knowledge (Prahalad & Hamel, 1990). Such knowledge resides in an evolving set of organisational assets, such as the employees, the structure, the culture and the processes of the organisation. Employee knowledge, and particularly tacit knowledge, has been identified to be the dominant asset, as it is decisive at all levels and has to be fully exploited (Nonaka, 1994). Such exploitation refers to the transformation of tacit knowledge to codified information, a process considered to be critical for organisational performance and success (Cohendet & Steinmueller, 2000).

For the above reasons, we argue that it is necessary to adopt a knowledge-based public policy and decision-making view in the development of the supporting technologies (Holsapple & Whinston, 1996). According to this view, public policies and decisions should be considered as pieces of descriptive or procedural knowledge referring to an action commitment. Moreover, any public policy and decision-making process should be viewed as a collaborative production of new knowledge, for example, evidence justifying or challenging an alternative, or practices to be followed (or avoided) thus providing a refined understanding of the problem.

Taking into account the above requirements, this chapter investigates whether and how argumentative collaboration for policy and decision-making can be effectively supported by an appropriately developed information system. The research method adopted for this purpose follows the 'Design Science Paradigm', which has been extensively used in information systems research (Hevner, March, Park, & Ram, 2004). We used this paradigm to develop a Web-based system for supporting:

N. Karacapilidis (✉)
Industrial Management and Information Systems Lab, University of Patras, Patras 26 504 Rio, Greece
e-mail: nikos@mech.upatras.gr

M. Cerreta et al. (eds.), *Making Strategies in Spatial Planning*, Urban and Landscape Perspectives 9, DOI 10.1007/978-90-481-3106-8_13,

(a) the collaboration required for public policy and decision-making; as well as (b) the creation, leveraging and utilisation of relevant knowledge. The proposed system allows for distributed (synchronous or asynchronous) collaboration and aims at aiding the involved parties by providing them with a series of argumentation, decision-making and knowledge management features.

The remainder of this chapter is structured as follows: Section 13.2 comments on literature related to the issue of argumentative collaboration; Section 13.3 presents the features and functionalities of the proposed system; while Section 13.4 describes its application in a real public policy problem. Finally, Section 13.5 discusses set of critical issues related to the proposed solution and draws conclusions.

13.2 Related Work

Designing software systems that can adequately address users' needs to express, share, interpret and reason about knowledge during an argumentative discourse has been a major research and development activity for more than 20 years (de Moor & Aakhus, 2006). Designing, building and experimenting with Information Systems for the development of specialised argumentation and decision rationale support systems has resulted in a series of computer-supported argument visualisation approaches (Kirschner, Buckingham Shum, & Carr, 2003). Technologies supporting argumentative collaboration include, among others, mailing lists, forums, group decision-support systems, as well as co-authoring, and negotiation support systems. There is also increasing interest in implementing Web-based tools supporting argumentative collaboration. These usually provide means for discussion structuring and user administration, while the more sophisticated ones allow for sharing of documents, online calendars, embedded e-mail, chat tools and so on.

The above approaches support argumentative collaboration at various levels, and have been tested through diverse user groups and contexts. Furthermore, all aim at exploring argumentation as a means of establishing a common ground between diverse stakeholders, to understand positions on issues, to bring to the surface assumptions and criteria and to collectively construct consensus (Jonassen & Carr, 2000). In the rest of this section, we present an overview of the existing software supporting argumentation that has been applied in different organisational and educational contexts. The primary aim of this overview is to highlight the features and functionalities of the existing argumentation tools, as well as to comment on their strengths and weaknesses in aiding argumentative collaboration.

Argumentation based on the exchange and evaluation of interacting arguments, which support opinions and assertions, has been extensively applied for collaborative decision support systems or for negotiation support in diverse organisational contexts.

gIbis (Conklin & Begeman, 1987), for instance, a pioneer argumentation structuring tool that has exhibited major impact on a series of other tools, was developed for the capturing of a design process rationale. This is a hypertext groupware tool that allows its users to create issues, assert positions on these issues, and make arguments in favour or against them. Sibyl (Lee, 1990), an extension of gIbis, is a tool for managing group decision rationale. This tool also provides services for the management of dependency, uncertainty, viewpoints and precedents, and can be viewed as a knowledge-based system. QuestMap (Conklin, 1996) is another approach based on gIbis's main principles that resembles a 'whiteboard', where all messages, documents and reference material for a project and their relationships are graphically displayed during meetings. QuestMap captures the key issues and ideas during meetings and creates shared understanding in a knowledge team. All messages, documents and reference material for a project are placed on the 'whiteboard', where the relationships between them are graphically displayed. Users end up with a 'map' that shows the history of an online conversation that led to key decisions and plans. Compendium (Selvin & Sierhuis, 1999) is a graphical hypertext system which can be used to gather a semantic group memory, when used in a meeting scenario. Compendium provides a participatory user interface for conceptual modelling frameworks and other diverse applications required by the community of users.

Other approaches, focusing on the representation of knowledge, include Euclid (Smolensky, Fox, King, & Lewis, 1987), a tool that provides a graphical representation language for generic argumentation, Sepia (Streitz, Hannemann, & Thuring, 1989), a knowledge-based authoring and idea-processing tool that supports the creation and revision of hyper-documents, Janus (Fischer, McCall, & Morch, 1989), which is based on acts of critiquing existing knowledge in order to foster the understanding of knowledge design, and QOC (Questions, Options and Criteria) which is another model to represent the rationale of reasoning in a decision-making process (MacLean, Young, Bellotti, & Moran, 1991).[1]

In the same context, Belvedere (Suthers, Weiner, Connelly, & Paolucci, 1995) is used for constructing and reflecting on diagrams of one's ideas, such as evidence maps and concept maps. It represents different logical and rhetorical relations within a debate and supports problem-based collaborative learning scenarios through the use of a graphical language. Finally, Hermes (Karacapilidis & Papadias, 2001), a tool supporting distributed, asynchronous collaboration by integrating features based on concepts from well-established areas such as Decision Theory, Non-Monotonic Reasoning, Constraint Satisfaction and Truth Maintenance, aims at augmenting classical decision-making approaches by supporting argumentative discourse among decision-makers.

In the context of argumentation theory, systems supporting the visualisation of argumentation have played a considerable educational role as they support teaching of critical thinking and reasoning skills. For instance, Araucaria (Reed & Rowe, 2001) provides an interface for the decomposition of text into argumentation premises and conclusions. It supports the contextual analysis of a written text and provides a tree view of the premises and conclusions. This software has been designed to handle advanced argumentation and theoretical concepts, which reflect stereotypical patterns of reasoning. These features, combined with its platform independence and ease of use, make Araucaria an interesting argumentation tool. The Reason!Able argumentation tool (van Gelder, 2002) also provides a well structured and user-friendly environment for reasoning. Through the use of an argumentation tree, a problem can be analysed or decomposed to its logically related parts, whereas missing elements can also be identified. Furthermore, Reason!Able provides the means for an elegant structuring of the tree diagram. Another educational software providing assistance in the creation and sharing of visual images of ideas is MindDraw,[2] a descendant of Spidermap. This software tool enables users to produce 'cause maps' (maps of causal relationships), thus supporting and encouraging self-reflection, inquiry and critical thinking. It is a special purpose, simple, point-and-click drawing tool that allows the creation, analysis and pictorial representation of ideas. MindDraw is a thinker's tool that is useful for students and learners of all ages, from primary school through graduate training and professional practice. Athena Standard and Athena Negotiator (Rolf & Magnusson, 2002) are two more examples of argument mapping software. Athena Standard is designed to support reasoning and argumentation, while Athena Negotiator is designed to facilitate analysis of decisions and two-party negotiations. It is directed at tertiary education, ranging from first year to postgraduate students or for elementary use by professionals. The above two systems are efficient argumentation structuring tools, but do not employ knowledge management features.

The above approaches have been thoroughly considered during the development of our approach and aided the conceptualisation, shaping and implementation of its currently integrated features and functionalities. For instance, the discourse graph of our tool is gIbis-like, while its reasoning mechanisms have exploited features of the above-mentioned argumentation tools. As noted earlier, majority of the existing argumentative collaboration systems focus mainly on the expression and visualisation of arguments. In this way, they assist participants to organise their thoughts and present them to their peers. However, their features and functionalities are limited (e.g., they pay almost no attention to knowledge management issues), they are tested, almost exclusively, in academic environments (i.e., not broadly used), they are not inter-connected with other tools, and they do not efficiently integrate the technological, social and pedagogical dimensions of collaboration. As acknowledged in de Moor and Aakhus (2006), traditional argumentation software approaches are no longer sufficient to support contemporary communication and collaboration needs. Our approach aims at filling this gap, by providing a list of features and functionalities described in the next section.

13.3 The Proposed Solution

Having followed an argumentative reasoning approach, we have developed a Web-based system that supports the multi-actor collaboration required for public policy decision-making, by facilitating the creation, leveraging and utilisation of the relevant knowledge. The overall framework of our approach extends the one conceived in the development of the Hermes system (Karacapilidis & Papadias, 2001), by providing additional knowledge management and decision-making features.

Discourses about complex problems in the public sector are considered as social processes and, as such, they result in the formation of groups whose knowledge is clustered around specific views of the problem. Following an integrated approach, our system provides public organisations, engaged in such a discourse, with the appropriate means to collaborate towards the solution of diverse issues. In addition to providing a platform for group reflection and capturing of organisational memory, our approach augments teamwork in terms of knowledge elicitation, sharing and construction, thus enhancing the quality of the overall process. This is achieved through its structured language for conversation and a mechanism for evaluation of alternatives. Taking into account the input provided by the individual public organisations, the system constructs an illustrative discourse-based knowledge graph that is composed of the ideas expressed so far, as well as their supporting documents. Through the integrated decision-support mechanisms, discussants are continuously informed about the status of each discourse item asserted so far, and reflect further on those items according to their beliefs and interests regarding the outcome of the discussion. In addition, our approach aids group sense-making and mutual understanding through the collaborative identification and evaluation of diverse opinions. Such an evaluation can be performed through either argumentative discussion or voting.

Furthermore, our system provides a shared Web-based workspace for storing and retrieving the messages and documents of the participants, using the widely accepted XML document format. Exploitation of the Web platform renders, among others, low operational cost and easy access to the system. The knowledge base of the system maintains all the above items (messages and documents), which may be considered, appropriately processed and transformed, or even re-used in future discussions. Storage of documents and messages being asserted in an ongoing discussion takes place in an automatic way, that is, upon their insertion in the knowledge graph. On the other hand, retrieval of knowledge is performed through appropriate interfaces, which aid users explore the contents of the knowledge base and exploit previously stored or generated knowledge for their current needs. In such a way, our approach builds a 'collective memory' of a public sector community.

The basic discourse elements in our system are issues, alternatives, positions and preferences. In particular, issues correspond to problems to be solved, decisions to be made or goals to be achieved. They are brought up by users representing a public organisation and are open to dispute (the root entity of a discourse-based knowledge graph has to be an issue). For each issue, the users may propose alternatives (i.e., solutions to the problem under consideration) that correspond to potential choices. Nested issues, in cases where some alternatives need to be grouped together, are also allowed. Positions are asserted in order to support the selection of a specific course of action (alternative), or avert the users' interest from it by expressing some objection. A position may also refer to another (previously asserted) position, thus arguing in favour or against it. Finally, preferences provide individuals with a qualitative way to weigh reasons for and against the selection of a certain course of action. A preference is a 'tuple' of the form (position, relation, position), where the relation can be 'more important than', or 'of equal importance to' or 'less important than'.

The use of preferences results in the assignment of various levels of importance to the alternatives in hand. Like the other discourse elements, they are subject to further argumentative discussion.

The above four types of elements enable the users of the system, who typically represent public organisations or other parties involved in a public policy or decision-making discourse, to contribute their knowledge on the particular social problem or need (by entering issues, alternatives and positions), and also to express their relevant values, interests and expectations (by entering positions and preferences). In such a way, the system supports both the rationality-related dimension and the socio-political dimension of the public policy and decision-making process. Moreover, the system continuously processes the elements entered by the users (by triggering its reasoning mechanisms each time a new element is entered in the graph), thus facilitating users to become aware of the elements for which there is (or there is not) sufficient (positive or negative) evidence, and accordingly, conduct the discussion in order to reach consensus.

The features and functionalities of the proposed system, as well as its applicability in supporting multi-actor collaboration for public policy and decision-making, are presented in more detail in the following section.

13.4 A Case Study

A real-life application of the system, for one of the most important, difficult and widely discussed public policy issues in Greece, was organised. The case concerned the establishment (or not) of non-state universities. Today in Greece, all universities are 'state' ones, established and supervised by the Ministry of National Education. According to the Greek Constitutional Law, higher education should be provided only by the State, and not by any private-sector enterprises. However, it has been proposed by some politicians and private companies that this status should be changed; initially, new 'state universities' should be established, not by the Ministry of Education, but by other public sector organisations, such as big municipalities, chambers of industry and commerce, the Church and so on. It has been also proposed that, as a next step, the Constitutional Law should be amended, so that it will allow higher education to be provided by private-sector companies as well. However, there are many parties and citizens who strongly object to the establishment of private universities. In this public policy issue many public organisations are involved (the Ministry of National Education, the Universities, the big Municipalities, the Chambers of Industry and Commerce, the Church, etc.), therefore extensive multi-actor consultation and collaboration is required among them, concerning this issue. In addition, there are private-sector stakeholders involved, namely, the owners of various existing private non-university level educational institutions. They are interested in establishing private universities (mainly in cooperation with foreign universities), providing the related Constitutional Law amendment are made. From the above, one can easily conclude that the public policy issue under consideration is quite complex, and diverse arguments both in favour and against all the proposed alternatives should be expected. Needless to say, the issue is of critical importance for many young people in Greece and their families.

Four groups of users participated in this application, each one representing a significant stakeholder in the issue: the Ministry of National Educational (three persons), university professors (four persons), the Chambers of Industry and Commerce interested in establishing non-for-profit universities (three persons), and owners of the existing private educational institutions (four persons). Participants were geographically dispersed and had access to the system via an Internet connection and a Web browser. They were all familiar with using computers and the Internet; all had previously participated (at least once) in an unstructured electronic forum on the Internet. They were trained by postgraduate students, who visited them in their own locations and introduced them to the basic functionality of the system. This training took on average less than an hour.

An instance of the argumentative discourse that developed during their collaboration appears in Fig. 13.1.[3] As shown, our approach maps the overall collaboration process to a discourse-based knowledge graph with a hierarchical structure. Each entry in the graph corresponds to an argumentation element (i.e., issue, alternative,

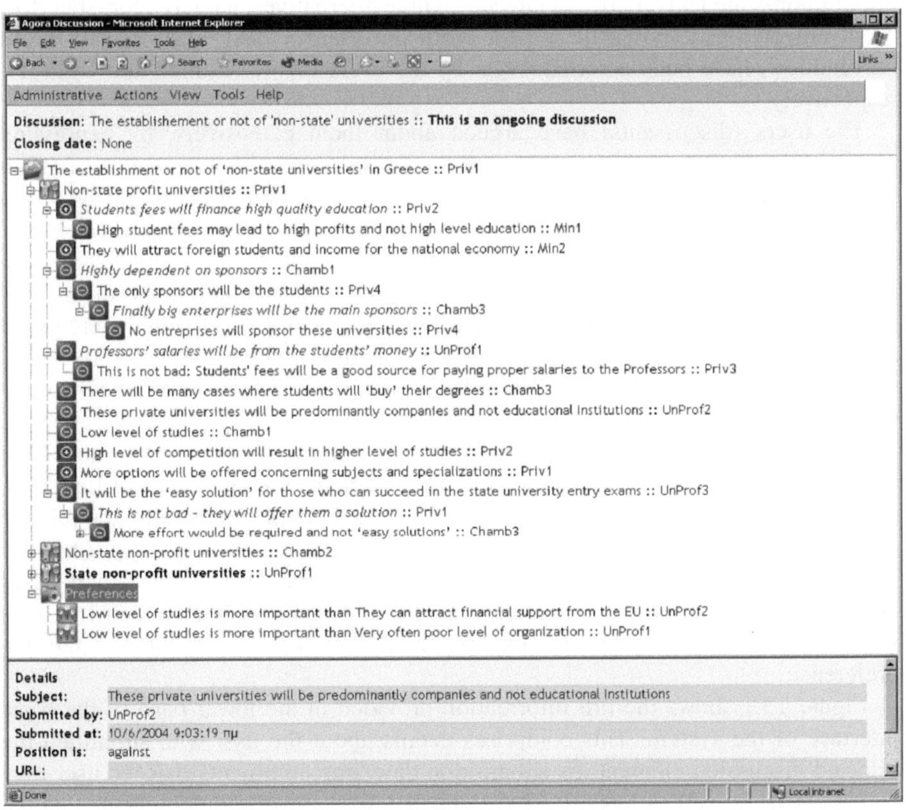

Fig. 13.1 An instance of the argumentative discourse

position or preference). Each element is accompanied by an icon that indicates the element type. There are also icons for folding/unfolding purposes, thus enabling users to concentrate on a specific graph's part; this is particularly useful in graphs of considerable length and complexity. Each entry in the graph may contain the username of the user who submitted it and the date of submission.[4]

In the application discussed in this chapter, the usernames used declare the type of the group the participant belongs to; for instance, the usernames Min1, Min2 and Min3 correspond to users representing the Ministry of National Education, the ones starting with UnProf correspond to university professors. The system may also support 'anonymous discourse', by not revealing the name of the user who entered an element.

According to literature (Beaudouin-Lafon, 1999; Lococo & Yen, 1998), such an approach may be useful in cases where more freedom in ideas generation is sought; also, it often allows users to evaluate each entry more impartially, without taking into account the hierarchical position, the social status and the other characteristics of the user who contributed it. The lower pane of the window shown in Fig. 13.1 provides more details about a selected entry of the discussion graph.[5]

In our case (Fig. 13.1), the overall issue under discussion is 'the establishment (or not), of non-state Universities in Greece'. Three alternatives, namely 'non-state-for-profit universities', 'non-state not-for-profit universities' and 'state non-for-profit universities', have been asserted so far by the users Priv1, Chamb2 and UnProf1, respectively.

The users (discussants) have argued about them extensively, by expressing positions speaking in favour or against them. For instance, 'They will attract foreign students and income for the national economy' is a position (asserted by Min2) that argues in favour of the first alternative, while 'Highly dependent on sponsors' is a position (asserted by Chamb1) that argues against it. All graph entries are subject to multi-level argumentation. For instance, 'Easy solutions are disastrous' has been asserted by UnProf4 to further validate the 'More effort would be required and not easy solutions' position (asserted by Chamb3), while 'No enterprises will sponsor these universities' to challenge the 'Finally big enterprises will be the main sponsors'.

As noted in the previous section, users may also assert preferences about the already expressed positions. As shown in the bottom of the main pane of Fig. 13.1, users UnProf2 and UnProf1 have expressed two preferences concerning the relative importance between the position 'Low level of studies' and two others (namely, 'They can attract financial support from the EU', and 'Very often [there is a] poor level of organisation'), arguing that the first position is (for them) of bigger importance. Users may also express their arguments in favour or against a preference.

Figure 13.1 shows the full information provided in the lower pane of the basic interface of the system. This comprises details about the user who submitted the selected discussion element, its submission date, any comments that the user may had inserted, as well as links (URLs) to related Web pages and documents that the user may have uploaded to the system in order to explain this element and aid his/her peers in their contemplation.

Further to the argumentation-based structuring of a discourse, the system integrates a reasoning mechanism that determines the status of each discussion entry, the ultimate aim being to keep users aware of the discourse outcome. More specifically, alternatives, positions and preferences of a graph have an activation label (it can be 'active' or 'inactive'), indicating their current status (inactive entries appear in red italics font). This label is calculated according to the argumentation underneath and the type of evidence specified for them. Activation in our system is a recursive procedure; a change of the activation label of an element is propagated upwards in the discussion graph. Depending on the status of positions and preferences, the mechanism goes through a scoring procedure for the alternatives of the issue (for a detailed description of the system's reasoning mechanisms, see Karacapilidis & Papadias, 2001).

At each discussion instance, the system informs users about the most prominent (according to the underlying argumentation), alternative solution (shown in bold font). In the instances shown in Figs. 13.1 and 13.2 (all items asserted under the first alternative are folded in Fig. 13.2, while items under the second and third

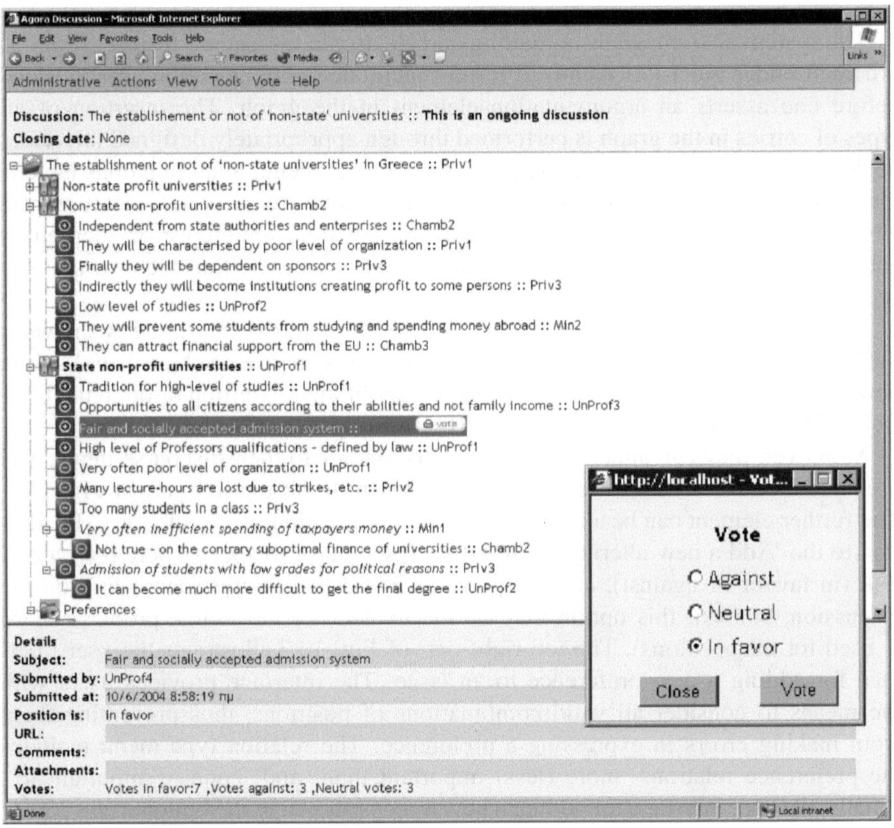

Fig. 13.2 Another instance of the argumentative discourse and the voting option

alternatives are unfolded – the opposite holds for Fig. 13.1), 'State non-for-profit universities' is the better justified solution so far. However, this may change upon the type of the future argumentation. In other words, each time an alternative is affected during the discussion, the issue it belongs to is updated, since another alternative solution may be indicated by the system.

Positions, preferences and alternatives may be evaluated also by voting. In such a case, the 'majority rule' is used in order to decide whether the item is active or inactive (i.e., whether it should be taken into account in the overall evaluation of the issue under consideration).

In order for an item to become subject to voting, the user who has asserted should take the appropriate action (the related option appears under the Vote menu). When an item is subject to voting, an indicative icon appears at the end of it. Any user may then vote about the validity of the item, having the options 'in favour', 'neutral' and 'against' (the related option also appears under the Vote menu, and the small window of Fig. 13.2 pops up). Such a case is shown in the discussion instance in Fig. 13.2, for the position 'Fair and socially accepted admission system', asserted by UnProf4. As one can see in the lower pane of the figure, 13 (out of 14) users have voted so far, while the results are 7 votes in favour, 3 votes against and 3 neutral votes.

The system also integrates e-mailing and electronic messaging features (options provided under the Tools menu) to further facilitate communication among users, before one asserts an argumentation element in the graph. The insertion of all types of entries in the graph is performed through appropriately designed interfaces deployed upon the user's selection under the Actions menu. Such functions include the opening of an issue, insertion of a new alternative (to an issue), insertion of a new position (in favour or against an existing position, preference or alternative) and insertion of a new preference (to an existing issue). Editing features are also provided.

The user interface for adding a new alternative to an existing issue is shown in the bottom left part of Fig. 13.3. As illustrated, users can give a subject (title), of the new alternative, but also provide more details about their assertion through the URL (related Web addresses), and comments (free text), panes.

Moreover, they can attach multi-media documents to their discourse items. The user interface for adding a new position is shown in the top left part of Fig. 13.3. The further element can be an alternative, another position, or a preference. In addition to the 'Add a new alternative' interface, users have to specify here the type of link (in favour or against), and the proof standard they prefer (depending on the discussion, context, this option may be inactivated; i.e., the same proof standard is used for all positions). The top right part of Fig. 13.3 illustrates the user interface for adding a new preference to an issue. The interface provides users with the means to consider all valid combinations of positions, thus preventing them from making errors in expressing a preference. The relation type menu includes the preference relations 'more (less) important than' and 'equally important to'. Finally, the user interface for adding a new issue is shown in the bottom right part of Fig. 13.3.

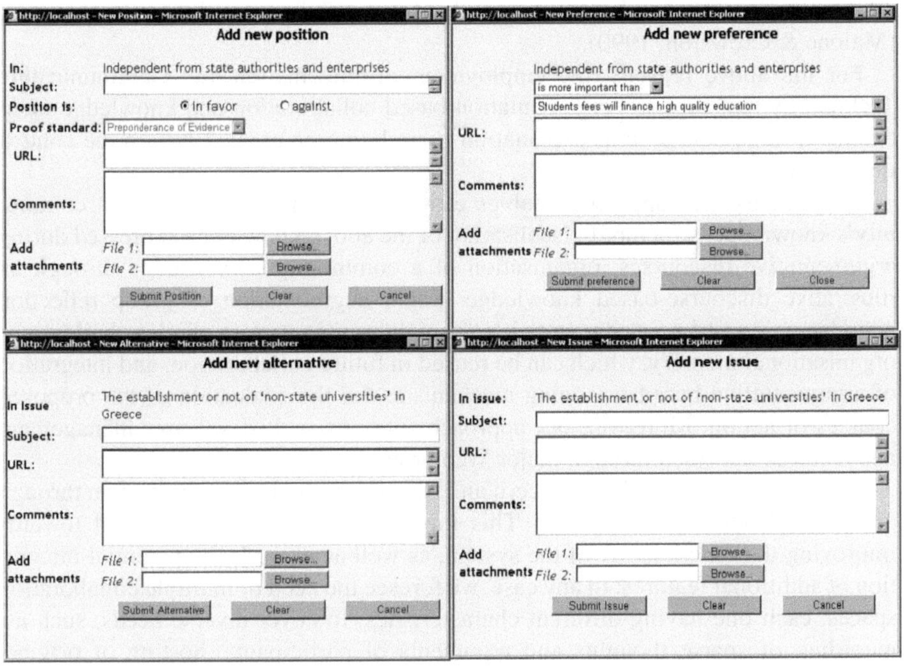

Fig. 13.3 User interfaces

13.5 Discussion and Conclusions

The proposed solution is a Web-based tool that attempts to assist and augment argumentative collaboration being held among multiple actors with diverse interests and backgrounds, by facilitating the creation, sharing, leveraging and utilising the relevant knowledge. The system follows an argumentative reasoning approach, which complies with collaborative principles and practices. As noted by many influential thinkers, argumentation is central to learning (Paul, 1989; Perkins, 1986; Resnick, 1987).

In a variety of contexts, argumentation is an essential element for effective learning, in that it enables people to develop their points of view and refine their knowledge. In an effective collaborative argumentation environment, participants focus on the same issues, and learn to negotiate conflicting opinions, until they accept or share the answer, solution and so on. (Veerman, Andriessen, & Kanselaar, 1998). Sharing information and creating common knowledge in argumentative discourse also contributes to trust development and enhances collaborative behaviour (Chesñevar, Maguitman, & Loui, 2000). Moreover, argumentation facilitates learning as it increases the coherence of organisational mental models by assuring their rationality, logical consistency, and by eliminating any internal contradictions (Rescher, 1970). Similarly, as it operationalises trust and power relations, argumentation has been proved to be an efficient coordination mechanism (Malone & Crowston, 1990).

For the above reasons, the employment of Information and Communication Technology that supports argumentation-based collaboration and knowledge management, 'argumentation as explanation' (van Eemeren et al., 1996) in the context under consideration, is crucial.

In summary, our approach enables easy expression and sharing of a community's knowledge, structured visualisation of the above knowledge expressed during argumentative discourses, organisation of a community's knowledge through an illustrative discourse-based knowledge graph, augmentation of group reflection and leveraging of knowledge creation through argumentation, efficient building of organisational memory, which can be reused in future collaboration, and integration of argumentation-based reasoning mechanisms for the evaluation of the proposed courses of action. Moreover, our approach supports multi-level user management and it can be accessed through major Web browsers.

Future research directions concern an extensive evaluation of the system through diverse real application settings. This input will be further considered towards improving the functionality of the system, as well as towards the potential integration of additional features. In any case, we foresee the need of multiple collaboration spaces, each one having different characteristics, to cover diverse needs, such as: recording of sparse thoughts and arguments of participants, hosting of original free-text dialogs, collection of original resources needed in the context of a specific session of collaboration, creation of new knowledge by elaborating original resources and so on. Such collaboration spaces should be tightly inter-connected, while the transition from one to another should be both transparent and user-friendly.

Notes

1. QOC provides the means to represent and integrate rationale of varying degrees of stability at the different stages of a design process.
2. See http://info.cwru.edu/minddraw/index.html
3. We asked participants to carry out this experiment in English.
4. Alternative forms in the appearance of each entry can be obtained through options provided under the View menu.
5. Users can select an entry by clicking on it.

References

Beaudouin-Lafon, M. (1999). *Computer supported collaborative work*. New York: Wiley.

Chesñevar, C., Maguitman, A., & Loui, R. (2000). Logical models of argument. *ACM Computing Surveys*, *32*(4), 337–383.

Cohendet, P., & Steinmueller, W. E. (2000). The codification of knowledge: A conceptual and empirical exploration. *Industrial and Corporate Change*, *9*(2), 195–209.

Conklin, E. J. (1996) *Designing organisational memory: Preserving intellectual assets in a knowledge economy*. Group Decision Support Systems Working Paper [http://www.gdss.com/wp/DOM.htm; accessed 22/01/2007].

Conklin, E. J., & Begeman, M. L. (1987) *gIBIS: A hypertext tool for team design deliberation*. In: Proceedings of the Hypertext'89 Conference. ACM Press, New York, pp. 247–252.
de Moor, A., & Aakhus, M. (2006). Argumentation support: From technologies to tools. *Communication of ACM*, *49*(3), 93–98.
Fischer, G., McCall, R., & Morch, A. (1989) *JANUS: Integrating hypertext with a knowledge-based design environment*. In: Proceedings of the Hypertext'89 Conference. ACM Press, New York, pp. 105–117.
Hevner, A. R., March, S. T., Park, J., & Ram, S. (2004). Design science in information systems research. *MIS Quarterly*, *28*(1), 75–105.
Holsapple, C. W., & Whinston, A. B. (1996). *Decision support systems: A knowledge-based approach*. St. Paul, MN: West Publishing Company.
Jonassen, D. H., & Carr, C. S. (2000). Mindtools: Affording multiple representations for learning. In S. P. Lajoiem (Ed.), *Computers as cognitive tools II. No more walls: Theory change, paradigm shifts and their influence on the use of computers for instructional purposes* (pp. 165–196). Mawah, NJ: Erlbaum.
Karacapilidis, N., & Papadias, D. (2001). Computer supported argumentation and collaborative decision making: The HERMES system. *Information Systems*, *26*(4), 259–277.
Kirschner, P., Buckingham Shum, S., & Carr, C. (2003). *Visualizing argumentation: Software tools for collaborative and educational sense-making*. London: Springer.
Lee, J. (1990) *SIBYL: A tool for managing group decision rationale*. In: Proceedings of the CSCW'90 Conference. ACM Press, New York, pp. 79–92.
Lococo, A., & Yen, D. (1998). Groupware: Computer supported collaboration. *Telematics and Informatics*, *15*(1–2), 85–101.
MacLean, A., Young, R. M., Bellotti, V., & Moran, T. (1991). Questions, options and criteria: Elements of design space analysis. *HCI*, *6*(3–4), 210–250.
Malone, T. W., & Crowston, K. (1990) *What is coordination theory and how can it help design cooperative work systems?* In: Proceedings of CSCW 90 Conference. ACM Press, New York, pp. 357–370.
Nonaka, I. (1994). A dynamic theory of organisational knowledge creation. *organisation Science*, *5*(1), 14–37.
Paul, R. W. (1989). Critical thinking in North America: A new theory of knowledge learning and literacy. *Argumentation*, *3*(2), 197–235.
Perkins, D. N. (1986). *Knowledge as design*. Hillsdale, NJ: Lawrence Erlbaum Associations.
Prahalad, C. K., & Hamel, G. (1990). The core competence of the corporation. *Harvard Business Review*, *68*(3), 79–91.
Reed, C., & Rowe, G. (2001). *Araucaria: Software for puzzles in argument diagramming and XML*. Technical Report. Dundee: Department of Applied Computing, University of Dundee.
Rescher, N. (1970). *The coherence theory of truth*. Oxford: Oxford University Press.
Resnick, L. B. (1987). *Education and learning to think*. Washington, DC: National Academy Press.
Rolf, B., & Magnusson, C. (2002) *Developing the art of argumentation: A software approach*. In: Proceedings of ISSA 2002 [http://www.athenasoft.org/; accessed 22/01/2007].
Selvin, A. M., & Sierhuis, M. (1999) *Case studies of project compendium in different organisations*. In: Proceedings of Computer-Supported Collaborative Argumentation for Learning Communities. Workshop held at Computer-Supported Collaborative Learning'99, Stanford.
Smolensky, P., Fox, B., King, R., & Lewis, C. (1987). Computer-aided reasoned discourse, or how to argue with a computer. In R. Guindon (Ed.), *Cognitive science and its applications for human-computer interaction* (pp. 109–162). Hillsdale, NJ: Erlbaum.
Streitz, N., Hannemann, J., & Thuring, M. (1989) *From ideas and arguments to hyperdocuments: Travelling through activity spaces*. In: Proceedings of the Hypertext'89 Conference. ACM Press, New York, pp. 343–364.
Suthers, D., Weiner, A., Connelly, J., & Paolucci, M. (1995) *Belvedere: Engaging students in critical discussion of science and public policy issues*. In: Proceedings of the 7th World Conference on Artificial Intelligence in Education (AI-ED '95), Washington, DC, pp. 266–273.

van Eemeren, F. H., Grootendorst, R., Henkemans, F. S., Blair, J. A., Johnson, R. H., Krabbe, E. C. W., et al. (1996). *Fundamentals of argumentation theory*. Mahwah, NJ: Lawrence Erlbaum Associates.

van Gelder, T. J. (2002). Argument mapping with Reason!Able. *The American Philosophical Association Newsletter on Philosophy and Computers*, *2*(1), 85–90.

Veerman, A. L., Andriessen, J. E. B., & Kanselaar, G. (1998) *Learning through computer-mediated collaborative argumentation* [http://eduweb.fsw.ruu.nl/ar-ja/PhD2.html; accessed 22/01/2007].

Chapter 14
Frames, Multi-Agents and Good Behaviours in Planning Rationales

Dino Borri

14.1 Organising Multi-Agent Plans

Spatial planning still lacks of robust scientific attention to knowledge and knowledge-in-action coordination in multi-agent environments (Faludi, 1973, 1987; Friedmann, 1987). Intuitively, this limitation is particularly invalidating, as the current generation of spatial plans aims at democratising its traditional expert and top-down approach and enhancing its knowledge contents and multi-logic potentials (Forester, 1989, 1999; Friedmann, 1987; Healey, 1997; Sandercock, 1998).

At the forefront of knowledge engineering, distributed and multi-agent intelligence, unfortunately, when paying attention to coordination of multi-agent micro-tasks in task accomplishment is still short in the elaboration of the integrated social thoughts that are prerequisites of the new generation of knowledge-based interactive spatial plans (Ferber, 1997).

However, in knowledge-based spatial planning engineering there is increasing awareness of the typical rational and computational complexity of Multiple Source Knowledge Integration (MSKI): problems like contradictions in beliefs and intentions, semantic redundancies and uncertainties, and other theoretical and practical inconsistencies definitely hamper the spreading new age democratic planning arenas, making their assumptions and tools largely ineffective (Borri, 2001).

The new strategic, interactive and strongly future-oriented and visionary socio-environmental planning, in which through cognitive sessions and forums a multiplicity of agents (stakeholders) interact to set and solve complex problems, is an interesting challenge to multi-agent coordination in knowledge engineering (Avlijas, Borri, & Monno, 2005).

Mainstream strategic planning has a typical iterative organisation, which consists of a number of stages, all of which are assisted by a knowledge engineer (as individual or group figure) performing roles of facilitator, mediator or reflexive agent: (i) preliminary organisation of the cognitive interaction (problem definition, selection of stakeholders to be involved), (ii) implementation of Computer Interaction

D. Borri (✉)
Department of Architecture and Town Planning, Polytechnic of Bari, Bari 70125, Italy
e-mail: d.borri@libero.it

M. Cerreta et al. (eds.), *Making Strategies in Spatial Planning*,
Urban and Landscape Perspectives 9, DOI 10.1007/978-90-481-3106-8_14,

(CI) (facilitation-mediation-negotiation assistance by the knowledge engineer, going from the traditional Delphi sessions to the recently adjusted ones characterised by recognisable agents who think and act in destructured and often conflictual ways, so being in tune with real processes of building of social knowledge) and (iii) finalised elaboration of the experimental outcomes in terms of problem-setting and/or problem-solving. An evident current methodological trend is increasing multi-agent protagonism and responsibility in all phases of social elaboration of MSKI – according to the new social knowledge style which pervades spatial planning – and enhancing endogenous knowledge potentials. Cognitive interventions by knowledge planners along these stages of multi-agent interaction can assume *active* modes, insofar as planners give definite structure to (i), make strong efforts to focus knowledge in (ii), and make interpretive more than descriptive use of the experimental outcomes in (iii), or *passive* modes, insofar as they facilitate problem emersion from stakeholders-agents who freely enter the arena in (i), do not focus knowledge in stage (ii), and make descriptive – more than interpretive – use of the experimental outcomes in (iii).

Another important notation about multi-agent cognitive experiments deals with their two alternative motivations: per se problem complexity requires support by and coordination of a multiplicity of cognitive agents; problem complexity is sufficiently small to allow that a cognitive agent can set-solve alone the problem, but cognitive democracy – more than one agent involved – is required because of the existence of a moral aspiration to it.

So, being sufficiently aware of MSKI complexity, we'll deal only with a limited number of aspects of strategic interactive planning, assumed as particularly interesting for their potential developments in the near future: emergence of new knowledge from the experiments and practical relevance of the cognitive experiments in problem-setting and/or solving.

14.2 Features and Outcomes of Multiple Source Knowledge Acquisition and Integration

Extensive experimentation of multiple source knowledge acquisition and integration (through a variety of methods, from the traditional ones to electronic forum-based methods, variously structured interviews, participant observation, etc.), during the last few years of fortune of participative and communicative turn in spatial strategic planning (Borri, Camarda, & Grassini, 2005a), have shown various important features and outcomes of Multiple Source Knowledge Acquisition (MSKA) and MSKI:

- general appreciation of these collective knowledge experiments by the communities that are involved: an appreciation which in some sense does not depend on the usefulness of the experiments and can be interpreted as a general aspiration to democracy in decisions and plans; an appreciation which contrasts with the tacit (but sometimes explicit) reluctance and distrust showed by the leading actors of

the play, hardly challenged by the serene and often successful progressing of the experiments (Borri, Concilio, Selicato, & Torre, 2005b);
- positive effects on the long-term from people involvement in plan-making, consisting for instance in: creation within the community at large of a sort of memory and ex-post-narrative reframing of plan-making experiences (often transmitted through story-telling to people who did not have the chance of participation), shared by politicians of different wings; promotion of a trustful social attitude towards planning efforts, in some way independently on the outcomes of these (Barbanente & Borri, 2000);
- rapidity and freshness of the whole process of participative knowledge elicitation, in contrast with the typical time-consuming and stale process of expert-rational (mostly monologic) knowledge elicitation;
- reliability of knowledge generated via participative experiments and especially via forum-based extensive and iterative cognitive interactions, because of potentials of public knowledge refinement and/or knowledge – often conflictual – debate: in fact, knowledge generated by this way substantially does not differ on fundamentals (apart from differences in format, negligible because they are easily manageable by the new available plurilogic and plurilanguage tools) from expert knowledge (Borri, Camarda, & De Liddo, 2004);
- rapid emergence of fundamentals of information and knowledge about situations and dynamics at hand, even if often in a raw format which needs refinement (Borri, Camarda, & Grassini, 2006);
- need of limiting exogenous interventions in the experiments by knowledge engineers, using these, instead, as mere organisers and listeners; in fact, for success' sake, knowledge elicitors should remain mute and sympathetic during the cognitive sessions, bypassing temptations of providing explanations, indications, judgements and so on: this to guarantee fluidity and neutrality to the experiments and to promote self-confidence within the involved cognitive agents (Barbanente, Borri, & Concilio, 2001);
- need of avoiding rhetoric and emphasis on theory by knowledge engineers, in order to provide cognitive experiments with friendly environments: this effort is at its best when paralleling working in common – in creative assemblies of social roles between experts and non-experts – to the derivation of cognitive frames from the matters at hand (Borri & Cera, 2005);
- relevance, in informal knowledge elicitation, of subjective assumptions and perceptions about past, present and future states of the world and originality – and appropriateness to cases at hand – of experiential statements presented by the agents: the informal knowledge that is elicited is mostly case-based, that is, it holds for the agents who enunciate it and for the others they know as sharing the same situation (Borri, Camarda, & De Liddo, 2005c);
- frequent getting stuck of the cognitive agents who participate in the informal knowledge elicitation experiments in specific whims – cognitive vicious circles – which apparently relate to recurrent and never-solved problems tormenting the agents and running wild because these fail in framing the tormenting problems within a hierarchy of values (Borri, Grassini, & Starkl, 2009);

- existence of a cognitive quantum (Soucek, 1997) in MSKA via informal knowledge elicitation, whose relation with local situations and dynamics deserves better exploration; in fact, informal MSKA proceeds through cycles concerning procedure and substance, according to a sort of *quantum*-development, so that extending knowledge elicitation efforts without introducing situation and/or organisational changes into the experiment setting is generally sterile; this sort of sets of cognitive spheres (Binetti, Borri, Circella, & Mascia, 2005), in which a multiplicity of cognitive agents interact in the light of the resources and constraints coming from theory-in-practice concepts on forms and structures, movements through cognitive frontiers and internal-external rules of behaviour, deserves consideration well beyond ambits and limits of political democracy and formal representativeness of agents (Borri, 2001).

Some other features and outcomes of MSKA and MSKI could be highlighted. But it is worthwhile to keep the list short, adding only few touches to the picture.

In the end, the symbolic value of participatory knowledge elicitation experiments at community level seems indisputable and dominant. It suggests relaxation of logical *aporias* of extended formal democracy and optimal organisation, in favour of the adoption of a reasonable and incremental approach: in these experiments, cognitions that are acquired do matter less than the outcome of self-confident and trustful cognitive environment on plan-making and its agents (politicians, technicians, citizenship organisations, individuals, etc.).

14.3 Making Research on Socio-Environmental Futures: The Frame Problem and the Generation of Multi-Agent Knowledge

Research on socio-environmental futures is a spill over effect of physical and emotional impacts of global socio-environmental change: many facets of this change are blind to normal science and call for integrated expert and non-expert, rational and non-rational (emotional) cognitive approach (Shakun, 1999).

Multi-Agent Knowledge Generation (MAKG), in the various field spanning from computer science (distributed knowledge) to policy science (learning organisations), is becoming a way to cope with this complex integration by the extraction of specific knowledge from the general knowledge bases – with their typical systemic resilient and generative lineaments – of a multiplicity of cognitive agents. In computer science, research on distributed knowledge has mainly dealt with agents' roles, non-monotonic and in general multi-modal logic, network architectures, language processing and so on; in policy science research on distributed knowledge has mainly dealt with stakeholder (cognitive agent) selection, agents' roles, decisions, problem-solving and setting, and in general organisational development, with increasing use of widely attended knowledge forums.

Some aspects of MAKG (see non-countability of relevant potential agents, hard reduction of the multiplicity of knowledge coming from these non-numerable agents to the oneness, and hard situation description in granular and sequential formats) are *de facto* intractable in conventional analytical terms, whereas, instead, they appear as more tractable when approached by the non-conventional analysis of situation calculus and cognitive framing: because of this reason, it is worthwhile to explore some hints offered to strategic planning MAKG by representational and computational framing.

The frame problem is a well-known problem in artificial intelligence (Minsky, 1986). Frames influence cognitive agents' abilities of building rationales when navigating through huge and in theory infinite problem spaces: in fact, situation frames are used for context-based and case-based pruning during navigation through dangerous and unfruitful regions of those spaces. More, apart from computer science, neuroscience is emphasising the role of non-rational and emotional knowledge – generally, organised in non-analytical formats – in decision-making performed by cognitive agents (Damasio, 1995), paving the way to a new understanding of individual and group knowledge formation for ill- and non-structured knowledge problems and forums.

Spatial planning – in particular strategic planning – do not adequately consider suggestions that come from the theory of framing: this ignorance or misconsideration limits the intelligence and performance of the current generation of plans; linear and sequential problem-solving architectures still dominate in the 'toy block' world even of the most intelligent plans.

A relevant problem in strategic planning forums, conceived as MAKG environments, is framing the forum's knowledge base and context: according to the above-mentioned general reference, this problem, familiar to cognitive and computer science, deals with agents' abilities of escaping cognitive paralyses and/or dispersions induced by slavish exploration of huge problem spaces (Fagin, Halpern, Moses, & Vardi, 1995; Minsky, 1986; Papadimitriou, 1994; Schank, 1982; Shanahan, 1997).

In fact, in strategic planning Cognitive Forums (CF), there is no other need than mere exogenous (activation of CFs by external facilitators) and/or endogenous (self-activation of CFs, which spontaneously find their internal facilitators) enunciations of subjects of discussion to catalyse frames spilling out from the participants' plural cognitions that are made of substance and procedure (ontologies and concepts, reported or experienced facts, procedural assertions): limits and potentials of this cognitive spillover are defined by the agents' previous cognitions and prevent the cognitive agents from inessential or harmful focusing on too much parochial (coming from the agents' non-expert side) or general (coming from the agents expert side, frequently inappropriate to the case at hand) parts of their previous cognitions (Kolodner, 1993; Minton, 1988; Sacerdoti, 1977).

In MAKG standard-type forums is usual that the organisers (knowledge engineers, acting as the intermediary-agents postulated by mainstream multi-agent cognition) provide the participants with frames – contingently structured – of the subjects that have to be elaborated: nothing of new, of course, as preliminary inquiries entrusted to competent agents on subjects to be submitted to group deliberation, on the basis of which group deliberation processes develop, are consolidated method and rite in many policy and decision-making and deliberation fields, including spatial planning.

Previous exogenous framing generally is at risk – when hyperformed – of self-fulfilling prophecies or – when hypoformed – of modest cognitive performances. Being evident that general models of framing are impossible for complex situations and that only case-based models (i.e., locally optimal models) can exist and be satisfying (incidentally, interesting theoretical and experimental research regards the *logically* proper relationship between frames and subjects), what is needed is dynamic framing paralleling forums implementation in the light of its context (regarding goals, action and means, and ends). In contrast with exogenous framing, previous endogenous framing – in completely bottom-up MAKG – can be innate (i.e., frames are part of the cognitive patrimonies of the forums' agents) or evolutionary syntheses of different frames activated during the forums by ecologies of multi-agent protocols and stories concerning the forums' subjects: inexistence of optimal frames and need of dynamic formation of frames during the multi-agent interaction exercise hold also in this case (Simon, 1982).

A relevant logical contradiction in the use of frames in MAKG forums is in knowledge focusing and selecting potentials that characterise framing and contrast with the aspirations to have free and creative forums (in fact, rational and emotional calculus should coexist in good visioning of futures) and to maintain democracy and neutrality – structural components of genuine collective cognition – in this kind of cognitive exercises (in fact, usually, the stakeholders who participate in the cognitive forums raise systematic criticisms against the whole set of framing contents and procedures).

Moreover, many impediments affect both formation of previous exogenous framing by single or collective intermediary-agents (lack of consolidated cognitive frames in exercises usually set up for ill-structured problems; difficult and controversial mediation among the multiple cognitions required by multi-agent framing; framing costs contrasting with the efficiency required from exercises) and formation of previous endogenous single or collective framing (difficulties in individuating and focusing the forums' subjects; difficult and controversial mediation among the multiple cognitions that structure the group mode of framing; and so on).

The problem of the indefinite and uncertain logical tractability of framing is particularly urgent in the starting phases of strategic planning forums, that is, when knowledge generation and dynamics are characterised by structural potentials of uncertainty (as many cognitive things can still happen, many cognitive directions can still be followed). But also the subsequent phases of the forum exercises – refinement, evolution, dialogue, convergence, divergence and so on – are strongly influenced by the framing mechanisms, in terms of (i) *internal adjustment framing*, oriented to mutual coherence and to development of agent-generated ongoing cognitions, (ii) *reflective return framing*, oriented to reflective and return insights by the agents on their own previous cognition in the light of the cognitive outcomes of the exercises and/or to starting new cognition cycles; (iii) more rarely, *dynamic creative framing*, paralleling implementation and evolution of the cognition exercise with generation of completely new frames (in a Schönian 'double loop' frame reflection).

The final phases of these cognition exercises, anyway, seem less suffering from framing scarce-definition and uncertainties, as they are streamlined through (regions of) problem spaces tractable by interacting cognitive agents who, in that final time, seem driven by a sort of inertial motion coming from the initial push towards a target located at the end of a cognitive path (Pearl, 1985; Schön, 1991).

If the above-mentioned problems are theory-in-practice framing logical *aporias* which hamper the routines of strategic planning MAKG forums, the cognitive agents practically muddle through, because of ability they have of starting, orienting and developing some personal cognitive cycle on the mere basis of a mechanism of (exogenous or endogenous) demand for and supply of cognitions they already have or can rapidly acquire by interacting with the real world (other participants in the exercise, knowledge engineers, available information or knowledge, etc.). Cognitive agents involved in MAKG exercises can even overcome the *aporias* of these by merely thinking to virtual realities based on concepts and procedures coming from personal theory-in-practice ontologies (even when being completely extraneous to a subject of exercise, the agent can build an hypothetical reality through creative imagination) (Cohen, 1995; Ishida, 1997).

However, in strategic planning forums, framing is useful but non-essential support (in terms of orientation, acceleration, delimitation, etc.) to the MAKG, especially when – as typical in the current strategic planning, strongly orientated to futures and problems exploration and representation instead than deliberation – what matters is problem-setting more than problem-solving: in problem-setting creative vantages compensate for the disadvantages in pertinence and investigation which come from scarce or null previous framing; by the way, in strategic planning forums increasingly characterised by multiplicities of cognitive agents, literature reports that abundance and variety of knowledge is potentially granted by the large numbers of cognitive agents involved (proud statement that should to be submitted to further experimentation) (Sutton & Barto, 1998; Watts, 1999).

Interesting conclusions come from the above-presented theoretical and experimental insights: first, exogenous framing is hampered by many theoretical and practical factors in forum-based MAKG exercises, so that standard problem-structuring methods for group knowledge elicitation often are inadequate to the task; second, exogenous framing is not strictly required as precondition of forum success, as agents, who participate in the knowledge forums and are aware of the subjects of these, spontaneously generate individual and/or group knowledge that is progressively refined during forum sessions; third, endogenous framing is highly context- and group-dependent, being at the same time conditioned by the prejudices that the agents have about the situation; fourth, aspirations to optimal (better: satisfying) preliminary and static framing should be replaced by efforts for creating interactive evolutionary framing able to develop in tune with forum development; fifth, simple computation and indexation support formal evolutionary models oriented to the understanding of the agents' knowledge interactions and fluxes; in the end, sixth, hybridisation and de-standardisation of methods of knowledge generation and representation are worthwhile in large group forums oriented to ill-structured or non-structured problems and issues.

14.4 Cooperation Versus Competition Dilemmas in Strategic Planning

The hard logical and computational tractability of the above-mentioned multi-agent spatial planning – in particular, see strategic planning, a form of planning currently à *la page* because of its democratic and visionary features – emerge from the operation of synthesising one individual choice from a plurality of social choices (an intriguing problem, defined by Arrow as insolvable if the conventional axioms of rationality are not relaxed) (Arrow, 1963).

The *reductio ad unum* of multiplicity is operation in some sense exoteric, requiring relaxation of rigour of classical rationality: in fact, the shift from unicity to multiplicity and vice versa – from One to Two and from Two to One (Shakun, 1999) – is sort of divine affair, that religion seems to manage better than science.

In spatial planning the attention to this hard but promising transformation is well exemplified by the appearance in the second half of twentieth century of a new form of interactive (also defined as communicative, dialogic, collaborative or transactive) planning, presented as complex and dynamic synthesis of a multiplicity of theories-in-action performed by a number of agents involved in socio-technical tasks. This move is crucial, after decades of domination of the so-called rational model of planning, based on the assumption that in thinking and acting any individual optimises her/his choice among alternatives (in fact, the choice would be made by an *homo oeconomicus* searching for maximisation of his/her individual profit). These two fundamental planning paradigms have proved to be able to coexist (Friedmann, 1987), even if both show increasing limits in front of the highly complex ecological rationality imposed to planning by the environmental challenge.

The cooperation-competition dilemma, with the whole range of possible fuzzy actions and regions evocated by the conflict (also moral) implied by its terms, deserves scientific reflection. On this terrain, one has to think to two main problems: (i) the assumptions and forms and rationality implied by the different possible worlds evoked by socio-environmental strategies (that are hyperdimensional and therefore intractable by mere economic rationality), (ii) the entities – in particular living beings – that are subjects and objects of strategic socio-environmental planning.

14.4.1 Rationales

Dealing with the first aspect, it is worthwhile to come back to the above-mentioned planning evolution: in the second half of the twentieth century the rationale of territorial planning has registered a shift from the prevalently abstract-procedural and normative models (Faludi, 1973) that had been mainstream for many decades, bivalued and based on a logic of maximisation of individual utility, to new prevalently practical-procedural and non-normative (contextual and adaptive) models (Forester, 1989; Friedmann, 1981), multi-valued and based on a logic of maximisation of collective utility.

But, after having considered this theoretical and historical frame of spatial planning models, we should admit that the rational – Abstract-Procedural and Normative (APN) – model of planning, whose origins are in systems theory and cybernetics, has been not *de facto* used in spatial planning, a part from the exception of some big planning experiments involving complex strategies and important public responsibilities (numerous examples can be found in the public planning in UK and a little bit less in the US). In fact, much more than in routine urban and regional planning, the APN model has mostly been practised in some planning macro-policies and in the evaluation field: this is so true that it is problematic even to find reported by literature evaluations of its pros and cons. The APN model is instead still largely influential in management science for optimisation planning of management science and in computer science for program routines. This model has a competitive feature insofar as it founds on action optimisation conceived from an interested individual's perspective (at least in the ethical sense of the individual agent's interest to survival: the case of the disinterested agent can be analysed as a virtual and moral case of existence of a constraint to full development of that individual interest to survival) (Borri et al., 2005b).

The rational – Practical-Procedural and Non-Normative (PPNN) – model of planning, whose origins are in philosophical theories of communicative action and speech acts and in psychology-management theories of group dynamics and organisational development, fundamentally inspired by the behavioural paradigms of social interaction and of technological change, has been largely used in spatial planning and in micro-planning in general (here the prefix micro does not have value of scalar dimension) has become object of variegated scientific insights, in the meanwhile being totally rejected by formal optimisation planning in management science and computer science because of the extreme complexity of the routines modelled on its functional and logical architecture. In spatial planning the PPNN model has been progressively extended from the small group ambits – the field numerically and suggestively defined by Lewin with the 2–12 interval (Lewin, 1948) – that are typical of site planning and community building to the large group ambits typical of comprehensive urban planning and now also – see strategic planning – of regional planning. This model has a cooperative feature insofar as it founds on agents' interaction and mutual listening.

However, the APN and PPNN models share the common feature of using systems theory and analysis to equip their routines with some typical systemic ideas: on the APN side the idea of rational drive and evolution in highly complex processes (systemic processes), on the PPNN side the idea of stronger resilience (robustness of dynamics) of the systems based on good variety of mechanisms and agents.

If the APN-type rational approach to planning substantially has not had any influence in spatial planning and particularly in its micro-practices, the PPNN-type rational approach seems to fail in the generation of non-local (i.e., found through exploration of wide spaces of optimal functionals) optimal or even satisfying (i.e., related to the only subset of the considered alternatives within the larger set of all the possible ones) solutions, while it is good for generating local satisfying solutions (compromises) through mediation or negotiation procedures; the PPNN approach,

30 years since its introduction in spatial planning, seems also weak in creative problem-solving, that is just in what should be its main trump in the light of its systemic heritage, given the assumption in systems theory that new proprieties emerge from interaction of elements.

Both fail in situations of high plurality of variables and agents involved and also in situations of semantic complexity (ambiguity, uncertainty), that is, where proper rationality – explorative and creative, definitely non-procedural both in abstractions and in practices – is needed. Socio-environmental phenomena are largely intractable by the two approaches, even if the PPNN approach provides some hints for the adoption of cautionary and compromise-oriented rationalities, highly required in socio-environmental change engineering and planning (the APN approach can provide this kind of rationality only in the dichotomic terms of a dilemma action/not-action) (Barbanente et al., 2001).

To overcome this limitation, completely new or hybrid approaches could be used: for example, the agents of a PPNN procedure could be forced to adopting an APN procedure as individuals (i.e., before interacting with other agents), even if this is merely hypothetical because such a complex planning architecture would require too much time for being usable in real behaviours, while, at the opposite, the insertion in APN of a logical procedure based on a social responsibility of the individual agent (dialogic, in the sense of assuming audience and acceptation) could be fine, should the analysis of its cost performances be encouraging.

14.4.2 Entities

Entities – especially living beings – involved in strategic planning cooperation-competition dilemmas put a problem of moral reason, insofar as they call for behaviours and argumentations based on principles external to local or contingent realities, finalised to conservation and transcendence of local and contingent realities by a superior order reality existing out of spatial and/or temporal boundaries (principles that come from ecological or religious axioms), on one side, and a problem of practical reason, insofar as they make explicit their own fight for surviving in those particular circumstances, out of any eschatological reflection and dimension, on the other side. Socio-environmental phenomena based on the agents' self-interests – the only interests by which the interests of a multiplicity of agents are pursued through a complex feedback chain and circular process – provide wide documentation of the proteiform and multi-logic characters of these dilemmas (Borri, 2002).

Sociology and in particular economic sociology deal in terms of rigid opposition with the two stances of cooperation and competition. Literature on underdevelopment as a phenomenon generated by the self-interest of dominant economies proposes local cooperation and self-organisation as antidote to economic and politic globalisation: the cooperative world should fight against the competitive world to defeat it, in a certain sense in the paradoxical terms of the Latin *Si Vis Pacem Para Bellum*. At the opposite, the competitive world tends to dominate on the cooperative one – assumed as expression of sterile refusal of development – through its own technologies and rules: in that world cooperation is functional to a competition seen as *conditio sine qua non* of infinite economic progress. In the end, the two worlds fail in rationally dealing with the problem of technology-based growth reduction that is timidly but clearly appearing at the strategic horizon of a new socio-environmental organisation (Latouche, 1991).

14.5 Concepts and Rationales of Plans

If complete information on a starting situation is provided a plan – in classic terms – "is a sequence of actions that leads the agent from the initial situation to a goal state" (Baral, Kreinovich, & Trejo, 2000, pp. 241–242). When the agent does not have complete information on the starting situation the above-mentioned sequence of actions is not the only one possible: in this case, if the agent can make observations during the execution of the plan, a preventive plan can be conditional and account for different sequence of actions in correspondence to the different possible initial states.

Often the agent's observations are limited or available only in some situations: these limited and sporadic observations can be thought as actions producing sense or knowledge, whose execution changes the knowledge that the agent has about the world more than changing the state of the world. So it is good when conditional plans incorporate these 'actions producing sense'. To be more explicit, it is convenient to remember the example made by Levesque (quoted by Baral et al., 2000, p. 242) of a traveller's plan to take a flight: the agent who at plan execution does not know the number of the boarding gate has to put an action producing sense as 'find the number of the boarding gate' before the sequence of actions that constitutes the plan per se (go to the gate and board). Within classic intelligent artificial planning the agent who performs the plan is seen as the only source of the transformation generated by the plan: so, since the completion of the plan the world will remain unchanged.

In the A language for action description proposed by Gelfond and Lifschitz in 1993, an action plan starts from a finite list of properties ('fluents') *fl, . . ., fn* which describe possible properties of a given state (a state being defined as a finite set of fluents) and from the assumption of the existence of complete knowledge on the initial state; in the A language, a finite set of actions describes any change of state (according to a *a causes f if fl, . . ., fm*-type formula); always in A, a 'D-domain description' is a finite set of 'value propositions' of *initially f*-type and a finite set of 'effect propositions' of *a causes f if fl, . . ., fm*-type; in A, in the end, a 'transition function R' describes the effect of an action *a* on a state *s* (Baral et al., 2000; Gelfond & Lifschitz, 1993, p. 242).

Notwithstanding the ever more diffuse positions against it, assumed as scarcely useful today in most real-world applications, according to mainstrem planning science, classical planning remains useful in many fields, from logistics to process planning and time and sequence programming (Giunchiglia & Spalazzi, 1999).

Further, in recent times classical planning has been supported by a number of contributions to the improvement of its intelligence and operationality: see case-based planning (Hammond, 1990), multi-agent planning (Durfee, 1988; Jennings & Wooldridge, 1998) and 'non-STRIPS'-type planning (Blum & Furst, 1995).

According to Yang, "classic planning mainly deals with generating plans to attain a set of pre-defined goals in situations in which the most relevant conditions of the external world are known and the changes that happen in this world do not affect the success of the plan" (Yang, 1997, p. VIII). Further: "[g]enerating a plan can be seen as a search in a space of nodes" (Yang, 1997, p. 39).

A planning system (a plan) can be seen as a set of basic activities: generation (of a plan), execution (action), sensorial activity with regard to the external world and control of planning agents in plan implementation and reparation of damages caused by failed actions (Giunchiglia & Spalazzi, 1999, p. 330); after half a century of approaching plans according to the cybernetic model, probably this is still the most influential conception of a plan for computer scientists, control engineers and spatial planners (Friedmann, 1987).

The generation phase of a plan, probably the most intriguing one in the whole set of phases of a plan, is usually based on methodologies and searches that can be typified in the following ways: searching in a space of states, for example, according to the well-known and effective way proposed by Fikes and Nilsson (1971), heuristic partial searching in the space of the plan à *la* Wilkins (Wilkins, 1988), searching case-based memories according to the already cited Hammond's mode (Hammond, 1990) and so on.

A plan can be generated with the aim of attaining a goal or reacting to an external event (reactive planning).

Usually, a goal is a condition posed with regards to intermediate and final states of actions (e.g., a set of desired behaviours): in classic planning, in most cases, a goal is a condition imposed on a final state, formally expressed as a conjunction of clauses à *la* Chapman (1987) or à *la* Fikes and Nilsson (1971).

As for the external world, classic planning usually assumes that the initial situation (initial state) is known, all the effects of the actions are known, and the world is not changing or is in negligible change (substantially, it refers to a close and static world): so, classic planning is clearly inapplicable to domains characterised by dynamics and unpredictability (a part from the social and environmental domains, whose intractability by classic rational planning is immediately evident, think to the domain of robotics or of network navigation).

In classic planning efficiency and effectiveness, procedure and substance, are usually distinct: according to Yang, an intelligent classic plan can be successful (not aborting during execution) and yet it can fail in attaining its goal (Yang quoted in Giunchiglia & Spalazzi, 1999, p. 330): in line with these insights, see also the Faludi's interpretation of spatial and environmental planning in terms of proceduralism and critical rationalism.

In both economic and philosophical literature, the theory of rational choice assumes that to attain a goal an agent has to evaluate alternative actions – immediately evident to her/him – referring to a probabilistic distribution of possible outcomes and to an utility function computed on these: in the simplest case, the agent has to combine probability and utility into a function of expected utility of her/his actions, in order to choose the action characterised by the maximum value of expected utility.

It is evident that fundamental practical situations remain intractable by this theory: desirability of a choice not immediately evident to an agent while this is acting to make the desirability measure clear to him/her, for example in a situation in which the benefit of an action is immediately evident while the cost is not evident; existence in the agents of previous sets of intentions (plans) (while the agents' resources are constrained) that inescapably frame the agents' reasonings and behaviours in whatever circumstance. In other words, substantially, any individual action of an agent cannot be evaluated in isolation from the other actions of that agent: in fact, it must be always evaluated against the background of a context of current plans (commitments on future activities) elaborated by the agent, plans that are only partially specified and specifiable in any given instant (or state of the world) (Horty & Pollack, 2001, pp. 199–200). Horty and Pollack make the example of a traveller who has to make the best choice between going to the airport by taxi and going by bus: if before going to the airport, for example, he/she has to participate in a business lunch with colleagues, then, location (should it be a university campus or the airport hotel) and relevance of that event will probably be essential in his/her choice.

The plans that are analysed by Horty and Pollack (2001) are – for the sake of simplicity – primitive (not subjected to hierarchies), complete (a plan is complete when any precondition of any pace is well defined and there is no risk of existence of causal links between precondition and effect), and deterministic in their outcomes (the outcomes of the actions on which they found). They are built, classically, of set of paces, temporal constraints on these paces (usually, in literature, mostly modelled in qualitative terms, that is with reference to ordinal logic of pace sequences; Horty and Pollack (2001) take into account also quantitative temporal constraints, referring to precise action time scheduling for start, conclusion and duration), and causal links that activate dependency relationships of a pace on another pace. In these plans, actions have – again in a classic mode – both preconditions and effects, expressible, for example, to be simple, in terms of propositional logic.

Horty and Pollack (2001) make an example of a plan made for 'buying a shirt in a shopping mall': a very simple plan in the real world, a plan which presents asperities in the artificial world. In the Horty and Pollack's situation, actions are very few and well definable, as they substantially refer to (i) going to the shopping mall, (ii) finding money and (iii) buying the shirt, preconditions and effects of actions are expressed in the form of true or false propositions, and the definition of the initial state (set of the propositions that are true or false at plan starting) is not required because all meaningful plans are characterised by dynamism and by propositions that can become true in various times, so that are not realistic plans that have only one initial state.

Horty and Pollack (2001) conceive choices (options) – consistently with the fact that they assume these as determined by contexts and in particular by the efforts (the actions) implied (their concept of 'strong compatibility' between plans and contexts) – as very plans per se, plans that are presented to the agents for acceptation or rejection (evidently, an extremely simplified binary model of alternatives: in more realistic terms, revision of plans will be searched, in order to come to intermediate options, context modification will be searched in order to adapt it to options, etc.). Further, Horty and Pollack think that the so-called 'admissibility filtering' – as a function of the agent's commitment to the action – is a crucial process, oriented to excluding consideration of options incompatible with a given commitment.

Hence, a realistic theory of rational choice must consider agents deciding in the context of their own existing plans – this way maintaining commitments only partially specifiable in whatever time – and having limited computational resources so to be able to assess the cost of a choice (of a decision or option) only on the basis of a desired level of rationality. Further, this modified – and not simply augmented – theory must consider that classic planning assumes actions as they were located in points of a semantic space, while a general approach to the plan is forced to conceive a plan in terms of something that specifies a set of possible futures (i.e., of futures consistent with plan implementation) with the negative consequence of the appearance of a ramification and pruning problem.

On this terrain, Horty and Pollack think also to the need of considering what is today usually neglected, in terms of a number of intriguing planning problems: problems of integration of various paces responding to particular contexts, simultaneous occurring of various actions (here planners belonging to the GraphPlan family, with their *mutex* relations, will be probably useful), incomplete and non-primitive plans (plans characterised by heavy time constraints – for instance, because of deadline or duration stringency – or non-instantaneous actions and hence structurally incomplete, insofar as they can be completed in different ways according to the evolution of the states of a dynamic world), pace hierarchy (i.e., problems asking for decomposable plans) and weak compatibility – or even incompatibility – between plans (options) and contexts (Horty & Pollack, 2001, pp. 218–219).

It is well known that planning problems, even very simple, are intractable – in the sense that they are complete in a problem space – by the classic model of a plan, that is, by a sequence of actions whose aim is going from a given initial state to a final state satisfying a given goal (Bacchus & Kabanza, 2000): only planners that use abstractions, or task hierarchies or other special mechanisms – generally based on heuristics or shaped in terms of open systems – to drive and control the searching of paces and solutions through potentially infinite spaces in which any other kind of search is usually blind can cope with real-world problems.

Bacchus and Kabanza cite the SOAR architecture, in particular, as the first one, at the end of the 1980s, in using a particular kind of expert heuristic mechanism of search driving and controlling (Laird, Newell, & Rosenbloom, 1987), followed, around the middle of the 1990s, by the PRODIGY architecture (Veloso et al., 1995): these intelligent planners are knowledge-based architectures for search controlling, which generally are based on domain-dependent (specific) knowledge, this way differing from search general heuristics based on domain-independent (general) knowledge; specific domain heuristics, in spite of their narrowness, are good means to make tractable otherwise intractable planning problems, in particular in the field of automatic planning (Bacchus & Kabanza, 2000, pp. 124–125).

Bacchus and Kabanza use first-order temporal logic to represent – in declaratory terms – knowledge inhering to search controlling and forward chaining of exploratory heuristic search in the plan space (TLPlan system). In doing this, they refer to what they assume as pioneer logical planners, that is, the ones by Green (1969), based on situation calculus, Rosenschein (1981), which makes use of dynamic logic and Bauer et al. (1991), which in its use of temporal logic resembles TLPlan. They warn against the asperities and computational risks that come from the fact that these prototypic logical planners assume coincidence between planning and theorem proving, as this forces the heuristic search to function in a proof space whose relationships with the typical structural space of a generic plan are unclear. They also show that their TLPlan can deal with conditional actions expressed in the Pednault's ADL language (Pednault, 1988, 1989), intermediate between STRIPS and situation calculus. Bacchus and Kabanza (2000) say that their forward chaining planner based on domain-dependent search heuristics performs well, is easily controllable and revalues forward chaining, that is what in AI planning – for reasons of efficiency and effectiveness – has been generally substituted by partial-order planners (whose search develops in a space of partially ordered plans) and more recently by planners whose search develops through graphs (GraphPlan system).

A forward chaining planner develops its search within a space of world states by examining the whole set of feasible sequences of action (these, apart from being per se plans, are prefixes of the whole set of plans that could derive from their expansion) that depart from the initial world and keep trace (memory) of the worlds that are generated by the implementation of the actions.

The individual worlds, in the time evolution of the sequences, are represented in form of typical STRIPS databases, so that any individual world is a complete list of the basic atomic formulas that hold in that world, any individual world is assumed as close (what means that any basic atomic formula that is not present in the database is false in that given world), the actions – starting from initial worlds – generate new worlds through specification of complete sets of further databases to be applied to the initial worlds, any sequence of actions generates a sequence of STRIPS-type complete databases.

As STRIPS databases are identical to the traditional relational databases and can be seen as first-order finite models against which it is possible to evaluate first-order logic formulas, this kind of plan produces a finite sequence of first-order models of the world.

Dealing with the terrible limitations which persist even in the very recent planners ascribable to the powerful generation inspired by domain-independent search, Bacchus and Kabanza (2000) note that the toy-world, a must in the experiments involving intelligent planners, still suffer a numerically ridiculous limit of block reconfiguration (ten blocks, roughly). This even if this world is based on the very simple assumptions that the blocks can form towers by bottom-up building and that in order to have the towers completed one should never interrupt the progression, that is, one should never remove a block which was put over another block.

The existing fastest planning systems that use domain-independent heuristics are Allen et al.'s and Selman's BlackBox (Allen, Kautz, Pelavin, & Tenenber, 1991; Selman, 1994), which codifies planning problems as problems of satisfaction, Koehler, Nebel, Hoffmann and Dimopoulos' IPP (Koehler, Nebel, Hoffmann, & Dimopoulos, 1997), which uses an optimised GraphPlan-type algorithm, augmented for dealing with the ADL language, and in general the HSP-type systems, as in the Bonet, Loerincs and Geffner's work (Bonet, Loerincs & Geffner, 1997).

The artificial intelligent planning systems, however, that have had the biggest impact in the planning practice are the HTN (Hierarchical Task Network) planners, which use domain knowledge in form of a scheme of task decomposition: compared to classic planners, they require large information on the domain and – a part from primitive operators – a set of tasks and decompositions of tasks; due to their architecture, they can bypass wide regions of the search space and limit their exploration of this only to the primitive sequences of action coming from some sequence of task decompositions. HTN planners eliminate a big number of primitive sequences of action potentially feasible in the space of search and prune – through a parsing algorithm – partially ordered plans based on primitive actions.

TLPlan has been used for a number of types of plans, always showing good performances, for instance for car pooling plans (Bacchus & Kabanza, 2000, p. 185). Bacchus and Kabanza, in order to illustrate the search heuristics and the TLPlan algorithm, describe the building plan of a tower made by small blocks starting from a casual configuration of these (initial state) and going towards a final configuration: it includes the concept of 'small block of a good tower' as expression of initial configuration of a small block that does not require to be moved for approaching the final configuration, because this partial configuration is per se, already, a definite part of the final configuration. The Bacchus and Kabanza's example fully expresses the distance which exists between the performance of current artificial intelligent planners and what would be required from these in real socio-environmental domains.

But also the simple world of the 24 h suitcase proposed by Pednault (1989) in order to illustrate his ADL language of action representation illuminates about the difference which exists between simple plans and complex plans. The Pednault's 24 h suitcase can move, have incoming and outcoming objects, and when contains objects and moves implies a movement of these objects as well.

Even the domain of a simple logistics (the one, for instance, whose world is made by trucks and aeroplanes – that is, by only two types of vehicles – and where trucks are used to carry goods within a city and aeroplanes to carry goods from airport to airport: here the problem starts with a set of objects in various locations in various cities and ends with the goal of redistributing those given objects within the space of their new locations) is the object of many well-known experimentations.

The world of tyres is another well-known domain of proof, originally ascribable to Russell and Norvig (1995): in it, there is a need of changing a deflated tyre through a sequence of actions that implies wheel lifting, bolt releasing and removing and so on.

The world of the list refers to listing a set of objects that have to be treated by different machines in order to get a set of operations and effects (for instance: shaping an object, painting it, polishing it, etc.).

The limited world of blocks is, for instance, the one of a table having limited dimensions and housing a cumbersome set of blocks, where moving blocks from place to place of the table is not easy.

These experimentation show that there are meta-strategies (meta-level strategies) applicable with small variations in different domains.

As for search heuristics in spaces of states or plans, it can be observed that in many real domains – for instance the socio-environmental ones that are centre stage in our analysis – planners explore only parts of (search) spaces of alternatives of action that would be otherwise infinite: they explore, in particular, those parts within which moving through seems familiar or promising to the planners, also in the light of the resource finiteness (see time resource) that constrains their potential of actions; in fact, for most planners and plans (and for most problems treated by these) exploration is instinctive, that is driven by an automatism selected through memory and among the numerous – to be precise, infinite – possible automatisms as the most convenient one for that planner and that plan; the infinity of the space of search is hence a mere abstraction which does not exist in the practical reality and in its bounded context, so that theoretical insights should address the way by which operational spaces of search are formed and function.

A relevant question: given a state constraints set describing some aspects of the world (i.e., given a system), how to integrate a representation of an agent's action in the world that could allow us to reason on the effects of the agent's action (first-order action) on the system and at the same time on the impact of the system's action(s) (second-order action) on the agent which affects the ability of this of performing that first order action? Here the problem is to reason on the so-called problems of 'structure' (a structure problem was posed for the first time by McCarthy & Hayes in 1969: how to characterise what does not change when an action is implemented), 'ramification' (a ramification problem – as Finger defined it in 1986 – deals with characterising the indirect effects of actions: it arises when a theory of action is integrated with a set of state constraints), and 'qualification' (a problem ascribed to McCarthy in 1977: how to characterise the preconditions of actions, a problem made worse by the existence of state constraints), that is on problems widely discussed but substantially never solved (McIlraith, 2000, pp. 88–89).

State constraints substantially play two roles in representing an action: they capture the relation between objects that exists in the world and hence the system's states that are in coherence; serve as ramification and qualification constraints, so implicitly defining the indirect effects of the actions and constraining, for this reason, further actions in their implementation.

The language of situation calculus (proposed for the first time by McCarthy in the 1960s as a scheme of logical representation for reasoning on the actions and for changing) in integration to a set of state constraints, for example, expressed in terms of first-order logic (McIlraith, 2000, p. 89), seems to fit well to solving the above-cited structure and ramification problems: it is a language based on actions, situations and a domain. In the case proposed by McIlraith as an example of 'diagnostic problem solving task' offered by the real world (McIlraith, 2000, p. 91) one has to reason about a system made of three potentially malfunctioning components: an energy generator, a pump and a water heater.

The McIlraith work adopts the vision proposed by a number of scholars (see Pednault, 1988; Reiter, 2001; Schubert, 1990) which assumes that successor state axioms and action precondition axioms provide effective solution, respectively, to structure and ramification problems and to qualification problems, just as they are axiomatic, monotonic and generally thrifty (McIlraith, 2000, p. 96).

Recently reactive planning, based on the stimulus-and-response principle, has been proposed as a valid alternative to classical planning, above all for rapidly changing domains: the 'universal plans' (Schoppers, 1987) are probably the most known example of a reactive plan of today.

A universal plan is a function which expresses the set of the states in the terms of the set of the applicable (possible) operators: hence a universal plan does not generate a sequence of operators that lead – as in the classical plan – from the current state to the final state but instead implies at any pace a decision – based on the state at the time of decision (current state) – on how to do the subsequent pace.

Quoting Ginsberg (1989a) universal plan expresses the set of the possible situations S as an arbitrary function of the set of the primitive actions A Ginsberg criticises the universal plans (Ginsberg, 1989a, 1989b) because they would imply exploration of huge spaces, but Schoppers (1987) counterargues that planning problems are generally well structured (an argument that is obviously a nonsense in socio-environmental domains) so that this structure can be exploited to create small and effective universal plans.

Selman (1994), among the few that have dealt with universal plans from a formal and complexity theory point of view, observes that the existence of small (having polynomial dimension) universal plans from which minimal plans could be generated would imply a collapse of the polynomial hierarchy, what is seen as almost completely false by mainstream literature (see, e.g., Papadimitriou, 1994).

From the above, a need comes of narrowing the concept of universal plan: for example, Jonsson, Haslum and Backstrom (Jonsson, Haslum, & Backstrom, 2000, p. 2) propose to limit universality to the small set of desirable properties consisting in plan solidity and in plan completeness.

In fact, for example, a 'A-completeness' problem (first and weakest level of completeness in the Jonsson et al.'s three levels scale) – a problem that when solvable should be solved by an universal plan in a finite number of paces – cannot be solved by a polynomial universal plan both in spatial dimension and in time dimension; but if the exigencies of time and of polynomial space are neglected universal plans can be built also for different completeness criteria (according to Jonsson et al., R-completeness is inverse of A-completeness: for a R-complete plan the universal plan in a finite number of paces can eventually warn us about the inexistence of a plan going from the current state to the final state, while Rplus-completeness is stronger than R-completeness because for a Rplus-complete problem the universal plan in only one pace can eventually warn us about the inexistence of a plan going from the current state to the final state).

To improve their performances there is an effort for adding probabilism to the universal plans, by linking these to casual sources (databases) and coherently redefining completeness in such a way that it can include the case: but even with this cunning universal plans remain inefficient in the face of problems of general plan (a not trivial problem, however, as it is devoid of restrictions and hence is a PSpace-complete) and can be applied – in space and time polynomial configurations – only to some classes of problems.

A performance comparison implemented in a variety of domains (the traditional toy blocks world, a traditional puzzle with movable tesseras, a tunnel divided in *n* sections, each illuminable independently from the other by switches located at the end of each section, in which a person could move only through the illuminated sections) between Stocplan, the Jonsson et al.'s probabilistic reactive universal planner, and Graphplan, the Blum and Furst's (1995) deterministic propositional planner which is considered today as one of the fastest existing planners of this type, does not point out relevant differences, even if Graphplan defeats Stocplan in a majority of cases (blocks and puzzle) (Jonsson et al., 2000, p. 3).

Acknowledgements Sections of this chapter refer to papers presented in various occasions. In particular Sections 14.1, 14.2 and 14.3 are a re-elaboration of an invited lecture that was presented in a seminar on multi-agents in planning organised by professor Lidia Diappi at the Polytechnic of Milan in 2002 and Section 14.4 is a re-elaboration of an invited lecture that was presented in a seminar on planning evolution organised by professor Corrado Zoppi at the University of Cagliari in 2007.

References

Allen, J., Kautz, H., Pelavin, R., Tenenber, J. (Eds.). (1991). *Reasoning about plans*. San Mateo, CA: Morgan Kaufmann.

Arrow, K. J. (1963). *Social choice and individual values*. New York: Wiley.

Avlijas, N., Borri, D., & Monno, V. (2005) *Facing the crisis in contexts in transition: Rethinking local development through experimentations of strategic visioning*. Paper presented at the Conference of the Regional Studies Association Regional Growth Agendas, University of Aalborg, Aalborg.

Bacchus, F., & Kabanza, F. (2000). Using temporal logics to express search control knowledge for planning. *Artificial Intelligence*, *116*(1), 123–191.

Baral, C., Kreinovich, V., & Trejo, R. (2000). Computational complexity of planning and approximate planning in the presence of incompleteness. *Artificial Intelligence*, *122*(1), 241–242.

Barbanente, A., & Borri, D. (2000). Reviewing self-sustainability. *Plurimondi*, *II*(4), 5–19.

Barbanente, A., Borri, D., & Concilio, G. (2001). Escapable dilemmas in planning: Decisions vs. transactions. In H. Voogd (Ed.), *Recent developments in evaluation* (pp. 355–376). Groningen: Geopress.

Bauer, M., Biundo, S., Dengler, D., Hecking, M., Koehler, J., & Merziger, G. (1991). Integrated plan generation and recognition. A logic-based approach. *Informatik-Fachberichte*, *291*, 266–277.

Binetti, M., Borri, D., Circella, G., & Mascia, M. (2005) *Does prospect theory improve the understanding of transit user behaviour?* In: Proceedings of the 9th Conference on Computers in Urban Planning and Urban Management (CUPUM), London.
Blum, A., & Furst, M. L. (1995) *Fast planning through planning graph analysis*. In: Proceedings of the International Joint Conferences on Artificial Intelligence (IJCAI-95), Montreal, pp. 1636–1642.
Bonet, B., Loerincs, G., & Geffner, H. (1997) *A robust and fast action selection mechanism for planning*. In: Proceedings of the 14th National Conference on Artificial Intelligence and 9th Innovative Applications of Artificial Intelligence Conference (AAAI-97/IAAI-97), pp. 714–719.
Borri, D. (2001). Planning in evolution. In N. Maiellaro (Ed.), *Towards sustainable building* (pp. 3–10). Dordrecht: Kluwer.
Borri, D. (2002). Intelligent learning devices in planning. In K. Alexiou & T. Zamenopoulos (Eds.), *Proceedings of the seminar on computational models in design and planning support*. London: Center for Advanced Spatial Analysis, University College London.
Borri, D., Camarda, D., & De Liddo, A. (2004). Envisioning environmental futures: Multi-agent knowledge generation, frame problem and cognitive mapping. *Lectures Notes in Computer Science, 3190*, 230–237.
Borri, D., Camarda, D., & De Liddo, A. (2005c). Mobility in environmental planning: An integrated multi-agent approach. *Lecture Notes in Computer Science, 3675*, 119–129.
Borri, D., Camarda, D., & Grassini, L. (2005a). Complex knowledge in the environmental domain: Building intelligent architectures for water management. *Lecture Notes in Computer Science, 353*, 762–772.
Borri, D., Camarda, D., & Grassini, L. (2006). Distributed knowledge in environmental planning: A hybrid IT-based approach to building future scenarios. *Group Decision and Negotiation, 15*(6), 557–580.
Borri, D., & Cera, M. (2005). An intelligent hybrid agent for medical emergency vehicles. Navigation in urban spaces. In P. van Oosterom, S. Zlatanova, & E. M. Fendel (Eds.), *Geoinformation for disaster management* (pp. 951–964). Berlin: Springer.
Borri, D., Concilio, G., Selicato, F., & Torre, C. (2005b). Ethical and moral reasoning and dilemmas in evaluation processes: Perspectives for intelligent agents. In D. Miller & D. Patassini (Eds.), *Beyond benefit cost analysis. Accounting for non-market values in planning evaluation* (pp. 249–277). Brookfield, VT: Ashgate.
Borri, D., Grassini, L., & Starkl, M. (2009) *Technological innovations and decision making changes in the water sector: Experiences from India*. Paper presented at the Palestinian Water Authority 2nd International Conference on Water Values and Rights, Ramallah.
Chapman, D. (1987). Planning with conjunctive goals. *Artificial Intelligence, 32*(3), 333–377.
Cohen, P. R. (1995). *Empirical methods for artificial intelligence*. Cambridge: MIT Press.
Damasio, A. R. (1995). *Descartes' error*. New York: Avon.
Durfee, E. H. (1988). *Coordination of distributed problem solvers*. Dordrecht: Kluwer.
Fagin, R., Halpern, J., Moses, Y., & Vardi, M. (1995). *Reasoning about knowledge*. Cambridge: MIT Press.
Faludi, A. (1973). *Planning theory*. Oxford: Pergamon.
Faludi, A. (1987). *A decision-centred view of environmental planning*. Oxford: Pergamon.
Ferber, J. (1997). *Multi-agent decision support systems*. London: Addison-Wesley.
Fikes, R. E., & Nilsson, N. J. (1971). STRIPS: A new approach to the application of theorem proving to problem solving. *Artificial Intelligence, 2*(3–4), 189–208.
Finger, J. (1986). *Exploiting constraints in design synthesis*. Ph.D. Thesis. Stanford, CA: Department of Computer Science, Stanford University.
Forester, J. (1989). *Planning in the face of power*. Berkeley, CA: University of California Press.
Forester, J. (1999). *The deliberative practitioner*. Cambridge: MIT Press.
Friedmann, J. (1981). *The good society*. Cambridge: The MIT Press.
Friedmann, J. (1987). *Planning in the public domain. From knowledge to action*. Princeton, NJ: Princeton University Press.

Gelfond, M., & Lifschitz, V. (1993). Representing actions and change by logic programs. *Journal of Logic Programming, 17*(2–4), 301–323.
Ginsberg, M. L. (1989a). Universal planning: An (almost) universally bad idea. *AI Magazine, 10*(4), 40–44.
Ginsberg, M. L. (1989b). Ginsberg replies to chapman and schoppers. *AI Magazine, 10*(4), 61–62.
Giunchiglia, F., & Spalazzi, L. (1999). Intelligent planning: A decomposition and abstraction based approach to classical planning. *Artificial Intelligence, 111*(1–2), 329–338.
Green, C. (1969) *Applications of theorem proving to problem solving*. In: Proceedings of the International Joint Conferences on Artificial Intelligence (IJCAI 69), Washington, p. 219.
Hammond, K. J. (1990). Case-based planning: A framework for planning from experience. *Cognitive Science, 14*(3), 385–443.
Healey, P. (1997). *Collaborative planning. Shaping places in fragmented societies*. London: MacMillan.
Horty, J. F., & Pollack, M. E. (2001). Evaluating new options in the context of existing plans. *Artificial Intelligence, 127*(2), 199–220.
Ishida, T. (1997). *Real time search for learning autonomous agents*. Dordrecht: Kluwer.
Jennings, N. R., Wooldridge, M. (Eds.). (1998). *Agent technology: Foundations, applications, and markets*. Berlin: Springer.
Jonsson, P., Haslum, P., & Backstrom, C. (2000). Towards efficient universal planning: A randomized approach. *Artificial Intelligence, 117*(1), 1–29.
Koehler, J., Nebel, B., Hoffmann, J., & Dimopoulos, Y. (1997) *Extending planning graphs to an ADL sub-set*. In: Proceedings of the 4th European Conference on Planning (ECP-97), Toulouse, pp. 275–287.
Kolodner, J. (1993). *Case-based reasoning*. San Mateo, CA: Morgan Kaufmann.
Laird, J., Newell, A., & Rosenbloom, P. (1987). SOAR: An architecture for general intelligence. *Artificial Intelligence, 33*(1), 1–67.
Latouche, S. (1991). *La Planéte des Naufragés: Essai sur l'Après-Développement*. Paris: La Découverte.
Lewin, K. (1948). *Resolving social conflicts: Selected papers on group dynamics*. New York: Harpres and Bros.
McCarthy, J. (1977) *Epistemological problems of artificial intelligence*. In: Proceedings of the International Joint Conferences on Artificial Intelligence (IJCAI-77), Cambridge, pp. 1038–1044.
McCarthy, J., & Hayes, P. (1969). Some philosophical problems from the standpoint of artificial intelligence. In B. Meltzer & D. Michie (Eds.), *Machine intelligence 4* (pp. 463–502). Edinburgh: Edinburgh University Press.
McIlraith, S. (2000). Integrating actions and state constraints: A closed-form solution to the ramification problem (sometimes). *Artificial Intelligence, 116*(1), 87–121.
Minsky, M. L. (1986). *Society of mind*. New York: Simon and Schuster.
Minton, S. (1988). *Learning search control knowledge*. Dordrecht: Kluwer.
Papadimitriou, C. H. (1994). *Computational complexity*. Reading, MA: Addison Wesley.
Pearl, J. (1985). *Heuristics: Intelligent search strategies for computer problem solving*. Reading, MA: Addison-Wesley.
Pednault, E. (1988). Synthesizing plans that contain actions with context-dependent effects. *Computational Intelligence, 4*(4), 356–372.
Pednault, E. (1989) *ADL: Exploring the middle ground between STRIPS and the situation calculus*. In: Proceedings of the International Conference on Principles of Knowledge Representation and Reasoning, Toronto, pp. 324–332.
Reiter, R. (2001). *Knowledge in action: Logical foundations for specifying and implementing dynamical systems*. Cambridge: MIT Press.
Rosenschein, S. (1981) *Plan synthesis: A logical approach*. In: Proceedings of the International Joint Conferences on Artificial Intelligence (IJCAI-81), Vancouver, pp. 359–380.

Russell, S., & Norvig, P. (1995). *Artificial intelligence. A modern approach*. Englewood Cliffs, NJ: Prentice Hall.
Sacerdoti, E. D. (1977). *A structure for plans and behaviour*. New York: Elsevier/North-Holland.
Sandercock, L. (1998). *Towards cosmopolis*. New York: Wiley.
Schank, R. C. (1982). *Dynamic memory: A theory of learning in computers and people*. Cambridge: Cambridge University Press.
Schoppers, M. J. (1987) *Universal plans for reactive robots in unpredictable environments*. In Proceedings of IJCAI-87, Milano, pp. 1039–1046.
Schubert, L. K. (1990). Monotonic solution of the frame problem in the situation calculus: An efficient method for worlds with fully specified actions. In H. E. Kyburg, R. P. Loui, & G. N. Carlson (Eds.), *Knowledge representation and defeasible reasoning* (pp. 23–67). Dordrecht: Kluwer Academic Press.
Schön, D. A. (1991). *The reflective turn: Case studies in and on educational practice*. New York: Teacher's College Press.
Selman, B. (1994) *Near-optimal plans, tractability, and reactivity*. In: Proceedings of the 4th International Conference on the Principles of Knowledge Representation and Reasoning, Bonn, pp. 521–529.
Shakun, M. (1999). Consciousness, spirituality, and right decision/negotiation in purposeful complex adaptive systems. *Group Decision and Negotiation*, *8*(1), 1–15.
Shanahan, M. (1997). *Solving the frame problem*. Cambridge: MIT Press.
Simon, H. A. (1982). *Models of bounded rationality*. Cambridge: MIT Press.
Soucek, B. (1997). *Quantum mind networks*. Split: FESB.
Sutton, R. S., & Barto, A. G. (1998). *Reinforcement learning*. Cambridge: MIT Press.
Veloso, M., Carbonell, J., Pérez, A., Borrajo, D., Fink, E., & Blythe, J. (1995). Integrating planning and learning: The PRODIGY architecture. *Journal of Experimental and Theoretical Artificial Intelligence*, *7*(1), 81–120.
Watts, D. J. (1999). *Small worlds, the dynamics of networks between order and randomness*. Princeton, NJ: Princeton University Press.
Wilkins, D. E. (1988). *Practical planning: Extending the classical AI planning paradigm*. San Mateo, CA: Morgan Kaufmann.
Yang, Q. (1997). *Intelligent planning: A decomposition and abstraction based approach*. Berlin: Springer.

Chapter 15
Knowledge Management and Strategic Self-Sustainability: A Human Systems Perspective

Milan Zeleny

15.1 Introduction

The Bata Zlin is an example of orienting a city not around a cathedral, not around a city hall, not around the main square or a tourist centre, but around the business and university core. Business provides jobs, university produces knowledge and a city becomes competitive. Nothing more is needed; all the rest is implied and will take care of itself. Supply jobs produce knowledge to attract more business – and therefore more jobs – and there will be plenty of cathedrals, tourist attractions, cultural centres and fewer municipal halls.

In this chapter we describe Bata Zlin and similar knowledge and digital cities. Then we attempt to create a methodology for transforming cities and regions into job-producing knowledge and innovation enterprises. Jobs and knowledge assure life, make a city come alive, make a city a living organism. A modern city is an enterprise. With this end in mind we present tools and definitions related to Entrepreneurial University, knowledge, innovation, quality, added value and strategy – all the building blocks of a modern knowledge and innovation city.

There is a city in Moravia called Zlin. Zlin has evolved through its long-term symbiotic relationship with business. In 1894, Tomas Bata founded his shoe factory in the town. From a sleepy town of less than 3,000 inhabitants in the 1900s, Zlin grew to a population of more than 21,000 by 1930. Bata became its mayor for four terms in a row and could design the town according to the needs of his employees and its inhabitants.

Tomas Bata died in a plane crash in 1932, and his half brother Jan A. Bata carried on successful development of the business and the city afterwards. In 1935, Jan Bata invited the famous Swiss architect Le Corbusier to design the layout for the whole town. Le Corbusier's plan represented a paradigm shift from his earlier conceptions of urban design. Here he abandoned an anthropomorphic, centralised city model in favour of the linear city format. In the end, the progressive Jan Bata rejected Le

M. Zeleny (✉)
Graduate School of Business Administration, Lincoln Centre, Fordham University, New York, NY 100123, USA
e-mail: mzeleny@fordham.edu

M. Cerreta et al. (eds.), *Making Strategies in Spatial Planning*, Urban and Landscape Perspectives 9, DOI 10.1007/978-90-481-3106-8_15,

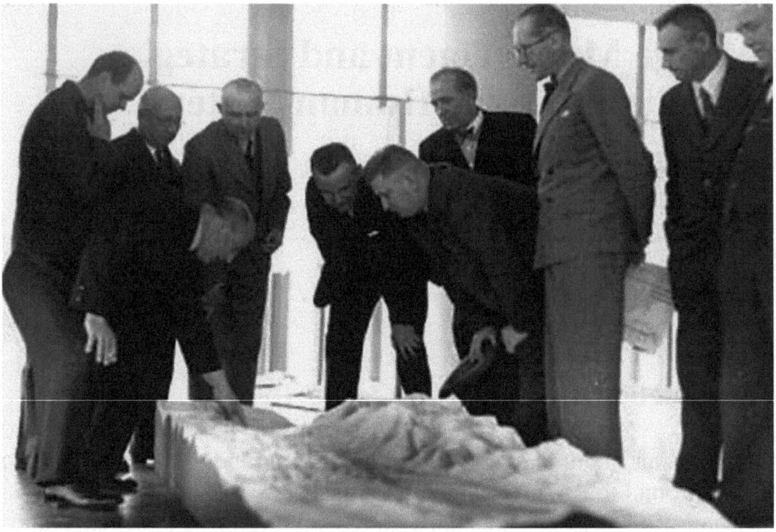

Fig. 15.1 Le Corbusier and Jan Bata looking at the plans for the city of Zlin

Corbusier's plans to bring 'the spirit of Louis XIV' to Zlin. He later decided to employ a local architect, Frantisek Lydie Gahura, a student at Le Corbusier's atelier in Paris (Fig. 15.1).

Although Gahura's plan for the city was never fully realised, it made Zlin the only city in Czechoslovakia (and perhaps in Europe) that is not built around a castle, a cathedral or the marketplace, but around the shoe factory. The functionalistic architecture prevailed: simple, box-like, red-brick houses are still among the best housing in the city and the Gahura style laid out during 1930s still asserts its presence. The unadorned brick, steel, concrete and glass surrounded by greenery in this garden city make Zlin a model of urban planning. The reconstruction and renovation work of the Bata industrial area, including the famous skyscraper (probably the first in Europe, 'Building 21'), and the new Convention and Cultural Centre of the University follow the style of the 1930s (Fig. 15.2).

By a strange twist of fate both Le Corbusier and Jan Bata died only a few days apart in 1965, as the announcement from *TIME* Magazine from Friday, September 3, 1965 confirms:

"Died (23.8.1965). Jan Antonin Bata, 67, Czech-born 'world shoe king' when he was boss (1932–1939) of the sprawling (now 80 plants in 67 countries), well-heeled (annual sales: some $400 million) producer of cheap shoes founded by Half Brother Thomas, but who in 1962 was relegated to an outpost in Brazil after Nephew Thomas Jr. of Canada's Bata, Ltd., won control of the family empire in a spectacular court fight; of a heart attack; in São Paulo, Brazil".

"Died (27.8.1965). Le Corbusier (real name: Charles-Edouard Jeanneret), 77, brilliant, Swiss-born, French architect of the reinforced concrete age; of a heart attack while swimming off Roquebrune-Cap-Martin, France".

Fig. 15.2 Bata industrial complex today

Another famous architect was Vladimir Karfik whom Jan Bata hired from the United States in 1930. Karfik built the famous 'Building 21', where Jan Bata installed his also famous office in the elevator. This first mobile office still functions today and attests to the innovative spirit of Zlin during the Bata times (Fig. 15.3).

Fig. 15.3 Karfik constructing the 'Skyscrape' and 'Building 21' today

By 1939, under the influence of Karfik, Bata became disenchanted with the mass produced functionalism of 'red-brick boxes' and together they planned major reconstruction and rebuilding of workers' housing in an individually-fitted style of the American modular design. Unfortunately, Jan Bata could not realise his grandiose plans, being forced into exile by the Nazi occupation in 1939.

One of the astonishing features of the city's architectural development was a synthesis of two modernist urban utopian visions: the first inspired by Ebenezer Howard's Garden city movement and the second tracing its lineage to Le Corbusier's vision of urban modernity. Zlin is a real garden city, full of trees, gardens and green acres.

There are many other architectural highlights in the city of Zlin. Among them, for example:

1. the Villa of Tomas Bata (finished in 1911), carried out by the famous Czech architect Jan Kotera, professor at Prague's Academy of Fine Arts; today the building houses the headquarters of the Thomas Bata Foundation;
2. Bata Hospital in Zlin (founded in 1927), designed by Frantisek Lydie Gahura;
3. the Grand Cinema (built in 1932), the largest cinema in Europe (it can seat 2,580 people) in its time, designed by Miroslav Lorenc (1896–1943) and František Lydie Gahura (1896–1958);
4. the Monument of Tomas Bata (built in 1933), designed by Frantisek Lydie Gahura; it has been the seat of the Philharmonic Orchestra of Bohuslav Martinu since 1955.

In 2008, the Tomas Bata University (TBU) built the University Centre, a part of the planned multi-functional complex 'Cultural University Centre in Zlin'. This complex consists of the new University Library building (5 floors), and the TBU Centre building (4 floors), that are designed as two independent structures, joined by a covered atrium. The urban-architectural concept of the complex was designed by the team of Prof. Eva Jiricna, studio AI Design Prague.

15.2 The Triad

As we can see in the example of Zlin, business can and should become the centre of city-regional design. The appropriate innovation strategy is created 'bottom-up' as an outcome of 'collective entrepreneurship' through cooperation between Business, Government and University, the 'BGU Triad', or 'triple helix' by Etzkowitz (2004) (Fig. 15.4).

The key event is the creation of an Entrepreneurial University, which takes initiatives together with government and industry to create a support structure for the formation of firms and regional growth.

The common objective of knowledge-based economic development efforts everywhere in the world is the creation of an 'Innovating Region'. An Innovating

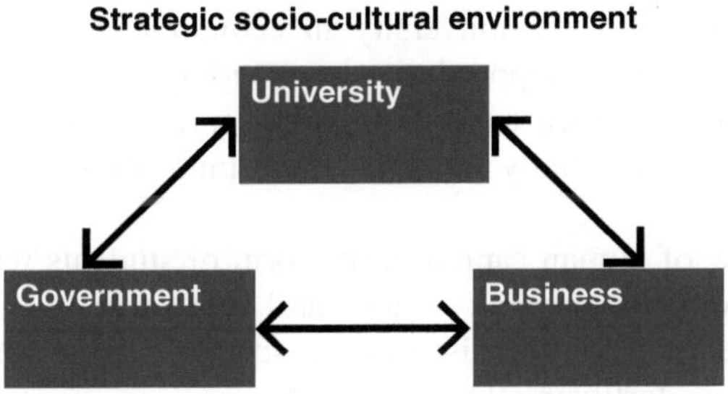

Fig. 15.4 The 'BGU Triad'

Region (or 'Knowledge-Wisdom City') has the capability to periodically renew itself through new technologies and firms generated from its BGU Triad cooperation.

1. *Business*, driving sphere; wealth producer, source of added value and competitiveness of the region; creation of employment opportunities, development of human capital in cooperation with the university sphere.
2. *Government*, enabling sphere, supporting factor; creates the optimal conditions for both driving spheres, that is, physical, institutional and social infrastructure, for effective functioning of cooperation between university and business spheres.
3. *University*, driving sphere; creation of human capital, production and transfer of information, knowledge and wisdom in cooperation with the business sphere.
4. Socio-cultural environment creates the social capital through cultural traditions, social institutions, values and preferences, behaviour and habits, trust and cooperation. It evolves relatively slowly and cannot be changed 'overnight'.

Among good examples of an 'Innovating Region' we find Linköping, Sweden. Traditional 'dyads' of university–government or government–industry are insufficient in the global era. They must be transformed into BGU Triad Transition to knowledge-based society as the basic premise of the Triad model:

1. the role of the university should be more prominent in innovation, on a par with industry and government in a knowledge-based society;
2. co-operative relationships among the three major institutional spheres: innovation policy is an outcome of interaction, not a prescription from government;
3. each institutional sphere also 'takes the role of the other'; Entrepreneurial University, assuming some of the traditional roles of industry and government, is the core institution of the Innovating Region.

A region with a traditional cluster of SMEs, rooted in a particular technological paradigm, is short-sighted and is in danger of decline once that technology has run its course (Zeleny, 2001).

A more active role for the university in economic and social development, rather than merely playing a supporting role providing human capital and research resources, is the defining characteristic of the Entrepreneurial University.

The university is an especially suitable site for innovation given:

1. high rate of flow of human capital in the form of students who are a source of potential inventors; the university is a natural incubator;
2. potential source of new interdisciplinary scientific fields and new industrial sectors, each cross-fertilising the other;
3. network of academic research groups and start-up firms, with alliances among large firms.

People representing the Triad functions include science park and incubator directors, the university, the municipality, the regional county council, private firms and small business support networks.

Among other examples we find Shanghai Municipal Government announcing in 2000 the Digital City Shanghai Strategy, and the Digital City Strategy in Waitakere, New Zealand.

To deliver the Digital City, four core concepts are needed:

1. *focused actions* – coordination of the delivery of solutions that add most value through rigorous evaluation and prioritisation of projects;
2. *smarter work* – by working together the efficient use of time, people and money is maximised;
3. *mobilised support for change* – ownership of the Digital City Strategy by City Council, businesses and universities to ensure that words are turned into actions and actions into results;
4. *stakeholders concerned* – Business, Government, Community agencies and groups, City Council, Education and University, Individual citizens.

Another concept is the Knowledge City,[1] approved by the University of São Paulo, now under construction with the support of banks, high-tech companies (IBM), NGOs and important media groups in Brazil which are developing a global network of knowledge cities. They have asserted that only if organisational culture and knowledge-producing cycles follow creative, innovative paths, evolution can take place.

Four main themes must be developed:

1. relation of global to local;
2. transformation of educational systems, problems of governing and governability;
3. relation of intellectuals and knowledge in the problem of educational change;
4. relation of knowledge management methods and educational reform.

In other words: knowledge creation, continuous innovation, competitive advantage.

Traditional observers view the organisation as a *machine* for information processing tradition. A *biological view* of organisation as a knowledge-producing organism is needed: create win/win solutions; learn from failures and setbacks of others; develop a culture of flexibility; ensure a humanistic style of leadership, guidance and teaching; understand the interdependent nature of relationships.

The Manchester City Council has created a concept called *Ideopolis: Knowledge City-Region*, that means:

1. national priorities create a framework for local priorities in a way that best meets local needs;
2. government policy needs to be more 'local';
3. regional institutions provide a framework that encourages Ideopolis to work together within the region;
4. Regional Development Agencies ensure that Regional Economic Strategies reflect the local needs;
5. Government Offices should help local institutions connect their policies;
6. more decision-making powers need to exist at a local level;
7. City-Region should have earned more autonomy where local leadership has proved effective;
8. the creation of City-Region institutions should be relevant to the local context, not a one-size fits all or so-called the 'best practices' approach.

It is self-evident that a human city cannot just be 'designed' as a piece of machinery or architectural layout. The city must become a *human city* (Fusco Girard, Forte, Cerreta, De Toro, & Forte, 2003) which, like Zlin, can become a self-sustainable living organism. We have a long way to go, but we now do have all the tools: the only missing piece is the will.

15.3 What Is Self-Sustainability?

Systems with limited or curtailed communication can be sustained and coordinated only through external commands or feedback; they are not self-sustaining. *Hierarchies of command are sustainable but not self-sustaining.* Any self-sustainable system must secure, enhance and preserve communication (and thus coordinated action), among its components or agents. It must also overlook their coordination and self-coordination competencies.

Consensual (unforced) and purposeful (goal-directed) coordination of actions is knowledge. Self-sustaining systems must be organised so as to continually 'produce themselves': their own capability of their own action coordination. Even though we often talk about sustainable systems, it is the *self-sustainability* of systems that is of real interest. The question is not how can *we* sustain a given system, but how can a system sustain *itself* in a given milieu?

Sustainability and self-sustainability are directly related to system organisation and its self-production (*autopoiesis*) (Zeleny, 1981). How systems are organised is much more important than how a system's individual agents think or what values they uphold. *Self-sustainable* systems are autopoietic and must therefore be organised for autopoiesis. Merely *sustainable* systems are heteropoietic because their sustainability does not come from within (from their own organisation) but from the outside: from planned, system-sustaining activities of external agents. *Non-sustainable* systems are allopoietic, that is, they are organised to produce things other than themselves. Allopoietic systems necessarily deplete their environment.

Heteropoietic systems can be sustainable as long as external agents sustain their system-sustaining efforts. Only autopoietic systems replenish their own environment and thus can become self-sustaining.

In summary, the presented view of sustainability can be characterised as follows: both sustainability and self-sustainability are time- and context-dependent system properties emerging from system organisation. System organisation must be continually produced or renewed via operating a common, shared resource system, optimally managed through the competition and collaboration of agents.

Continued functioning of the organisation requires continued coordination of action, that is, continued production of knowledge. Most systems can be sustained over long periods of time through an external supporting agent that disburses ideas, efforts, money or resources. Once this external agent withdraws its support, a system's sustainability can be directly challenged. *Externally Sustainable Systems do not have to be internally self-sustainable.* Any relationship (External Agent → Sustainable System) can be transformed into a *self-sustainable meta-system* (External Agent ↔ System). While an external agent can in principle make any system sustainable, only an integrated agent-system can become self-sustainable: through making the external agent an internal part of the system.

Many human and social systems are temporarily self-sustaining: they emerge, build up, persist, degrade, decay and disappear. Societies emerge, flourish and collapse. Cycles of self-sustainability are spontaneously self-organised, then amplified through human action, but ultimately depleted and dissolved.

A good example[2] where such transformation of self-sustainability into non-sustainability can be simply demonstrated is the case of three remote and isolated Pacific islands: Mangareva, Pitcairn and Henderson. Because of their remoteness, they could be simply *sustained* by external imports, but as an interacting trio had to become *self-sustainable* through their own internal flows of resources, in the end turning *non-sustainable* and collapsing (Fig. 15.5).

None of the three islands could support human population without external inflows of resources. Only Mangareva received such inflows, hundreds of miles away from Pitcairn and Henderson. Yet, all three islands supported an ancient Polynesian population. By the seventeenth century the Mangareva population was decimated, the other two populations collapsed and the islands became uninhabited. Here is the story of their bio-cycle.

Mangareva was capable of supporting a large human population of several thousand: plenty of water, forests, fish and shellfish, oysters, plantations of sweet

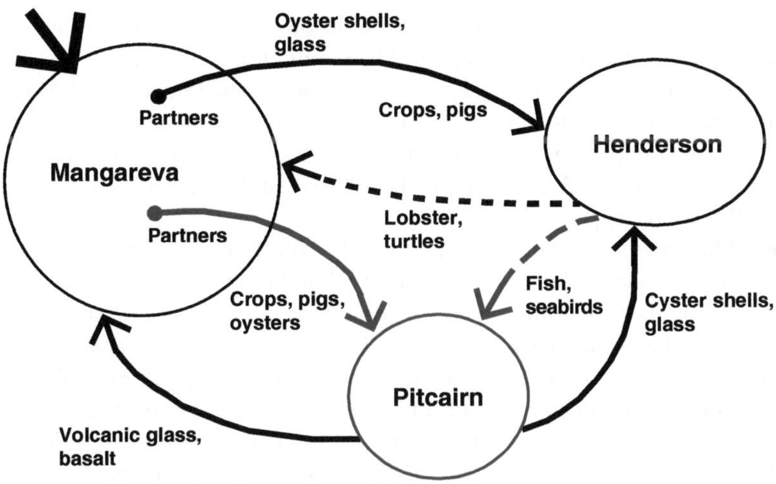

Fig. 15.5 Self-sustainable interactions and their dynamics

potatoes, yams, taro, breadfruit, bananas and so on. It had crops and pigs, but it did not have high-quality stone for making sharp-edged tools and adzes for shaping wood needed for houses, canoes and other necessities.

Pitcairn had what Mangareva needed: plenty of volcanic glass and fine-grained basalt, perfect for making adze. Only some hundred people could have been supported on Pitcairn. It needed crops, pigs and oysters.

Henderson had conditions for only marginal human existence. It is not volcanic, has no rocks or basalt, no sources of fresh water other than rain, only small trees and cave-based housing. But it had lobsters, crabs, green turtles, million-sized colonies of seabirds, large land pigeons. Yet only a few dozen people were permanently sustained.

So the pattern emerged: as the population of Mangareva grew it started importing volcanic glass and basalt for adzes from Pitcairn, while exporting oyster shells for fishhooks and tools to Pitcairn and Henderson. Volcanic glass and basalt from Pitcairn and Mangareva went also to Henderson, in exchange for sea turtles, parrot feathers, fruit doves, red-tailed tropic birds and other luxuries. Mangareva was also a source of crops and pigs for Pitcairn and Henderson. Exchanges of marriage partners and skilled individuals were especially crucial for small populations of Pitcairn and Henderson. The mutually advantageous, life-sustaining flows of resources are depicted as a cyclical network. After a period of time, the inflows and outflows of resources were balanced and stabilised, populations settled and self-sustainability established.

This balance of self-sustaining trade flows was indispensable for life on the islands. It is a precarious balance, susceptible to *external perturbations which have to be compensated through internal adjustments*. However, even small perturbations can be amplified when such compensations fail; the systems move into disequilibrium and an accelerating degradation of trade settles in. By about A.D. 1500 all trade had stopped in the region. The populations collapsed.

This first happened in Henderson when canoes from Pitcairn and Mangareva stopped arriving for the luxury items, bringing necessary metal, stones and oyster shells. Pitcairn collapsed next, about 100 years later, when canoes from Mangareva did not come anymore. Populations of Henderson and Pitcairn were trapped, and doomed. So, what happened?

When Mangareva got their glass and basalt from Pitcairn – and produced their adzes and axes – the building boom started. New houses, gardens and canoes, all required mining more and more wood, leading to fast and severe deforestation. The population swelled to many thousands, but the topsoil was carried away and ground was denuded. Canoes were not built (there were no big trees), and fishing yields declined. There were too many people and too little food. The links with Pitcairn and Henderson collapsed (no canoes, no need for adzes). The population of Mangareva slid into civil wars, permanent hunger and all forms of cannibalism.

With the collapse of Mangareva, the whole biocycle of trade network not only with Pitcairn and Henderson, but also with the Marquesas, Societies and Tuamotus, disintegrated. Imported rats killed off the seabird populations of Henderson; humans overharvested the shellfish. Pitcairn became massively deforested from desperate attempts to establish gardening and build canoes. With no exchanges of marriageable people, incest and inbreeding further worsened the fate of the survivors. Nothing worked, the bio-cycle was broken.

Self-sustainable systems must maintain their ability to coordinate their own actions, producing *knowledge.* Self-sustaining systems must be knowledge producing, not only labour or capital consuming entities.

A *knowledge-based city* revolves around business and university. In order to become a self-sustainable city/region, the concept of university must change. It must stop generating just information and start producing active knowledge. That is, it must produce businesses, firms, jobs and knowledgeable graduates. It must become an ‘Entrepreneurial University’.

15.4 What Is Entrepreneurial University?

We are entering an era of re-assessment of business programmes, shifting from description of action (functional, ‘scientific’ model) towards action itself, that is, an *entrepreneurial model.*

It is being realised globally that management should become a *profession* and schools of management *professional schools*, like schools of medicine and law. Professions are always more about knowledge and wisdom, less about information, always more about doing and less about describing.

Professions work with an *accepted body of knowledge* (not information), *certify and guarantee* acceptable practice; they are *committed to the public good*, and rely on an enforceable *code of ethics*.

Bennis and O'Toole recently wrote, "[t]he problem is not that business schools have embraced scientific rigor but that they have forsaken other forms of knowledge" (2005, p. 102).

Every business school should run its own business, as proposed by Polaroid's E. Land. This need for practice, innovation and entrepreneurship takes us to the notion of the *Entrepreneurial University.*

The Entrepreneurial University not only produces knowledge (rather than information), but engages in a new mission of *capitalisation of knowledge*. It produces not only graduates and alumni, but also firms and companies. It becomes an economic actor in the regional and possibly – through a network – also in global economic and social development. This new mission puts the university into direct cooperation with the state and corporate sectors, forming the 'Triad of Cooperation'.

In the vision of the triune EU network of alliances, the university should change from being a 'conservatory' of information and knowledge, and their producer and transmitter, to *an entrepreneur*, which would be pre-eminently and prominently *positioned* to assume global leadership in translating this vision into reality.

The University–Industry–Government is the proper triad for successful regional development. *New firms and their capitalisation* is the proper output of a professional, entrepreneurial school. One-way, linear outflow without feedback is replaced by a self-sustaining cycle of knowledge and wisdom.

The Entrepreneurial University still produces graduates and publications, of course, but 'packages' them in firms and companies to take the created knowledge out with the newly minted entrepreneurs.

The trend is towards global alliances and networks in business and economic cooperation. It is moving away from self-absorbed islands of bureaucracy and political roller-coasters. Education, entrepreneurship and innovation are the next frontiers. Knowledge becomes the key to regional and national success in a global society.

15.5 What Is Knowledge?

Knowledge is the purposeful coordination of action. Achieving its purpose is its sole proof or demonstration. Its quality can be judged from the quality of the attainment (its *product*), or even from the quality of the coordination (its *process*) (Zeleny, 1987, 2002, 2005).

What do we mean when we say that somebody knows or possesses knowledge? We imply that we expect one to be capable of coordinated action towards some goals and objectives. Coordinated action is the test of possessing knowledge. *All doing is knowing, and all knowing is doing.*

Every act of knowing brings forth a world. We 'bring forth' a hypothesis about the relationships and test it through action; if we succeed in reaching our goal, we know.

Bringing forth a world of coordinated action is human knowledge.

Bringing forth a world manifests itself in all our actions and all our being. Knowing is effective (i.e., coordinated and 'successful') action. So, knowledge is *not* information. *Everybody in the world is now informed, only some are knowledgeable, just a few are wise.*

While information allows us to do things right (*efficiency*), knowledge already aspires to also do the right things (*effectiveness*). Doing the right thing, especially in business, requires not only knowing how, but also knowing why. *Explicability* of purpose is an essential ingredient of its effectiveness in attainment. *Wisdom is about explicability and ethics* of our doing (Maxwell, 1984).

Many informed people know what to do, quite a few knowledgeable experts know how to do it, but only a few *wise persons* know why it should (or should not) be done.

The last row of the taxonomy table (Fig. 15.6) already aims at wisdom:

- wisdom is *knowing Why* things should or should not be done – locally, regionally and globally – and is, and will remain, in short supply.
- *asking Why* is fundamentally different from *asking How*.

Whenever we explore a coordinated process in the sense of *What* or *How* (what is to be done, how to be sequenced, how to be performed, etc.), we already accept and fixate that process. The process is becoming *a given* subject to learning or mastering, but not subject to exploration or change.

It is only when we start asking *Why* (why to do it at all, why this operation and not another, why this sequence, etc.), we question the very structure of knowledge (coordination of action), and introduce the possibility of change. The *Whys* and the *Why Nots* are the most important questions in business and management and they should not be taken as givens. Only asking questions can lead to innovations.

In the global economy, frequent or continuous strategic change will become the norm of competitiveness. Doing the same, given thing better and better (continuous improvement), will be inadequate for strategic success. One has to *do things differently* (not just better), and *do different things*, not just the same ones. Such an important mode of strategic thinking cannot be learned and mastered by asking *How*, but mainly by asking *Why*.

	Technology	Analogy (Baking Bread)	Effect	Purpose (Metaphor)
Data	EDP	**Elements:** H2O, yeats, bacteria, starch molecules	**Muddling through**	**Know-Nothing**
Information	**MIS**	**Ingredients:** flour, water, sugar, spices + recipe	**Efficiency**	**Know-What**
Knowledge	**DSS, ES, AI**	**Coordination of the baking process result, product**	**Effectiveness**	**Know-How**
Wisdom	**WS, MSS**	**Why bread? Why this way?**	**Explicability**	**Know-Why**

Fig. 15.6 Taxonomy of knowledge

15.6 What Is Innovation?

There are clearly many definitions and concepts as well as many popular images of what innovation entails. Generally speaking, innovation is such quantitative or qualitative improvement of a product, process or business model which significantly adds value to the customer/user, business or both. Because added value is realised only through market transaction, innovation is created at the moment of customer/user purchase.

Alternately, we can specify innovation as the change in the hardware, software, brainware or support network of a product, system or process that *increases the value for the user or customer*.

From this definition it should become clear that not every *invention* (a discontinuous, qualitative change) is an innovation, and also not every *improvement* (a continuous, quantitative change) is an innovation. Innovation adds value and value can only be realised after the purchase. While invention can exist even without a customer, innovation, in order to exist, must be valued and purchased by a customer.

The value is being added through the *Innovation Cycle* (Fig. 15.7).

It is clear that innovation is a process, a self-reinforcing and continually repeating cycle of activities. It starts with *Understanding* (U) what a customer wants and how the resources are to be used to satisfy him. Then a corresponding *Design* (D) solution is prepared and its value-adding (and money-making), potentials evaluated. If they are found to be significant, the design is *Implemented* (I). The actual service delivery is achieved through its actual *Operation* (O).

The U-D-I-O Cycle shown in Fig. 15.7 is a simplified interpretation adapted from Jackson's *The Escher Cycle* (2004). This is a self-reinforcing learning cycle which must be continually repeated if any *learning from operating* is to take place.

The cycle must be *effective*, that is, delivering the right answers to the right questions, not just *efficient*, that is, delivering the right answers to possibly wrong questions, and thus developing *wrong* services and products quickly and cheaply; this would be the worst possible outcome.

Because innovation must add value, we can also conceive quality in the same vein and context. The only difference is that quality does not have to pertain only to innovation but to a standard product or process as well. In fact, most products must have been innovations at some point.

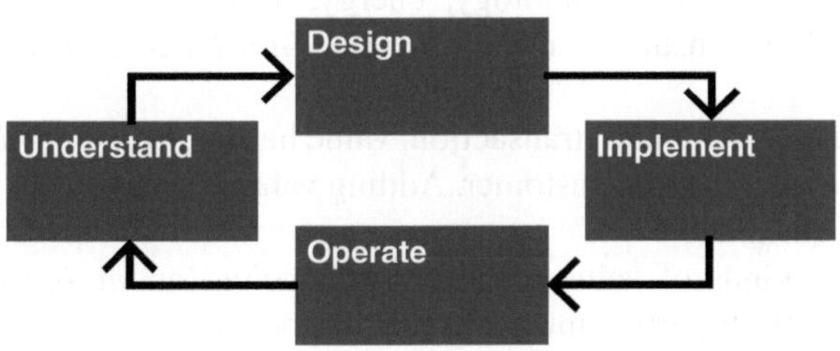

Fig. 15.7 Innovation U-D-I-O Cycle

15.7 What Is Quality?

It is clear that the notion of quality must be closely associated with the notion of added value. In fact, quality and value seem to be inseparable, although value is more encompassing, while quality is often stripped to minimal technical and efficiency standards.

Quality is widely considered the optimal balance between effectiveness and efficiency. A quality product, process or service provides the right customer balance between doing the right thing (*value*) and doing things right (*cost*). The value the customer receives for his money, that is quality.

The price (*money*) is therefore a constitutive part of quality. A 'high quality' item cannot be priced above the maximum price a customer is willing to pay because then the added value is reduced to zero and the item is not purchased. It has 'low quality' for me, it does not deliver value, and I don't want it.

Quality therefore does not exist per se. It is realised and thus created in the act of purchase (more precisely in the subsequent use), and through the transaction. So, the notion of quality is intimately associated with the customer/user and his act of purchase. Quality is not stored in warehouses; it only emerges through the act of purchase and subsequent use.

We can of course recognise and acknowledge quality in items we do not purchase, but that is recognising the quality for others, not for ourselves. Such 'quality' is meaningful only as a point of reference or benchmark, not as a living aspect of our own economic behaviour. In business, it does not matter what people think, say or imagine; the only thing that matters is *what they do*.

Unsold or unsalable goods, products or services *cannot be* of high quality, by definition. Quality is derived from customers' preferences and realised through the purchase. It is not 'built in' by engineers and 'stored' in a warehouse, waiting for the customer to recognise the engineer's sincere imagination and vision.

15.8 What Is Added Value?

Knowledge is measured by the value our coordination of effort, action and process adds to materials, technology, energy, services, information, time and other inputs used or consumed in the process. *Knowledge is measured by added value*.

In any business (and human) transaction, value has to be *added to both* participating sides: the provider *and* the customer. Adding value is what makes the transaction satisfactory and sustainable.

There are two kinds of value to be created: *value for the business* and *value for the customer*. Both parties must benefit: the business, in order to make it; the

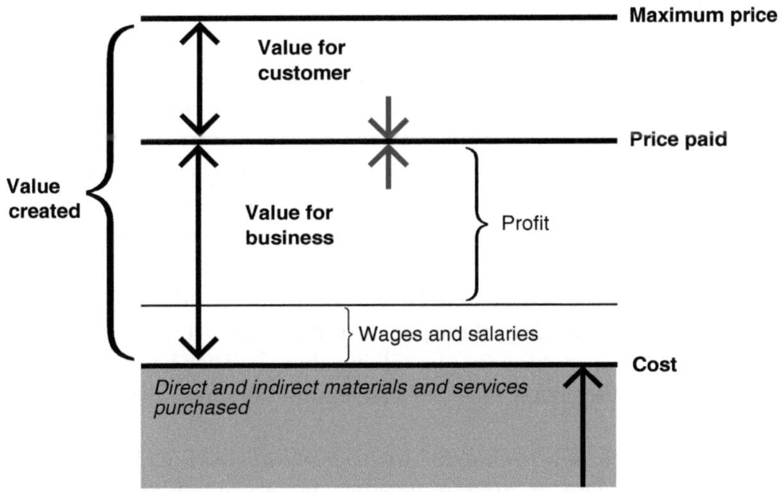

Fig. 15.8 Adding value for the customer (Zeleny, 2007, p. 38)

customer, in order to buy it. In the global age it is precisely this business-customer *value competition* that is emerging as the hardest and the busiest battleground.

In Fig. 15.8 we attempt to explain the process of creating new value. This is crucial for the identification and assessment of innovation.

First, the customer pays for the service or product: the *price paid.* The producer subtracts the *cost incurred*, including all direct and indirect materials and services purchased. The difference is the *added value* for the business. This added value can also be interpreted as the *value of knowledge* engaged in producing the service or product. In order to pay wages and salaries, the production process and its coordination must generate this added value. Added value is the only source of corporate wages and salaries and profits.

If the added value does not *cover* the wages and salaries, then these must be correspondingly lowered. If no value has been added, then the value of knowledge is zero and no payment can be attributed to it. The business must add enough value in order to *cover* at least its workers and managers, their salaries and wages. If even more value has been created, then *profits* can be realised, up to the price received.

The customer, of course, must be willing and ready to pay more for the service/product than he actually pays. The *maximum price* the customer would be willing to pay must exceed the price the producer has asked for. The difference is the added *value for customer.*

If there is no value for customer – the maximum price is lower than the price to be paid – then the customer will not buy the service or product. In a competitive market, the customer pays money only for the value received, that is, the value for the customer.

15.9 What Is Strategy?

Strategy is what a company does. And what a company does is its strategy. It is plain and simple. Every company has a strategy as long as it is doing *something.*

There is no question what its strategy is; the only question is its effectiveness: if it brings forth the desired results. Assorted corporate mission and vision statements are *not* strategy and have little to do with it. They are simply descriptions of intentions, desires and plans – just words substituted for action. The dichotomy between the action and its description is often palpable. The gap between knowing what to do and actually doing can be excruciatingly real (Pfeffer & Sutton, 2000).

The old-fashioned strategists of the pre-information era were so effective because they did not have a proper space for just *talking* about it. They had to deliver: engage in action and deliver the goods. One can also admire the action-based strategy of the animal world.[3]

We have the sketch of a *traditional strategy paradigm* (Fig. 15.9). There could be some additional details in different versions, but essentially this is how the strategy is still pursued by many companies. Some institutions spend years hammering out their mission and vision statements and defining, testing and measuring the goals, before encountering the 'Cloud Line' and the problem of *Implementation.* The 'Cloud Line' is a real phenomenon: those above it do not see below, the strategists do not understand the problems of operations. Those below it do not see above, the doers do not understand what is being asked of them and how the strategy

Fig. 15.9 Traditional 'strategy paradigm' and its 'Cloud Line'

is to be implemented. One needs only to contemplate this scheme in order to realise that no viable strategy can ever emerge from such procedure, except by chance. Everything about the Cloud Line is just *symbolic descriptions* of the intended future action. Everything below the Cloud Line is only *pure action*, no descriptions. These are two separate domains: descriptions of action and action itself. They can and do differ; very rarely do they meet – unless the description refers to the 'actual' action, present or past, not the intended action of the future. Strategy describes the future; the two domains can rarely intersect. So there arises an eternal problem of implementation. How does one transcribe the descriptions of action into action itself? How does one implement a strategy? Most executives say that they agree with the descriptions, they are fine above the Cloud Line and sometimes they do not understand why it should be so difficult to implement their descriptions (mission and vision statements).

The very notion of implementation is typical of the gap between doing and talking. If we accept that corporate strategy is about doing and not about saying, then the notion of implementation becomes mute and uninteresting. If the strategy is what the company does, then there is nothing to implement: the strategy is already enacted. What we want is not implementing a description but changing the strategy itself: changing from one type of action to another. The purpose is the change of action, the change in strategy. One of the end-products of such a change can be a description, a mission-vision statement *derived from the action itself*, not from executive musings. That way, the strategy and its change take place in one domain, in the domain of action. Let us outline the steps and proper sequencing of the strategic process.

First, we have to create a detailed map of key corporate activities to find out what the company *is* doing, to reveal its actual strategy that is embedded in action. Remarkably, many corporations do not know what they are doing (their own activities and processes); they do not know their own strategy. They only know what they say, through their mission statements. Here we can get some help from Porter (1996) and his idea of *activity maps*. We have adopted a short example of an activity map from Porter (1996). According to Porter, strategy is its activities (Fig. 15.10).

Activity maps show how a company's strategic position is contained in a set of customised activities designed to deliver it. In companies with a clear strategic position, a number of higher order strategic themes can be identified and implemented through clusters of tightly linked activities. The activity map presents high-order strategic themes in black circles and their corresponding activities in grey circles.

Second, after creating coherent *activity maps*, one has to analyse the activities and evaluate their performance by comparing them to benchmarks of competitors, industry standards or stated aspirations. First, one has to ask a series of questions, like:

- Is each activity consistent with the overall positioning, the varieties produced, the needs served and the types of customers accessed?
- Are there ways to strengthen how the activities and groups of activities reinforce one another?
- Could changes in one activity eliminate the need to perform others?

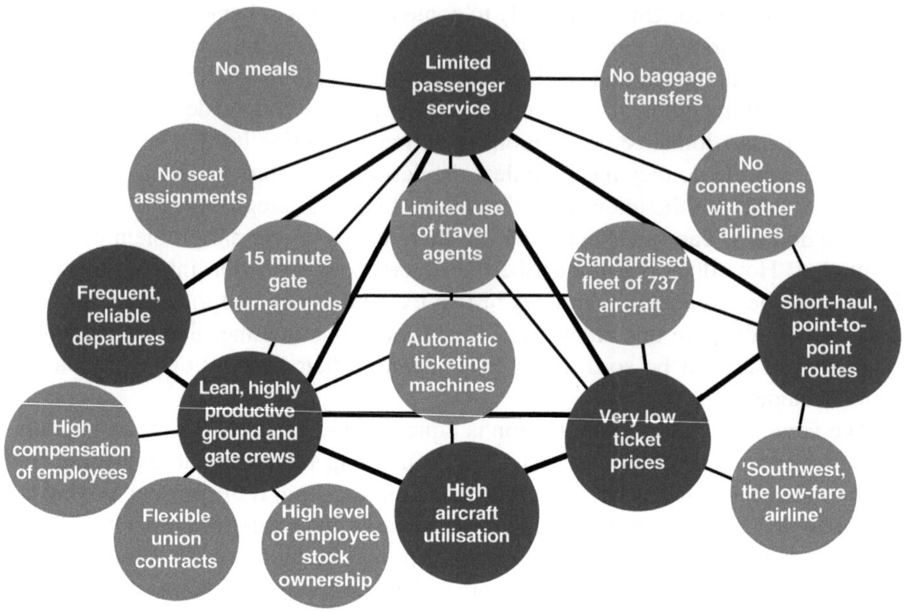

Fig. 15.10 Activity map according to Porter (1996)

Answering such and similar questions already leads to changes in activities themselves, leading to reformulation and redrawing of the map. Each implied change can be immediately enacted and the actual corporate strategy firmly established.

The next task is *benchmarking*. One has to be careful here: the purpose of benchmarking is *not* to match strategies or performances of the others. One does not want to become 'like them', similar or the same. The imperative of the global era is *differentiation*. We want to know how the others are doing not in order to become the same, but to become different.

This is somewhat similar to the *strategic diversification* of H. I. Ansoff (1965), concentrating on the different directions in which a business might branch out or expand from where it is today. The firm is defined by the customers or markets it serves and the products or services it sells. It is not defined by its missions and visions, or by its symbolic statements.

In order to establish the directions in which some activities should be changed, one has to make comparisons with customer desires, competitor performances and industry standards and corporate aspirations. In the next step, another tool is needed: the *value-curve maps* or their earlier version – *radar (or spider-web) diagrams*.

Third, so-called *value-curve maps* are produced in order to differentiate one's activities from those of competitors. *Differentiation, not 'catching up' or imitation* is the key to effective competitiveness and strategy.

Below we present such a generic value-curve map (Fig. 15.11). On the horizontal axis we list criteria or attributes while on the vertical axis we record their performance levels. In concordance with Zeleny, Kasanen and Östermark (1991)

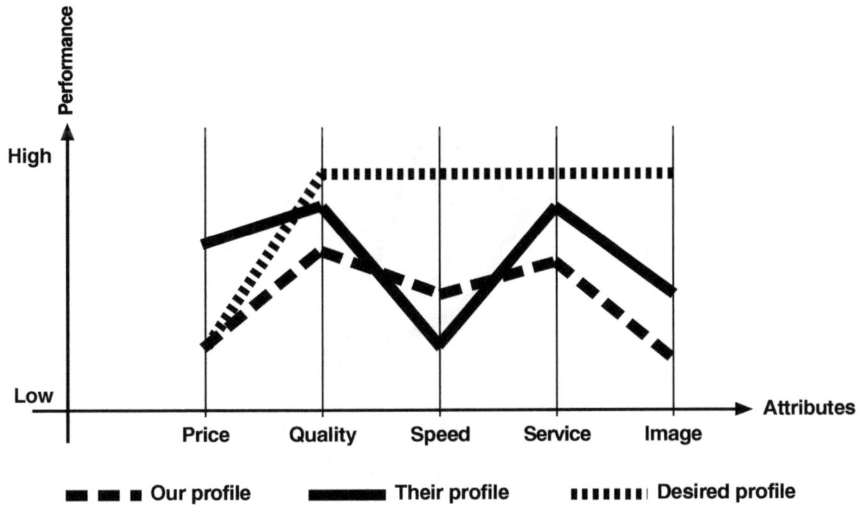

Fig. 15.11 Search for differentiation via value-curve map

we refer to the individual patterns as *profiles*. Thus, we can have our profile, their profile and the desired profile.

Individual attributes or criteria correspond to Porter's themes. So, the task of differentiation is to identify not just the performances we want to achieve on existing criteria, but to develop a set of new criteria (attributes or themes), which would differentiate us from the competition or standards. Once we identify such new attributes, we can develop the corresponding sets of activities (from activity maps), which have to be removed, changed or added (generated). This way we start changing the activity map and its corresponding activities in a directed and purposeful way of desired differentiation.

We can transcribe the value-curve map used by Kim and Mauborgne (1997, 1999) into the informationally identical form of the radar (or spider-web) diagram of Zeleny et al. (1991) (Fig. 15.12). In the radar diagram one can see, in a compact form, the domination, non-domination and relative positioning of individual profiles.

The purpose of these maps and diagrams is to uncover the spaces and niches for which a strategic entry would be desirable. Companies do not succeed or fail because of their mission statements, but because of how well they *fit with their environment*. To fit well means, like everywhere in nature, to create a *niche*, to differentiate itself from others and to compete 'head on' as little as possible.

One has to *find a space in the market*, not just mindlessly emulate what all the others are doing. In order to identify such 'open spaces', we often have to create them. Traditional benchmarking leads to standardisation, commoditisation and 'sameness'. The name of the game is *differentiation*.

Below we present an actual example of strategic positioning adapted from the work of Kim and Mauborgne (1997, 1999) (Fig. 15.13). It is the example of a hotel

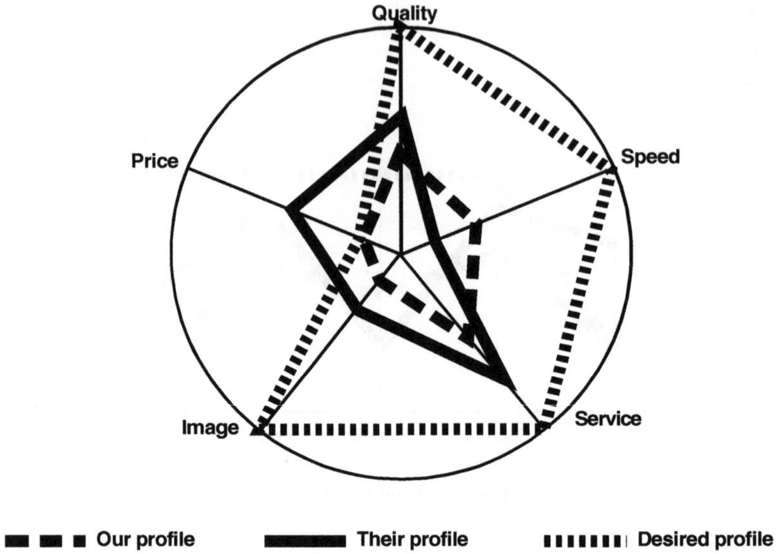

Fig. 15.12 Radar diagram of the search for differentiation

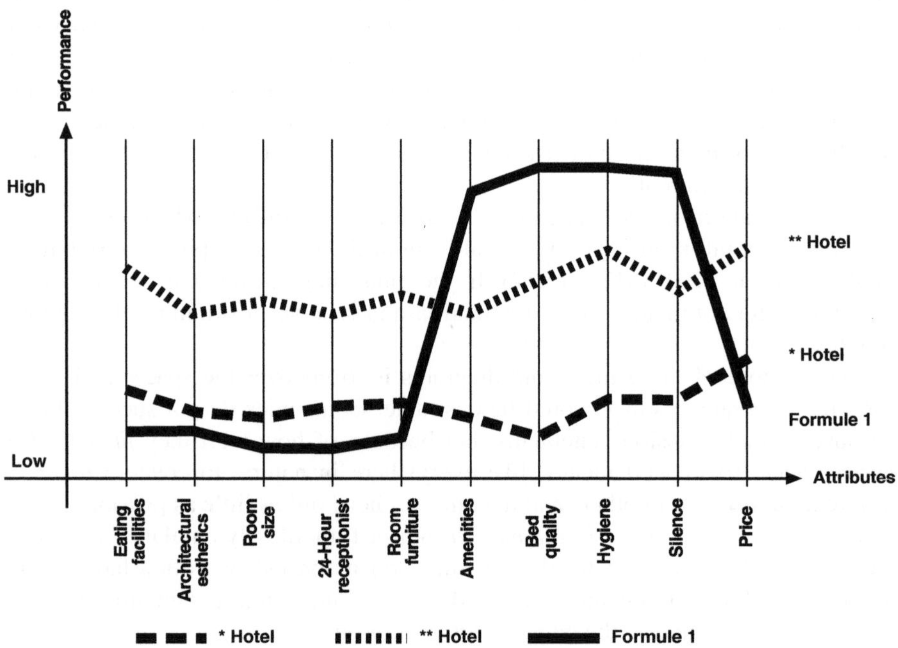

Fig. 15.13 Formule 1 example by Kim and Mauborgne (1997)

chain *Formule 1* and its differentiation from traditional one-star and two-star hotels. Observe that comparing these two kinds of hotels with respect to ten attributes leads to virtually parallel *value lines*, adding little value to the customer.

It is extremely difficult to enter such a competitive and well-covered market of identical, self-copying profiles at any interesting environmental level, other than strictly local. The competition is intense and 'bloody' – the *Red Sea Strategy*. The purpose is to create a less crowded, more differentiated space where the competition is based on complementarity, cooperation and differentiation. Those who are complementary can better cooperate and enter alliances than those who compete head-on. One should seek, in the terminology of Kim and Mauborgne (1997), the less intense and more complementary *Blue Ocean Strategy*.

Formule 1 chain is an example of a successful innovation, which has created its own new space and a significant new value for customers. They chose not to compete along traditional 'hotel' dimensions (got rid of the 'piano music in the lobby'), and focused on *bed quality*, *hygiene*, *silence* and *price*. In these four key customer-driven attributes or themes they easily surpass their 'industry standards'. Their innovation adds value.

The radar-diagram version of the Formule 1 example is in Fig. 15.14. Observe that the dimensionality is well preserved and individual action patterns can be well developed.

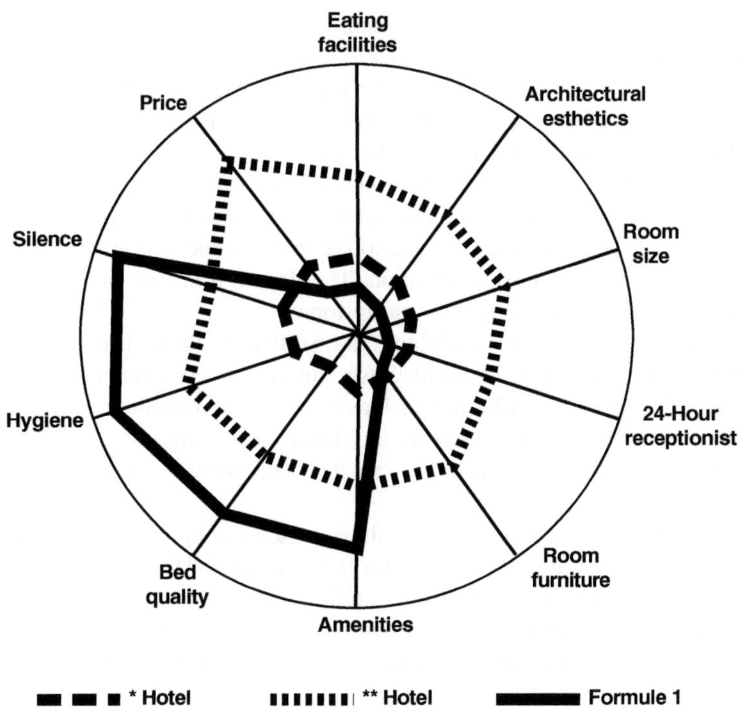

Fig. 15.14 Radar-diagram version of Formule 1 example

The next, the fourth step is about identifying the activities which are to be changed, dropped or added, thus creating a new activity map, a new strategy.

Fourth, identified selected activities are changed – in order to fill the opportunity spaces revealed by value-curve maps – as being most effective for successful differentiation. The rest of action space is *conserved*.

It is important to realise that the notion of change (which activities should be changed) is best handled not through attacking the change directly, but indirectly: identifying which activities should *not* be changed first. What should be conserved in the existing strategy opens new spaces of change in the new strategy (Fusco Girard, 2007).

What is central in evolution or any history is not what has changed, but what has been *conserved*. The study of change in human systems cannot be about what changes but what persists unchanged and remains conserved. The same applies to corporate strategy.

The structures change, the organisation remains. Some life forms disappear but living systems go on. Companies go bankrupt but business continues. Departments are cancelled and formed but organic corporations live on. Individuals come and go but institutions persist.

The conservation of system organisation is the true contents of history. When some pattern of relations is being conserved, there is a space opened for all other relations to change around. There is no change without conservation.

When we say that a particular company, like the Bata company, has existed since 1894, we mean that something has been conserved – that which we perceive as constituting the *identity of the company*. Because of that preserved pattern, the company has a history. All the rest could and did change.

Finally, we are ready to enter the last stage of the strategy formation process.

Fifth, after a newly changed action space (and its activity map) has emerged and become reliably functional, the descriptive mission and vision statement can be drawn for the purposes of communication. The description now *actually* describes the action and the action reflects the description.

Through the *wisdom systems* (Zeleny, 2006), that is, through exploring corporate action via wisdom cycle of inquiry, we can effectively change the action and consequently the strategy, without ever leaving the action domain. Corporate strategy remains the doing, even though we are doing something else. There is no need to implement or execute the 'strategy' (a set of statements) – it has already been enacted.

Executives are supposed to 'execute' their strategic statements. Traditional strategies are hard to execute as they are probably created 'above the Cloud Line', far removed from the corporate doing. Often they should not be executed at all. Effective (forced) execution of incorrect or impossible to implement strategies is likely to damage the corporation and its strategic self-sustainability and resilience.

15.10 Conclusion

The old-fashioned 'Competitive Strategy' has doubtful use in the global society. It does not address cooperation, alliance or complementarity; it is not 'Cooperation Strategy' and it does not cooperate with the customer/user. There is even a notion that customers somehow 'undermine' strategy. Such customer-free thinking makes strategy often unsustainable.

Such an approach is based on traditional tradeoffs; it does not recognise tradeoffs-free economics and thus cannot add value to *both* the customer/user and business. Every company *has* strategy but some companies lack missions, visions and similar declarations. They need flexibility and continuous adjustment of action, not engraved and near-perfect mission statements.

Most companies, firms, cities and regions need *good* strategies; the bad ones they already have.

Notes

1. The City of Knowledge is a research project founded in 2001 at the Institute of Advanced Studies of the University of São Paulo (IEA-USP), Brazil.
2. Adapted from Jared Diamond's Collapse, Chapter 3, The Last People Alive.
3. Observing a pride of lions in action teaches us more about strategy through mutual adaptation and readjustment than any flashy PowerPoint presentation of a symbolic description (lions are lucky that their space for PowerPoint presentations is so limited because they would have died of inaction and empty roars a long time ago).

References

Ansoff, H. I. (1965). *Corporate strategy*. New York: McGraw-Hill.

Bennis, W. G., & O'Toole, J. (2005). How business schools lost their way. *Harvard Business Review, 83*(5), 96–104.

Etzkowitz, H. (2004). The evolution of the entrepreneurial university. *International Journal of Technology and Globalisation, 1*(1), 64–77.

Fusco Girard, L. (2007). Urban system and strategic planning: Towards a wisdom shaped management. In Y. Shi, D. L. Olson, & A. Stam (Eds.), *Knowledge and wisdom: Advances in multiple criteria decision making and human systems management* (pp. 316–340). Amsterdam: IOS Press.

Fusco Girard L., Forte B., Cerreta M., De Toro P., Forte F. (Eds.). (2003). *The human sustainable city. Challenges and perspectives from the habitat agenda*. Aldershot: Ashgate.

Jackson, F. (2004). *The Escher Cycle*. Mason, OH: Thomson Learning.

Kim, W. C., & Mauborgne, R. (1997). Value innovation: The strategic logic of high growth. *Harvard Business Review, 75*(1), 103–112.

Kim, W. C., & Mauborgne, R. (1999). Creating new market space. *Harvard Business Review, 77*(1), 83–93.

Maxwell, N. (1984). *From knowledge to wisdom*. Oxford: Blackwell.

Pfeffer, J., & Sutton, R. I. (2000). *The knowing-doing gap*. Boston, MA: Harvard Business School Press.

Porter, M. E. (1996). What is strategy? *Harvard Business Review*, *74*(6), 61–78.
Zeleny M. (Ed.). (1981). *Autopoiesis: A theory of living organisation*. New York: North Holland.
Zeleny, M. (1987). Management support systems: Towards integrated knowledge management. *Human Systems Management*, *7*(1), 59–70.
Zeleny, M. (2001). *Information technology in business*. London: Thomson.
Zeleny, M. (2002). Knowledge of enterprise: Knowledge management or knowledge technology? *International Journal of Information Technology and Decision Making*, *1*(2), 181–207.
Zeleny, M. (2005). *Human systems management: Integrating knowledge, management and systems*. Singapore: World Scientific Publishers.
Zeleny, M. (2006). Knowledge information autopoietic cycle: Towards the wisdom systems. *International Journal of Information Technology and Decision Making*, *7*(1), 3–18.
Zeleny, M. (2007). The mobile society: Effects of global sourcing and network organisation. *International Journal of Mobile Learning and Organization*, *1*(1), 30–40.
Zeleny, M., Kasanen, E., & Östermark, R. (1991). Gestalt system of holistic graphics: New management support view of MCDM. *Computers and Operations Research*, *18*(2), 233–239.

Chapter 16
Bricolaging Knowledge and Practices in Spatial Strategy-Making

Grazia Concilio

16.1 Introduction

"[T]here is no 'one best or one single way' to carry out strategic spatial planning. The most appropriate approach depends to a large extent on the challenges faced, the particular (substantive and institutional) context of a place and the values and attitudes of the main actors of the process" (Albrechts, 2006, p. 1150). Consequently, practices and approaches to strategic spatial planning are widely investigated and reveal several diverse traditions and approaches (Albrechts, 2004, 2006, 2007, 2008; Carmona, 2009; Healey, Khakee, Motte, & Needham, 1997; Pugliese & Spaziante, 2003). Literature focuses on processes for developing and formulating strategies, including strategic analysis. Implementation is not investigated with a comparable emphasis and, when this is the case, it is scarcely analysed from the insight of cognitions and practices, and reveals to be much more complex than strategic analysis and strategy formulation in strategy-making.

The complexity of strategy implementation and the failure often associated with implementation can be related to diverse causes. Considering strategy-making as being explicitly concerned with the recognition of the need for a significant change, we can identify for such a failure at least two causes being significant for the discussion in this chapter. The first cause is related to the belief that a needed change can be translated into a whatever endstate spatial strategy. It is already recognised that strategic planning cannot be conceived as oriented to an end-product but rather as a "complex governance processes, through which concepts of spatial organisation are mobilized" (Healey, 2007, p. 527) by and for a "strategic enabling of means-based activity" (Tewdwr-Jones, 2002, p. 278). *The strategy, as an end-product, refers to a fixed form of the future.*

The second cause is related to the adoption of a pre-determined solution as approach to strategic spatial planning (Hillier, 2007). Gunder and Hillier (2007) already criticised the essence of strategic planning: the plan as a statement of what

G. Concilio (✉)
Department of Architecture and Planning, Polytechnic of Milan, Milan 20133, Italy
e-mail: grazia.concilio@polimi.it

M. Cerreta et al. (eds.), *Making Strategies in Spatial Planning*,
Urban and Landscape Perspectives 9, DOI 10.1007/978-90-481-3106-8_16,

the city, the territory "ought to become", or "what ought to happen" (ibid, p. 467). *The strategy is a guide towards an already known future.*

This chapter tries to look at strategy-making by abandoning the vision of strategy as an end-product and also considering that the future is changing along the path from present to future which means that it cannot be known in advance. Therefore the concept of strategy itself needs to be re-framed in order to adapt strategy-making to the dynamics of future.

The attempt to re-conceptualise strategy-making starts from considering modes for complex organisations to develop a strategy by using, producing and appropriating knowledge and practices while composing knowledge and practices in a coherent whole towards the needed change.

The increasing complexity of spatial systems and the speed-up of their dynamics make spatial strategy-making knowledge intensive processes. Increasingly, knowledge is considered the most strategically important resource and learning the most strategically important capability necessary for complex organisations to manage complex issues. Therefore, in strategy-making processes, the way knowledge is managed is crucial to the effectiveness of the processes themselves. Knowledge management becomes crucial when trying to overcome the traditional vision of strategy and to reconceptualise it as a dynamic framework within which an organisation coordinates its activity throughout a needed change.

In the first part, this chapter explores the connection between knowledge and action in strategy-making, recognising that knowledge and action are linked together by a mutual framing dependency. Being embedded in social relations, routines and day-to-day practices, knowledge cannot be moved towards the planning action; it is rather action that needs to be developed inside those spaces of the organisation where knowledge is available for use, that is, is actionable. Referring to strategy-making, such spaces are identified as *strategic episodes* through which organisations appropriate knowledge and practices while testing them against a needed change. Strategy is seen as a dynamic entity evolving together with the organisational structure and is described as the dynamic product of a *bricolage* activity resources for the *bricolage* are knowledge and practices explored and internalised by the organisation with respect to a needed change.

In the second part, the 'story of a strategy' is described and analysed: it refers to the planning experience carried out in Torre Guaceto, a Natural Reserve in southern Italy. This experience shows clearly that strategy is not a predetermined entity and that the organisation does not know a priori what its future will be. The strategy, in Torre Guaceto, is a *bricolage* product of diverse resources: knowledge and practices developed in very particular organisational spaces, defined as *strategic episodes*, where an organisation is forced to re-think itself against and towards a needed change.

Finally, the chapter considers the possibility to look at *strategic episodes* as spaces for the micro-foundation of strategy and opens a small perspective for further research towards other micro-foundational aspects or spaces in strategy-making.

16.2 Bridging Knowledge to Action in Strategy-Making

16.2.1 Knowledge and Action in Planning: A Gap to Overcome

Traditional planning relies on two different categories of knowledge: expert and non-expert (this last having different characterisation: lay, local, common, etc.); expert knowledge has been long considered an object to be owned (by planners) and used together with non-expert knowledge which had to be captured and/or acquired in order to be used. Such a vision of the relation between planning and knowledge is based on two main assumptions: (1) knowledge is additional; (2) knowledge can be moved out of the relational contexts and (world of) practices that produce and share it, and transferred to the planning arena in order to be used by planners. Knowledge is seen as being stable, reducible to a synthetic body (not conflictual, not competitive among components), ready and actionable for action in any place, at any time.[1]

More and more, in the last decades, knowledge in planning is recognised as multiple (Sandercock, 1998) and embedded in social relations; it is 'situated in social context' (Fuller, 2002). It is multiple because it has a variety of sources and takes a variety of forms. It is embedded in social relations and gives shape to the related activity infrastructure (knowledge is the capacity to act, and this is a capacity that 'emerges' from the relationships that exist within organisation; Boer, van Baalen, & Kumar, 2002; Hendon, 2000). In this sense, knowledge in planning is coherent with the 'community view' of knowledge as described by Jakubik (2007). The community or social view assumes that knowledge is not stable, but rather a dynamic and evolving entity and "that it is created in social interactions: knowledge is a social construct" (Jakubik, 2007, p. 14).[2] This vision of knowledge is centred on both process and context and assumes that knowledge is constructed within organisations also through processes of dialogue and interactions, and that knowledge is imbued with routines, standards and with day-to-day practices (Brown & Duguid, 1991).

The notion of knowledge as multiple and embedded in social relations gives rise to other acknowledgements. *Knowledge is not additive* (Evans & Marvin, 2006): it is not the result of multiple knowledge combination, it is synergic (Maisseu, 2006). Knowledge is the outcome of continuous, complex, hidden negotiations of languages, visions, world views, meanings, beliefs, claims, values and learning, communicating, reflecting and inquiring modes. *Knowledge is not stable*: it is rather transformative, it is a mutant entity ('knowing') continuously or discontinuously adapting and adjusting; a consequence of the openness and dynamics of the relational context activated by that knowledge and/or using, producing it. *Knowledge cannot be packed*: it never becomes an end-product; it is strongly related to the evolving nature of the relational context which shares, produces and uses it and therefore cannot be moved out from it. *Knowledge is not always actionable*: it is not always ready for use; it is acknowledgeable only through those practices that use it even keeping it in its tacit dimension[3]; knowledge is actionable only when action can make use of it albeit the embedded nature of that knowledge.

As they have just been described, the characteristics of knowledge are challenging for planning activities. The wide reliance on deliberation and communication approaches, as possible answers to this challenge, is showing more and more its weakness (Rydin, 2007): "bringing actors (expert and non-expert) together into the planning action is not enough" (Rydin, 2007, p. 55) and still represents the attempt to move knowledge from its relational context to the context of the planning action.

Making of knowledge a resource ready for action requires that action becomes the frame (with borders of space and time) in which knowledge is mobilised and activated for the action itself. Consistently, with the concept of *Ba* proposed by Nonaka and Konno (1998), action needs to be conceived as a 'shared context' sustaining the knowledge system and keeping knowledge *actionable*.

With respect to such a complexity of knowledge dynamics, planning requires to reconceptualise action as 'situated' (Lave & Wenger, 1991; Wenger, 1998). The 'knowing in action' (Amin & Roberts, 2008) perspective shows some potential with respect to the situatedness challenge. Amin and Roberts consider 'knowing in action' as a situated practice which:

1. handles the variety of knowledge dynamics;
2. takes into account differences in knowledge and modes of knowing;
3. makes use of portions of knowledge which already exist in and are or can be shared by the relational context through a sort of *bricolage* activity (Lanzara, 1999);
4. takes into account cognitive mechanisms taking places at the periphery of the action context;
5. becomes part of the dynamics of the cognitive organisation of the relational context;
6. is concerned about context dependency of participation and communication rules.

The 'knowing in action' perspective gives emphasis to the situated condition of the planning action which is relevant to preserving the context dependency of knowledge dynamics and to making the planning action part of that dynamics.

The 'knowing in action' perspective also requires a shift of the planning focus to knowledge. In planning, as well as in other domains (mainly the business domains), the crucial issue is no longer that of finding, collecting and making available for use the necessary knowledge (no longer the traditional knowledge management perspective); the most crucial issue is to recognise knowledge, distributed and/or concentrated, explicit and/or tacit, already existing and/or being produced, diverse and/or similar, belonging to individuals and/or to organisations, as an evolving and collective whole, framing the planning action. At the same time, planning action affects the knowledge infrastructure of the relational context: it asks for new knowledge by activating reflection and learning mechanisms (Schön, 1983), it uses knowledge from outside, it produces, collects, shares and manages data and information which affect knowledge.

Bridging knowledge to action in planning means to recognise that knowledge and action are linked together by a mutual framing dependency. In order to make knowledge an effective resource for the planning action and, the other way around, the planning action a resource for knowledge to become *actionable*, planning action cannot be just any complex process of collecting, sharing and using knowledge; it needs to be reconceptualised in strict inter-dependency with knowledge dynamics and the related relational context and needs to intrinsically include the goal to make knowledge *actionable*.

How does this inter-dependency affect strategy-making?

16.2.2 Reframing Strategy and Strategic Action

Generally speaking, a strategy is supposed to lead an organisation through changes and shifts to secure its future wellness and sustainability. Consistently with these general conceptions of strategy, change management becomes a crucial issue of the strategy implementation process. As it is well known, changes are not obvious consequences of decisions, regardless how consistent they may be with the overall strategy. Many problems arise, and many of these are largely associated with knowledge ability and management and with learning mechanisms. This is also evident in spatial strategy-making: Healey observes that "spatial strategy-making activity is taking place in a (...) context in which 'knowledge ability' and learning capacity are emphasised by policy-makers" (Healey, 2008, p. 861). The need for bridging knowledge to action augments its importance in strategy-making.

If we keep on conceptualising strategy-making as a linear sequence of two main activities, strategy formulation and strategy implementation (and change management), the knowledge-action gap stays un-resolved (Angehrn, 2005; Pfeffer & Sutton, 2000). Strategy-making needs to be reconceptualised within the perspective of bridging knowledge to action and making knowledge and action reciprocally shaping entities. This implies that: strategy-making has to be thought of as a process in which strategy is identified and formulated throughout the 'change management' activity; and implementation loses its whole significance. Zeleny with his idea of strategy (2008) makes a significant contribution in this regard.

The problem of implementation is described by Zeleny, 2008; (see also Chapter 15, this book) as the *Cloud Line* problem. The 'cloud line' is a real phenomenon well known in nature: from above the cloud line you cannot see below. Zeleny transfers the concept to strategy-making and observes the same phenomenon: operators of strategy implementation do not understand what is being asked and how the strategy has to be implemented (ibid, p. 66). "Everything above the cloud line is just a symbolic description of the intended future action. Everything below is only pure action, no descriptions. These are two separate domains: (1) description of action and (2) action itself. They can and do differ; very rarely do they meet – unless the description refers to the 'actual' action, present or past, not the intended action of the future" (ibid, p. 66). Till strategies prevail in describing the future, the eternal problem of implementation remains unresolved. How can a strategy be implemented?

Zeleny suggests that strategy be reconceptualised as it is about doing and not about saying, thus making of implementation an 'uninteresting' issue. He considers that the notion of strategy abandoned the 'mission-vision' paradigm and reinstated action in the centre: *an organisation's strategy is what the organisation is doing and not what it is saying* (ibid, p. 65).

"What we want is not implementing a description but changing the strategy itself: changing from one form of action into another" (ibid, p. 66). Strategy-making starts with action, with current action, not with the identification of a mission or vision formulation; its product is a mission or vision derived from the action itself.

According to Zeleny, the problem originates from keeping knowledge and action as different and distinct concepts. Viewing knowledge and action as mutually framing entities, as envisaged in the previous paragraph, is consistent with the Zeleny's idea of knowledge as action: "knowledge is a purposeful coordination of action" (ibid, p. 66).

Knowing is acting and acting is knowing. When this gap is bridged a different approach to strategy-making is possible. The problem in spatial strategy-making, and in general in spatial planning, is that the relation between knowledge and action has been often looked at as a gap. Shifting the point of view to the knowledge-action relation implies that strategy-making can be reconceptualised and looked at as a complex activity of knowledge management by coordinating action towards a necessary change.

When coordinating action in complex systems such as spatial organisations, the action space is not completely known a priori; action is carried out within high uncertainty and many risks are envisaged. In such systems, action coordination asks for an exploratory approach in order to guarantee the systems from irreversible organisational, social and environmental consequences. Some authors discussed this problem in terms of micro-action or micro-decision (Barbanente, Borri, & Pace, 1993; Zeleny, 2002). For similar reasons, although concerning the business organisational world, Johnson, Melin and Whittngton (2003) discuss the issue of micro-strategy. Having an exploratory approach to action coordination requires that an empirical value be assigned to knowledge and practices until these are acknowledged of any strategic shared value for the organisation and also for its related environment. Strategy-making has to be conceptualised as complex framing of 'empirical spaces' where cognitions and practices are explored thus enabling the new strategy to take shape together with the new developed knowledge ability.

The idea of 'strategy-as-practice' (Jarzabkowski, 2003, 2004; Johnson, Langley, Melin, & Whittington, 2007; 2003; Whittington, 1996) is consistent with the conceptualisation of strategy just discussed. The 'strategy-as-practice' idea is derived from the need to look at strategies with a deeper focus "on the processes and practices constituting the everyday activities of organisational life and relating to strategic outcomes" (see www.strategy-as-practice.org quoting Johnson et al., 2003, p. 3).

The notion of strategy developed by the 'strategy-as-practice' approach depicts strategy as an activity undertaken by people who are components of the organisation (Carter, Clegg, & Kornberger, 2008) asked to respond the needed changes. According to Whittington (2004), the innovation of 'strategy-as-practice' is to treat strategy as important organisational practice. Trying to clarify the distinction between actions and practices in the 'strategy-as-practice' approach, Carter et al. (2008) observe: "we should forget (for a moment, at least) the word strategy and see which practices produce endurable or recurring events that eventually turn into 'things' or 'events' that are then addressed as 'strategy'. Hence, we have good reason to assume that strategy does not exist independently of a set of practices that form its base. In fact, strategy might happen (...) in different circumstances and different contexts; however, only a small percentage of actions that occur will be called 'strategic' because they revolve around a set of practices that constitute what is formally acknowledged to be strategy. From this perspective, a strategy as practice approach would research those practices that constitute the (...) 'strategy'" (Carter et al., 2008, p. 92).

Following Carter et al.'s definition and considering that practices and knowledge are strictly inter-related and reciprocal, we can consider strategy-making as the search for those practices and knowledge which are consistent with the needed change. In a certain sense, strategy can be looked at as an exploratory learning process where practices and their related knowledge undergo an 'appropriation' process: knowledge and practices become properties of the organisation and are kept as new actionable resources for the organisation itself and for its strategy.

16.3 Strategic Episodes in Strategy-Making

Strategy-making needs to have an appropriative nature. It has to be developed around one or more values/needs for change and the whole organisation needs to develop an appropriation of that/those value/s by empirically testing knowledge and practices consistent with that/those value/s. These tests can be intended as laboratories of knowledge and practices activated by *strategic episodes*. *Strategic episodes* are defined by Hendry and Seidl (2003). Hendry and Seidl look at 'episode' as providing a mechanism by which a system can suspend its routine structures and thus initiate a reflection on and change of these structures.[4] They define *strategic episodes* within the idea that strategic changes need modification of communication structures for new strategic discourses (Hendry & Seidl, 2003, p. 185).

For the purpose of the present discussion, I will consider *strategic episode*s assigning a larger meaning, reducing the communicative dimension crucial in the Hendry and Seidl's definition, keeping their idea that a strategic change is a change of the context from which the organisation is observed[5] and shifting the concept towards the spatial strategic action. A *strategic episode is a any condition for the routine knowledge and practices constraints in spatial management and transformation to be suspended and alternative knowledge and practices to be explored.*

By assigning significance to *strategic episodes*, an organisation can distance itself from itself thus allowing itself to observe itself and, from this position, to start a change. The activated exploration represents one, possibly additional, step forward into the strategy-making activity and can be seen as the place in which knowledge and practices are specified, transformed and finalised thus becoming an internalised and shared property. The exploration, in fact, enables the appropriation of knowledge and practices which therefore become resources embedded in social relations and able to shape action. In a sense, knowledge and practices are explored and transformed till they become collective *actionable* resources. Only in this way, knowledge and practices can be seen as responsible for strategic changes, in the organisation.

Strategy-making is explicitly concerned with the recognition of the need for a significant, often radical change. Processes of change can obviously be activated unintentionally, incrementally or through organisationally distributed bottom-up processes. More often they are auspicated by a managerial or institutional intentionality. If we agree to abandon the image that strategy is a starting point (thus overcoming the idea to run strategy-making by first developing visions and formulating strategy), a key issue for starting change is starting managing knowledge by coordinating action along strategic episodes. Action coordination in strategy-making can be coherently conceptualised as capturing opportunities for strategic episodes and activating or managing them as a coherent whole towards the needed change.

Activating strategy-making, both when it is a bottom-up or top-down approach, requires capturing of 'strategic episodes'. Strategic episodes enable the appropriation of knowledge and practices that are tested and specified/developed hopefully in line with the needed change. The knowledge and practices which are acknowledged to be consistent with and appropriate for the needed change can be referred to as composing the strategy.

16.4 The *Bricolaging* Character of Strategy-Making

16.4.1 Actionable Knowledge and Practices in Strategy-Making

The concept of 'actionable knowledge' is well known in the domain of knowledge management and it is considered as the knowledge that is *ready-to-use*.

Actionable knowledge, as opposed to information or other types of knowledge, refers to knowledge that is useful in guiding behaviour in that it tells us how to create or produce something we believe has external validity (Argyris, 1993, 1996 quoted in Adams & Flynn, 2005). For example, knowing that the use of chemicals in agriculture affects ground water depending on specific draining characteristics of soil is information with external validity: it can be used for choosing one or more fertilisation methods in agriculture among diverse alternatives. Knowledge that informs cultivators how to quantify chemicals depending on soil's draining characteristics in order to avoid groundwater pollution is actionable knowledge because it provides the link between the general knowledge and setting specific knowledge to make draining mechanisms knowledge externally valid.

Actionable knowledge and practices are reciprocally shaping and cannot be disjointed: Chris Argyris clarified that actionable knowledge is not only relevant to the world of practice, it is the knowledge that people use to create that world (1993).

In strategy-making, this implies that an organisation needs to be the owner of actionable knowledge and related practices in order to be able to conduct the organisation itself towards the needed change. The appropriation process requires the acknowledgement of general knowledge or information with external validity as well as the test/development of actionable knowledge for linking that external validity to practices.

Therefore actionable knowledge in strategy-making can be defined as an organisational cognitive property developed throughout strategy-making and, at the same time, shaping the strategy itself through its related practices.

Knowledge is made 'actionable' for strategy-making, when strategy-making is considered to be an exploratory 'social/organisational activity' enabling the appropriation of knowledge, that is, making actionable knowledge produced and/or revealed, tested and therefore shared as a common good.

As they have been defined above, strategic episodes enable the appropriation of knowledge ready for shaping action towards the needed change, that is, for shaping strategy.

Actionable knowledge for strategic changes is a product of the system itself but not necessarily within evolutionary mechanisms. Strategic episodes, by suspending the routine cognitive and practice mechanisms, can activate practices and actions which are inconsistent with the pre-existing mechanisms but at the same time consistent with the new values empirically explored within strategic episodes. The mechanism can be continuous (Weick & Quinn, 1999) or episodic (Ford & Ford, 1994), or evolutionary/revolutionary (Weick & Quinn, 1999), that is, in continuity or discontinuity with the pre-existing structure of practices, but nevertheless manageable by the organisation because the knowledge supporting change has been internalised by the organisation itself through an appropriation process.

In a certain sense, strategic episodes can be considered as *Ba* (Nonaka & Konno, 1998) environments where knowledge and practices are created and transformed into available and actionable resources for organisations.

16.4.2 Bricolaging Knowledge and Practices

The question is: what makes and how to make actionable knowledge and practices, that have undergone an appropriation process throughout strategic episodes, compose a strategy?

Relying on the definition of strategy-making presented in this chapter, strategy is not a starting point; it is framed throughout the strategy-making process and has a retrospective dimension: in order to acknowledge and frame the strategy we need to look backward, on what has already taken place, searching for actionable knowledge and its related practices, possibly consistent with the needed change. It is not feasible that all the available actionable knowledge and practices appear promising with regards to the needed change: those being promising have to be sorted out of a chaotic set currently composing the whole organisational action and framed together. It is a *bricolage* activity and does not have an end. Strategy is dynamic: it evolves together with the dynamics of activities carried out by the organisation; the more experimental these activities are and the more empirical nature they have, the more the strategy is changing.

The dynamic view of strategy has been envisaged and analysed by many authors (Johnson et al., 2003; Regnér, 2008; Whittington, 2003). Regnér, among others, emphasises the view of strategy "as something immanent in purposive action that draws on tendencies and predispositions, rather than as individual purposeful action, as traditionally conceived" (Regnér, 2008, p. 575).

Generally speaking, we could say that *bricolaging* can be intended as 'creating order out of whatever resources are at hand'. In this sense, "*bricoleurs* act in chaotic conditions and put order out of them" (Weick, 2001, p. 110). Guiding a strategy-making process means managing actionable knowledge through the coordination (based on the exploration and capture of capabilities) of action in order to *bricolage* a coherent whole towards a needed change. Regnér defines this approach to strategy as *inductive strategy-making*: "strategy [is] developed through [. . .] exploratory activities involving trial and error, informal contacts and noticing, experiments and heuristics" (Regnér, 2003, p. 77); in these conditions new knowledge and practices that can enable significant changes are created and developed. There is a great focus on capturing opportunities from available resources, which is the basic assumption in the concept of *bricolage*.

In strategy-making, *bricolage* refers to a creative and adaptive management of knowledge/practices resources towards a needed change: it can be seen as a practical adaptation/composition of knowledge and practices.

Many authors already observed the use of knowledge artefacts as an activity grounded in the *bricolage* involved in everyday strategy-making (Chia, 2004; Jarzabkowski, 2004; Whittington, 2003; Jarzabkowski & Wilson, 2006; Wilson & Jarzabkowski, 2004). Great emphasis is given to the idea that knowledge artefacts are already existing, thus augmenting opportunities for easy use and reducing the demand for learning. "*Bricolage* is inherent in the practical use of knowledge, utilizing those knowledge artefacts that are at hand (. . .) about future strategy. Practitioners act upon future strategy without accurate foresight. Strategy artefacts assist in this process not as rational tools for diagnosing future action, but as tools that may be fashioned to effect current actions in ways that may bring about future actions. Rather than seeking new knowledge, in *bricolage* the use of an existing, well-known tool that is readily to hand is likely because such tools may be more easily fashioned to the (. . .) intent. Strategists continue to draw upon established artefacts (. . .) because these have technical, cultural and linguistic legitimacy that makes them easily appropriable" (Jarzabkowski & Wilson, 2006, p. 361).[6]

Considering actionable knowledge and practices developed in strategic episodes as resources to be *bricolaged* into a strategy means that the strategy adaptively comprises things that the organisation is already familiar with (it has already undergone an appropriation process) and that are ready to become routine, because they have been already tested against the needed change by the organisation.

16.5 The Story of a Strategy

16.5.1 The Experience of Torre Guaceto in Italy

The case we present here refers to one of the planning activities carried out by the Park Agency of the Torre Guaceto wetland, a Natural Reserve in Southern Italy.[7] Torre Guaceto is located in the Apulia region, on the Adriatic coast, about 15 km north of the city of Brindisi. Among others, including greater natural value, the Natural Reserve covers a large area used for agricultural activities: mainly olive trees and vegetables cultivation. The agricultural area was included in the Natural Reserve because its environmentally oriented management is committed to the protection of wildlife environments.

The Natural Reserve is managed by the Park Agency which is responsible for the Land Use Management Plan (LUMP). The LUMP is considered one of the available means to develop a change in local agricultural practices towards natural production; therefore, according to national laws, it includes regulations for the agricultural practices.

In 2000, the Park Agency started working on the LUMP and, in early 2002 presented the plan to the agricultural community. With respect to agricultural practices, the LUMP prescribed a shift from current practices towards biological ones. The way of shifting agricultural practices from the standard ones to natural was not explicitly defined in the LUMP. The underlying strategy of the LUMP was centred on the idea that changes in practices are possible if you change rules. Obviously, the constraints imposed on land use practices activated strong reactions by the agricultural community.

Faced with a conflict the Park Agency decided to adjust the norms by introducing less restrictive rules but this effort was not enough to reduce the conflict, and consequently the LUMP was adopted without an agreement with the agricultural community, although in its less restrictive version.

A short time later, the Park Agency was involved in a wetland project for introducing participatory practices in wetland management. This was considered an opportunity to manage the conflict, but the structured participatory protocol tested by the project resulted in the escalation of conflict. Eventually, attempts to communicate and interact with the agricultural community were abandoned by the Park Agency.

Some months later, the Park Agency hired a consultant to contact and interview the farmers cultivating land on the Natural Reserve in order to develop a financial and economic program of the Reserve. The interviews were conducted in a face-to-face fashion. Farmers began to cooperate and communication between the Park Agency and the agricultural community resumed: the Park Agency realised that communication needed to be managed within a more dialogic and individualistic mode. In the same period, the Park Agency was invited to join an Interreg project[8] under which incentives for testing innovative and environment friendly practices were available to farmers who wanted to participate. However, only eight farmers accepted to be involved in the project and test new practices for olives cultivation and olive oil production ('The Park Gold' project within the Interreg project).

The project 'The Park Gold' was a great success. Although very small in terms of the number of participants and land, it gained a symbolic value for the agricultural community: many other olive trees farmers expressed interest in being involved, although the Interreg project could no longer support them. Those who had joined the 'The Park Gold' project decided to set up the Torre Guaceto Association of Biologic Olive Oil Producers. The Statutory Rules of Association contain mainly prescriptions for olive oil production, and these rules are much more restrictive than those related to the same production and initially contained in the LUMP.

The Park Agency assigned strategic value to this result especially considering the main goal of shifting towards biologic production the cultivations in the area and decided to announce to the agricultural community that the olives farmers joining the new Association could benefit from using the Park Label on oil packaging.

The success of 'The Park Gold' initiative triggered additional experiments devoted to innovative biological methods of vegetable cultivation. The first one was a special cultivar of tomato (*fiaschetto*) that, in the past, was grown in Puglia in dry-cultivation. Dry-cultivation makes products more resistant and less demanding of chemicals. Farmers started different parallel tests in different areas of the Natural Reserve using different protocols to find out which protocol works best for strengthening the product and increasing productivity. The development of *fiaschetto* production has been and still now is supported by Slow Food[9], an international organisation founded to counteract fast food, fast life and the disappearance of local food.

In order to commercialise the *fiaschetto* effectively, some tests were set up to transform the *fiaschetto* into tomato sauce to be produced in very traditional manner (the way the sauce was produced in the local families' tradition) by involving women (the wives of the cultivators) and also a national organisation (Libera Terra[10]) which in Puglia is managing the (properties of) ancient farms confiscated from the local mafia.

Similar tests are currently carried out with lettuce and other vegetables. All the experiments are collaboratively designed by the Park Agency and the agricultural community. New practices for agricultural production are being developed and transformed into new routines.

Crucial to the discussion in this chapter is the following: the Park Agency abandoned the idea to handle a strategic change by carrying out the adoption and implementation of the LUMP and developed a sort of a strategic ability to coordinate action along strategic episodes and *bricolaging* knowledge and practices towards the needed change.

16.5.2 Analysing Strategy-Making in Torre Guaceto

16.5.2.1 Strategic Episodes

Strategic episodes can be identified with reference to four different fields of practice: ‘spatial plan design’, ‘communication’, ‘agricultural practices’, ‘community and organisational management’. In particular, as it is visualised in Fig. 16.1, strategic episodes can be clearly identified in the last three fields of practice and refer respectively to: the exploration of communication practices (the participatory models proposed within the framework of the Wetland project and the face-to-face communication model adopted by the consultant hired by the Park Agency to develop a financial and economic program); the exploration of biological agricultural production practices (olive oil cultivation, *fiaschetto* tomato cultivation and transformation, traditional vegetable cultivation); and the exploration of alternative dynamics of the organisation (involvement of external actors).

It is evident from the story that the identified events are strategic episodes clearly approached with an empirical approach by components of the organisation who felt

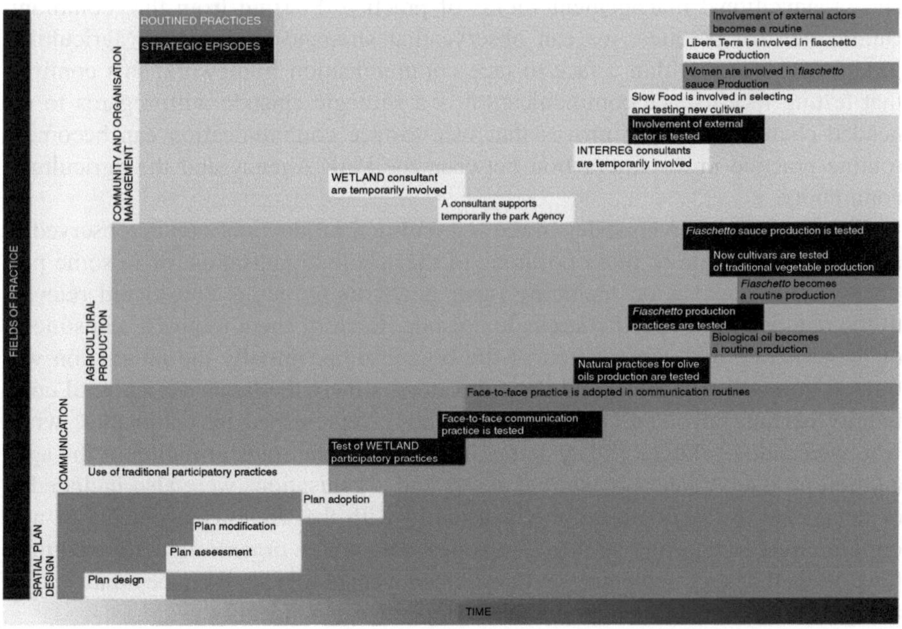

Fig. 16.1 Strategy-making in Torre Guaceto

themselves engaged in the need for change. Strategic episodes represent an intentional opportunity for the organisation to look at itself 'from outside' and reflect on itself with regards to the needed change.

An interesting episode, not indicated in Fig. 16.1, is the one related to the spatial plan design. The activity carried out at the time when the LUMP was being developed, did not have any empirical end, at least not determined by any intentionality. The failure of any attempt by the Park Agency to begin a dialogue with the agricultural community made the Park Agency search for opportunities to engage cultivators, even if only few, in collaborative initiatives. Moreover, this episode made the Park Agency abandon the idea that the LUMP could be of any value in moving towards a change. The Agency also learned/realised (and learn) that the agricultural community would never accept any modification of cultivation practices outside of practice itself.

The episode described above does not constitute a strategic episode as defined it in the early paragraphs; still it represents a key episode through which the Park Agency started developing the strategic ability to recognise opportunities for activating strategic episodes, conceptualised as laboratories of practices. The idea that the strategic episodes identified above are really strategic is also indicated by the fact that they are not only referred to as changes in the agricultural practices (being the main focus of the needed change) but also as other dimensions of the whole organisation of the Torre Guaceto wetland.

Strategic episodes in the 'agricultural production' field of practice take place in connection with strategic episodes related to the 'communication' and 'community and organizational management' fields of practice. Starting from the 'communication' field of practice, we can observe that strategic episodes in 'agricultural production' occur within a face-to-face communication framework: this confirms that testing face-to-face communication is a strategic episode with regards to the needed change; in fact it proves that face-to-face communication can become a routine practice in the interaction between the Park Agency and the agricultural community.

Similarly, strategic episodes in the 'agricultural production' can be observed as related to one single or more portion/s of the whole organisation, or to some portions of the organisation including other new-coming actors considered relevant to the outcome of a specific strategic episode. Each of them required adjustments of the structure and composition of the organisation: initially the interaction was limited to the Park Agency and the cultivators; slowly the larger agricultural community became involved (women were mainly represented by cultivators' wives who were acknowledged as key actors in the traditional transformation of the agricultural products); later on external actors and organisations were also included in or connected to the organisation consistently with the acknowledgements of new opportunities for starting strategic episodes. The whole organisation stopped thinking of itself as a closed entity and developed a capability to re-conceptualise itself as a pulsating entity, able to maintain strong geographic identity.

This analysis shows that the key issue for activating change in Torre Guaceto was starting coordinating action along strategic episodes. This experience of strategy-making shows/indicates that action coordination can be easily recognised as capturing opportunities for strategic episodes and activating or managing them as a coherent whole towards the needed change.

16.5.2.2 Actionable Knowledge and Practices

The Torre Guaceto strategy-making reveals what I mean by 'actionable knowledge' and how this knowledge is produced by strategy-making and, reciprocally, how it provides structure to strategy-making for governing the Torre Guaceto territory.

In the Torre Guaceto experience knowledge isn't *actionable* until the entire organisation is in agreement as to what that knowledge really is. The knowledge becomes actionable through collaboration in experimental forms of land use and is kept actionable by turning related practices into routines of the organisational activities. In terms of Argyris' definition (1993), actionable knowledge in this experience is the knowledge that the Torre Guaceto organisation used to create its new strategy, that is, its new world of practices.

In this sense, the activation of change started when the action coordination along strategic episodes enabled consistent (with the needed change) knowledge management or, reciprocally when the coordination of action along strategic episodes started to be structured by knowledge management.

It is possible to identify actionable knowledge and practices relevant for the Torre Guaceto strategy-making. Looking at the organisation as a whole (the Park Agency, the cultivators and their families the agricultural community, external actors or consultants), some examples of knowledge and practices which shape strategy-making in Torre Guaceto can be identified:

1. the knowledge activated to keep communication and collaboration active and used by the Park Agency in face-to-face communication routines;
2. the knowledge embedded in organic olive oil production practices;
3. the knowledge embedded in *fiaschetto* organic production practices;
4. the knowledge activated to involve external actors in strategic episodes.

These examples of knowledge and practices are knowledge artefacts provided to a new shared context, an empirical context, and developed collaboratively within the shared context itself. They frame strategy-making and are, at the same time, products of strategy-making itself: such knowledge artefacts found their way to practice and within practice: they are the actionable forms of knowledge.

It is interesting to observe that these knowledge artefacts are not only useful to the routine they are embedded in and responsible for: they are also used, although differently, while developing and carrying out other, subsequent strategic episodes. These knowledge artefacts are characterised not only by a dimension specifically related to the practice but also by a more general dimension related to the modes (communication and collaborative mechanisms) and conditions (success or failure of the strategic episodes) necessary to develop the artefact till it becomes actionable.

This second body, more abstract, is responsible for the *bricolaging* activity, that is, the way knowledge is managed by action coordination.

16.5.3 Discussion

The Torre Guaceto experience shows how a strategy-making process works as *bricolaging* knowledge and practices towards a needed change.

Initially, the Park Agency presented its own vision of the future by the means of the LUMP, imagining that the change could be achieved by LUMP norms implementation. The change could start when the Park Agency stopped describing its future vision and started managing and *bricolaging* knowledge by coordinating action through strategic episodes and towards a coherent change. In the Torre Guaceto experience, strategic episodes represented a lens through which actors could look at their environment and consider different ways to interact with it. In this way there was no need to extract knowledge from its relational context because that relational context is the context carrying on the strategy-making through practices within an empirical approach.

Strategy-making became a social practice, not necessarily intentional, carried out throughout the 'collaborative exploration of practices'. At the same time strategy is operationalised by recognising and *bricolaging* 'actionable knowledge' and related practices within a sort of knowledge governance framework. Here, by knowledge governance I refer to what is widely shared (Foss & Michailova, 2009): learning, creating and managing knowledge crucial for the future of the organisation that becomes a space where modes for creating and modifying the organisational reality are discovered.

The Torre Guaceto organisation gained a cognitive and relational dimension where knowing is *knowing in action*, that is, running the multiplicity of opportunities and modes for building its own world of practices within its specific spatial, environmental and organisational constraints.

16.6 Conclusions: Towards the Micro-Foundation of Strategy

This chapter explored spatial strategy-making as a process based on knowledge management and coordination of action towards needed changes, thus trying to reduce the gap between knowledge and action.

The knowledge–action gap has been analysed within two different domains: spatial planning and strategy-making. With regard to the first domain the analysis of this gap shows that knowledge has long been considered as a resource for planning to be captured and made available to the planning action. The idea that knowledge is not additive, not stable, not available to be packed and not always actionable, suggests that knowledge be kept within the relational context, where it is used, produced and shared and modes and opportunities for creating spaces where knowledge is actionable be explored.

With regard to strategy-making, the analysis shows that while keeping conceptualising strategy-making as a linear sequence of two main sequential activities, that is, strategy formulation and strategy implementation, the knowledge–action gap still represents a problem. Reconceptualising strategy-making as a process in which strategy is identified and formulated throughout the 'change management' activity makes knowledge and action reciprocally shaping entities, and the gap a non-existing issue.

Adopting a knowledge/practice-based perspective in strategy-making in order to make the gap between knowledge and action not a critical issue with regard to the planning effectiveness, strategy-making is proposed as an exploratory knowledge management process keeping knowledge within the relational context and continuously re-aligning it and the strategy itself within an empirical approach. When moving knowledge out from its socio-relational context, it is not possible to consider it in its actionable dimension and the social structures and interactions appropriate to the strategy cannot be formed.

Taking the above into consideration strategy-making is described as a *bricolaging* activity, capturing actionable knowledge and related practices as resources shaping the strategy and being shaped by the strategy-making process. When strategy-making is conceived as producing an end-product, it destroys the intrinsic idea of the *bricolage* concept: various elements and components are used and adjusted into the *bricolage* products when they are recognised as effective and consistent with the context evolution and requirements. The *bricolage* product evolves together with the context producing/using it. *Bricolaging* in strategy-making can be one of the possible ways to make strategy a vehicle for enabling knowledge to be activated as a resource for action.

Actionable knowledge and related practices are considered knowledge artefacts of the strategy-making process and, in particular, of *strategic episodes* activated as empirical spaces where the organisation that is making strategy can suspend its routine practices and explore new ones which are eventually recognised as consistent with the needed change.

Strategic episodes, as spaces where new actionable knowledge and related practices are developed as a dynamic whole coherent with the needed change, can be considered spaces for the *micro-foundation of strategy*.

The term micro-foundations is well known in economics and refers to the micro-economic analysis of individuals' behaviour that underlines macro-economic theory.[11] More generally, "[m]icro-foundations refer to the micro-level activity that underlies a macro-level phenomenon" (Stoker, 2008, p. 3). Strategic episodes represent only one of the possible micro-foundational spaces for strategy. Micro-foundations of strategy can be related to any opportunity or space indicating how actors, activities, practices and organisational structures are related towards strategic outcomes (Regner, 2008) consistent with the needed change.

Considering strategy-making as an exploratory process, the search for micro-foundations can be conceived as a process aimed at identifying any/all possible micro-organisational-level mechanisms which bring about aggregate organisational outcomes, which are key for the strategy. In strategy-making as described in this chapter, the organisational outcome, that is, the phenomenon at the macro-level can be related to the organisational dynamics observed through changes in practices. Micro-foundations can be intended as providing basic understanding of micro-mechanisms in the organisation that helps to guide the *bricolaging* activity towards the strategy.

Notes

1. "When dealing with tangible resources, it is possible to manage those resources by distributing them efficiently according to functions and goals. Knowledge is not a tangible resource. It is rather intangible, boundary-less, dynamic. If it is not actionable at a specific time in a specific place, it is of no value for action" (Nonaka & Konno, 1998, p. 41).
2. See also Searle, 1996.
3. Zeleny (2008, p. 66) considers that "there is no other knowledge than tacit". See also Zeleny, 1987 with regard to this concept.
4. Hendry and Seidl refer great part of their reflection on episodes and strategic episodes to the Luhmann's model of social systems change (Luhmann, 1990).
5. In these conditions, novel combinations of routines are made possible by reflections of actors on existing routines (Feldman, 2000).
6. The author wants to make the reader aware that Jarzabkowski and Wilson (2006) refer to actionable knowledge assigning to this a different meaning than that assigned by this chapter. Jarzabkowski and Wilson assume actionable knowledge as distinct from theoretical knowledge; it includes tools, techniques, models and methodologies developed by theory. Although this difference their quotation helps in clarifying the implications of a *bricolage* approach.
7. The planning experience of Torre Guaceto has been already analysed in other publications by the author (Celino & Concilio, 2006; Celino, Concilio, & De Liddo, 2008).
8. The Interreg project 'TWReferenceNET: Management and sustainable development of protected transitional waters' is designed to improve and reinforce conservation of natural heritage in protected transitional ecosystems and to enlarge their sustainable fruition. The project is financed by the Community Initiative INTERREG III B (2000–2006) CADSES.
9. Slow Food is an international non-profit, eco-gastronomic member-supported organisation founded in 1989 to counteract fast food and fast life, the disappearance of local food traditions and people's dwindling interest in the food they eat, where it comes from, how it tastes and how our food choices affect the rest of the world (http://www.slowfood.com).
10. See: http://www.liberaterra.it.
11. See Barro, 1993.

Acknowledgements I sincerely thank Adele Celino, my best friend and colleague: this chapter refers to many reflections and ideas developed in our collaborative work. I also thank Valeria Monno who supported the writing of this chapter by offering insightful criticism.

References

Adams, S. M., & Flynn, P. M. (2005). Actionable knowledge: Consulting to promote women on boards. *Journal of Organizational Change Management, 18*(5), 435–450.
Albrechts, L. (2004). Strategic (spatial) planning reexamined. *Environment and Planning B, 31*(5), 743–758.
Albrechts, L. (2006). Shifts in strategic spatial planning? Some evidence from Europe and Australia. *Environment and Planning A, 38*(6), 1149–1170.

Amin, A., & Roberts, J. (2008). Knowing in action: Beyond communities of practice. *Research Policy*, *37*(2), 353–369.

Angehrn, A. A. (2005). Learning-by-playing: Bridging the knowing-doing gap in urban communities. In A. Bounfour & L. Edvinsson (Eds.), *Intellectual Capital for Communities: Nations, Regions, Cities* (pp. 299–316). Burlington: Elsevier Butterworth-Heinemann.

Argyris, C. (1993). *Knowledge for action: a guide to overcoming barriers to organizational change*. San Francisco, CA: Jossey-Bass Inc.

Argyris, C. (1996). Actionable knowledge: design causality in the service of consequential theory. *Journal of applied behavioural science*, *32*(4), 390–406.

Barbanente, A., Borri, D., & Pace, F. (1993). Micro-problems and micro-decisions in planning by artificial reasoners. In R. Klosterman & S. P. French (Eds.), *Proceedings of the Third International conference on computers in urban planning and urban management* (pp. 17–33). Atlanta: City Planning Program Georgia Tech.

Barro, R. J. (1993). *Macroeconomics*. New York: Wiley.

Boer, N. -I., van Baalen, P. J., & Kumar, K. (2002) *The importance of sociality for understanding knowledge sharing processes in organizational contexts*. Report Series Research in Management [http://publishing.eur.nl/ir/repub/asset/179/erimrs20020308162214.pdf; accessed June 2008].

Brown, J. S., & Duguid, P. (1991). Organizational learning and communities-of-practice: Toward a unified view of working, learning, and innovation. *Organization Science*, *2*(1), 40–57.

Carmona M. (Ed.). (2009). *Planning through projects: moving from master planning to strategic planning – 30 Cities*. Amsterdam: Techne Press.

Carter, C., Clegg, S. R., & Kornberger, M. (2008). Strategy as practice? *Strategic Organization*, *6*(1), 83–99.

Celino, A., & Concilio, G. (2006). Managing open contents for collaborative deliberation in environmental planning. In F. Malpica, A. Oropeza, J. Carrasquero, & P. Howell (Eds.), *Proceedings of the 4th international conference on politics and information systems: technologies and applications* (PISTA 2006) (pp. 155–160). Orlando: International Institute of Informatics and Systemics.

Celino, A., Concilio, G., & De Liddo, A. (2008). Managing knowledge in urban planning: Can memory support systems help? In M. S. Ackerman, R. Dieng-Kuntz, C. Simone, & V. Wulf (Eds.), *Knowledge management in action* (pp. 51–65). Boston: Springer.

Chia, R. (2004). Strategy-as-practice: Reflections on the research agenda. *European Management Review*, *1*(1), 29–34.

Evans, R., & Marvin, S. (2006). Researching the sustainable city: Three models of interdisciplinarity. *Environment and Planning A*, *38*(6), 1009–1028.

Feldman, M. S. (2000). Organizational routines as a source of continuous change. *Organization Science*, *11*(6), 611–629.

Ford, J. D., & Ford, L. W. (1994). Logics of identity, contradiction, and attraction in change. *Academy of Management Review*, *19*(4), 756–785.

Foss, N. J., & Michailova, S. (Eds.). (2009). Knowledge governance. Themes and questions. In N. J. Foss & S. Michailova (Eds.), *Knowledge governance: Processes and perspectives* (pp. 1–24). Oxford: Oxford University Press.

Fuller, S. (2002). *Knowledge management foundations*. Boston, MA: Butterworth-Heinemann.

Gunder, M., & Hillier, J. (2007). Planning as urban therapeutic. *Environment and Planning A*, *39*(2), 467–486.

Healey, P. (2007). *Urban complexity and spatial strategies: Towards a relational planning for our times*. London: Routledge.

Healey, P. (2008). Knowledge flows, spatial strategy-making, and the roles of academics. *Environment and Planning C*, *26*(5), 861–881.

Healey P., Khakee A., Motte A., Needham B. (Eds.). (1997). *Making strategic spatial plans. Innovation in Europe*. London: UCL Press.

Hendon, J. A. (2000). Having and holding: Storage, memory, knowledge, and social relations. *American Anthropologist, 102*(1), 42–53.

Hendry, J., & Seidl, D. (2003). The structure and significance of strategic episodes: Social systems theory and the routine practices of strategic change. *Journal of Management Studies, 40*(1), 175–196.

Hillier, J. (2007). *Stretching beyond the horizon. A multiplanar theory of spatial planning and governance*. Aldershot: Ashgate.

Jakubik, M. (2007). Exploring the knowledge landscape: Four emerging views of knowledge. *Journal of Knowledge Management, 11*(4), 6–19.

Jarzabkowski, P. (2003). Strategic perspectives: An activity theory perspective on continuity and change. *Journal of Management Studies, 40*(1), 23–51.

Jarzabkowski, P. (2004). Strategy as practice: Recursive, adaptive and practices-in-use. *Organization Studies, 25*(4), 529–560.

Jarzabkowski, P., & Wilson, D. (2006). Actionable strategy knowledge: A practice perspective. *European Management Journal, 24*(5), 348–367.

Johnson, G., Langley, A., Melin, L., & Whittington, R. (2007). *Strategy as practice: research directions and resources*. Cambridge: Cambridge University Press.

Johnson, G., Melin, L., & Whittngton, R. (2003). Micro strategy and strategizing: Towards an activity-based view. *Journal of Management Studies, 40*(1), 3–22.

Lanzara, G. F. (1999). Between transient constructs and persistent structures: Designing systems in action. *Journal of Strategic Information Systems, 8*(4), 331–349.

Lave, J., & Wenger, E. (1991). *Situated learning: Legitimate peripheral participation*. Cambridge: Cambridge University Press.

Luhmann, N. (1990). Anfang und Ende: Probleme einer Unterscheidung. In N. Luhmann & K. Schorr (Eds.), Zwischen Anfang und Ende: Fragen an die Pädagogik. Frankfurt a.M., Suhrkamp, 11–23.

Maisseu, A. (2006). Gestalteconomy: The economic bases of knowledge management. *International Journal of Nuclear Knowledge Management, 2*(2), 174–198.

Nonaka, I., & Konno, N. (1998). The concept of "Ba": Building a foundation for knowledge creation. *California Management Review, 40*(3), 40–54.

Pfeffer, J., & Sutton, R. I. (2000). *The knowing-doing gap. How smart companies turn knowledge into action*. Boston, MA: Harvard Business School Press.

Pugliese T., Spaziante A. (Eds.). (2003). *Pianificazione Strategica per le Città: Riflessioni dalle Pratiche*. Milano: FrancoAngeli.

Regnér, P. (2003). Strategy creation in the periphery: Inductive versus deductive strategy-making. *Journal of Management Studies, 40*(1), 57–82.

Regnér, P. (2008). Strategy-as-practice and dynamic capabilities: Steps towards a dynamic view of strategy. *Human Relations, 61*(4), 565–588.

Rydin, Y. (2007). Re-examining the role of knowledge within planning theory. *Planning theory, 6*(1), 52–68.

Sandercock, L. (1998). *Towards cosmopolis*. London: Wiley.

Schön, D. (1983). *The reflective practitioner*. New York: Basic Book.

Searle, J. R. (1996). *The construction of social reality*. London: Penguin.

Stoker, G. (2008) *The microfoundations of governance: Why psychology rather than economics could be the key to better intergovernmental relations*. Paper presented at Zhejiang University [http://www.soton.ac.uk/ccd/people/stokerg.html, accessed June 2008].

Tewdwr-Jones, M. (2002). *The planning polity: Planning, government and the policy process*. London: Routledge.

Weick, K. (2001). *Making sense of the organization*. Oxford: Blackwell Publishing.

Weick, K., & Quinn, R. (1999). Organizational change and development. *American Review of Psychology, 50*, 361–386.

Wenger, E. (1998). *Communities of Practice: Learning, Meaning, and Identity*. Cambridge: Cambridge University Press.

Whittington, R. (1996). Strategy as practice. *Long Range Planning*, *29*(5), 731–735.
Whittington, R. (2003). The work of strategizing and organizing: For a practice perspective. *Strategic Organization*, *1*(1), 119–127.
Whittington, R. (2004). Strategy after modernism: Recovering practice. *European Management Review*, *1*(1), 62–68.
Wilson, D. C., & Jarzabkowski, P. (2004). Thinking and acting strategically: New challenges for interrogating strategy. *European Management Review*, *1*(1), 14–20.
Zeleny, M. (1987). Management support systems: Towards integrated knowledge management. *Human Systems Management*, *7*(1), 59–70.
Zeleny, M. (2002). Knowledge of enterprise: Knowledge management or knowledge technology? *International Journal on Information Technology & Decision making*, *1*(2), 181–207.
Zeleny, M. (2008). Strategy and strategic action in the global era: Overcoming the knowing-doing gap. *International Journal of Technology Management*, *43*(1–3), 64–75.

Part IV
Value-Based Approach in Strategic Thinking

... the need to discuss the valuation of diverse capabilities in terms of public priorities is (...) an asset, forcing us to make clear what the value judgments are in a field where value judgments cannot be – and should not be – avoided. Indeed, public participation in these valuational debates – in explicit or implicit forms – is a crucial part of the exercise of democracy and responsible social choice. In matters of public judgment, there is no real escape from the evaluative need for public discussion. The work of public valuation cannot be replaced by some cunningly clever assumption. Some assumptions that give the appearance of working very nicely and smoothly operate through concealing the choice of values and weights in cultivated opaqueness.

Sen, A. K. (1999). *Development as freedom* (p. 110). Oxford: Oxford University Press.

Chapter 17
Creative Evaluations for a Human Sustainable Planning

Luigi Fusco Girard

17.1 Introduction

'Development' can be defined as contrary to 'envelopment'. It is 'release' from constraints and exploitation of existing potentials and energies; it is 'promotion' of quality of life; it is fulfilment of weal for people and society; it is fostering *happiness*.

An 'economicistic' interpretation of development has mixed up means with goals: means have been transformed into goals. Now, after multi-decennial growth of per capita incomes, the *happiness* level shows to be steadily stable along time (Layard, 2005; Frey & Stutzer, 2002). The development of 'hard components' has not been abreast with that of *soft* values. Large economic richness, for example, has been produced together with large ecologic poverty.

When discussing the development/environment binomial, we are generally referring to the *sustainable development* model that, in a richer meaning, can be stated as *human sustainable development*; within this model, evaluation and strategic planning approaches are developed which are strongly integrated, and based on and oriented to the exploitation and promotion of relations, co-relations and inter-relations.

The technology eco-effectiveness approach in urban planning is a necessary condition to human sustainable development; nevertheless it is not sufficient. It is not even sufficient for re-activating a co-evolutionary man/nature relation and therefore for properly managing the natural system. New modes for thinking and developing priorities are needed also based on values, meanings and sense.

Human sustainable development is based on the triad sustainability–resilience–creativity; it finds its foundations in the cultural dimension and its assumptions in *values* (Fig. 17.1).

L. Fusco Girard (✉)
Department of Conservation of Architectural and Environmental Heritage, University of Naples Federico II, Naples 80132, Italy
e-mail: girard@unina.it

M. Cerreta et al. (eds.), *Making Strategies in Spatial Planning*,
Urban and Landscape Perspectives 9, DOI 10.1007/978-90-481-3106-8_17,

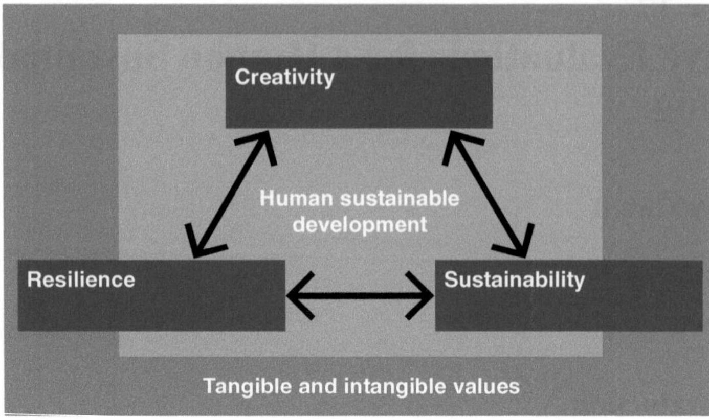

Fig. 17.1 The human sustainable development: three relevant components

What *values*?

It is well known that nowadays values are the *hard* values of economy.

This implies that an extraordinary simplification of choices is made: intangible, immaterial, non-monetary, *soft* values are all neglected.

All the services produced for free by the ecosystems, for example, are neglected although they enable diverse human activities and climate stability (self-regulation of water, carbon and oxygen cycles, etc.), and supply the necessary support to life. Similarly all common goods and social cohesion, community, co-membership and identity values are neglected.

What kind of choices/actions can result from such an omission?

Reducing nature to simple physic/quantitative values can riskily mean using violence towards natural and social environments.

Reflecting the triumph of the economic culture, the current culture separates the ecological from the social system. These systems are, on the contrary, strongly inter-relater and inter-dependent: one system's dynamic deeply affects the dynamic of the other and vice versa.

Indeed, conservation and re-production of tangible (*hard*) and intangible values (*soft*) and related capitals are the key elements for implementing human sustainable development of the city.

Human sustainable development is based on relational values: relations among components of the same subject, among different subjects, among subjects and the ecosystem.

Relational density stimulates coordination ability, develops systemic behaviours and improves communities' resilience (i.e., its ability to keep its identity through changes however chaotic or turbulent).

Much more than the hierarchic model, the network organisational model fosters multiple relations/co-relations which are sources of communication/cooperation and also stimulus for innovative/creative exchange.

New energies are released by networks. Reciprocally, creativity creates new networks. Labour multiplies relations and inter-dependencies: it is a 'bridge' towards others.

This paper focuses on tangible and intangible values and on possible ways to manage these values in order to improve choices in planning, avoiding damages on socio-cultural heritage, environmental heritage, intangible heritage and so on.

Evaluation should be conceived as having a relational nature, as a process based on interpretation and comparison and able to activate and develop relationships (among persons, among persons and their environment, etc.) (Fig. 17.2).

Indeed, interpretation and comparison are key elements of critical thinking and therefore of cultural resilience. Creative evaluations combine *hard* and *soft* values, quantity and quality, specific interests and common good in innovative way. This is the fundamental characteristic of creative actions/projects/plans.

All over the world conflicts are more and more concentrated in places with high symbolic, cultural, social, spiritual and environmental values and characterised by economic pressures and interests. New towers or malls are planned to be built near (or into) city centres or in coastal areas, generating strong conflicts between economic and non-economic interests/values (e.g., Saint Petersburg, Vienna, Prague).

How can we manage *soft* and *hard* values? How can we compare benefits produced by buildings (towers, skyscrapers, malls having high economic value), with costs associated to negative changes in urban landscape? The value of urban landscape is more important than that of a single building, but it is expressed only in *soft* or qualitative terms. Can economic evaluation of urban landscape balance the economic impact of its transformation? Some answers are explored, considering both technical/positivistic and social/constructivist approaches.

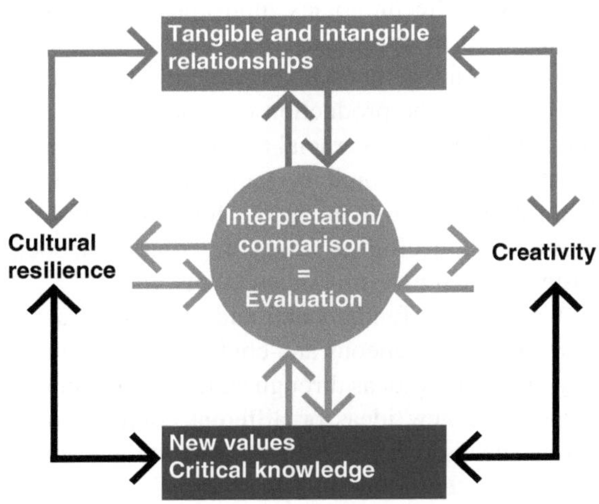

Fig. 17.2 Interaction among relationships and values

17.2 Human Sustainable Development: A Strategic Perspective

Being strictly related to urban and regional planning, evaluation is asked to supply operational answers able to support the growth of urban resilience against many diverse threats (cultural, economic, social, ecological, geographic, environmental, etc.).

Cities like Copenhagen Malmö and London are developing operational programmes and plans to face the consequences of climate change and to foster ecologic resilience. But we need to go beyond the ecologic resilience: we need to foster cultural resilience of cities, that is, the capability to keep its own identity throughout the society changes, however deep and radical they are. Such a resilience is not founded on organisational, technological or economic innovations; it is rather founded in a widespread culture: in the culture's ability of 'keeping together' and being 'gluing' on the basis of shared values. Finally, such a resilience is founded in the ability of fostering and promoting relations.

Within this perspective, cultural resilience is strictly related to creativity: they are together essential to foster human sustainable development.

Facing the distortions of the current urban development and its consequent (often dramatic) impacts, what can be done concretely?

First of all an innovation of the design ability is required, looking for innovative solutions. Creativity needs to be 'released' and a new and creative commitment of regional and urban planning is required.

Creativity cannot be limited to advanced technologies: creativity, for example, is a fundamental resource not only of urban planning but also of everyday choices.

Technological innovations becomes essential to protect environment, to guarantee conditions of *good life to everybody* and to respond to the never-ending human need for *happiness* (Sen, 2002); *happiness* being dependent on the intensity of relations of the individual with himself/herself, among the individual and its social, natural and ecologic environments.

Innovative technologies, more and more ecologically efficient, 'factor 10' or 'factor 4', help in optimising: they enable the reduction of environmental resources consumption together with the multiplication of results; they decouple the economic welfare from the production of negative environmental impacts and of ecological poverty. Therefore a crucial role has to be assigned to the energy sector that, fundamental to the economic system, asks for more and more effective and widespread innovations affecting the modes for producing architectures and cities. Environment is protected, ecological crises are contrasted, social and economic sustainability is fostered when investing on creativity.

Creativity is fundamental and reflects the ability to synthesise among elements however heterogeneous and conflicting they are. It depends on *human capital* (able to generate new ideas through education and knowledge) and on *social capital* (able to generate new ideas or different combinations of ideas – useful to respond to diverse needs – by fostering ideas communication and exchange).

The concentration and the integration of these two forms of capital determines the potential for endogenous development. Fostering creativity/innovation depends on knowledge and does not refer to a single individual but rather to the collective genius.

City planning needs creativity as resource (Albrechts, 2005; Hall, 2004; Healey, 2002; Kunzmann, 2004, 2005; Ache, 2000; Sartorio, 2005; Stein, 2005) to implement human sustainable development through the following steps:

- to build a more attractive 'image' of future, a strategic *vision* founded on the 'spirit of city', that reinvents the role of the city thus fostering inhabitants creativity; the vision expresses the creativity of a city as much as it combines in a synthetic way its identity with changes, old with new values, rationality with emotions, conservation with development, wealth destruction with wealth production and so on;
- to identify the most attractive paths in order to implement this image or vision over the time;
- to implement these paths concretely by means of real projects, with new rules, financial and economic incentives.

Creativity in urban planning has to reinvent city organisation founded on ecological principles and on the interpretation of human beings as a 'multi-dimensional entities'. Creative urban planning should promote a new relation between individual and society. Technological innovations can be useful in this perspective and also guarantee city-nature co-evolution (McHarg, 1969; Soleri, 2006). Urban planning can be able to make city more liveable and sustainable if it is founded on 'places' as spaces where existing relations, bonds and sense of belonging can be multiplied and rebuilt.

Urban planning is paying growing attention to new technologies. In the past, technologies have shaped the physical space of the city. Nowadays new technologies are first of all related with issues like efficient and renewable energy, green chemical, nanotechnologies and so on: these new technologies are becoming the inspiring principle of new urban morphology (Nijkamp, Button, Baycan Levent, & Batey, 2008).

Urban planning challenged by human sustainable development perspective should rebuild human/social ecology: social bonds, sense of community, social capital and cultural resilience.

The new urban planning is focused on production and regeneration of public spaces, as 'places', specific areas of identity, social relationships and life. Indeed, places are spaces characterised by an extraordinary diversity (among forms, typologies, morphologies, cultures, traditions, etc.). In places strong relationships exist among independences of use values, instrumental values and market values.

The relations existing within an individual, among individuals and among individuals and the ecosystem are deeply affected by the character of physical spaces. Each space is expression of a specific 'atmosphere' able to communicate and transfer values: spaces 'talk' to people who live there (de Botton, 2008; Ruskin, 1849).

'Places' are particular spaces in which tangible and intangible values are strictly intertwined. They are complex systems in which natural, man-made, human and social capitals are characterised by strong relationships. They fuel connections, relationships, emotions, links among material and immaterial elements, stones and human beings.

A particular relationship between people and stones is maintained in places, thus determining their particular identity: the spirit of places (Norberg Schulz, 1979). This relationship has economic, social and also emotional nature. It is due to the specific combination of different material and immaterial elements, ancient and new architectures, stones and people. Places are spaces of social aggregation, of communication, of community, of multiple values: historic, artistic, cultural and social ones.

They become fundamental elements for urban resilience capacity: city maintains its identity through its places.

The regeneration of cultural heritage in ancient city centres becomes a 'creative' urban initiative, as long as it is carried out by interpreting the spirit of places and transforming it into a new-built environment able to improve economic competitiveness – through the production of new goods/services sold outside the area – to reduce unemployment and poverty, to foster the capacity of better living 'together'.

Integrated and metabolised spirit of places becomes engine of local development.

17.3 Complex Values and Evaluation Processes

17.3.1 Towards 'Good' Evaluation Processes

The achievement of economic, ecological and social objectives into a win-win perspective requires different evaluation processes that go beyond economic and financial goals, to be able to grasp all the concerned *hard* and *soft* values (like landscape, symbolic, environmental values, etc.) (Fusco Girard, Cerreta, De Toro, & Forte, 2007).

A transdisciplinary/interdisciplinary evaluative approach is required. It could reduce the theory/practice gap in the planning domain by improving decision-making, promoting stakeholders' involvement, supporting the implementation of human sustainable development in the cities and, consequently, improving urban governance.

Ex ante, in itinere and ex post evaluations should be proposed to overcome traditional trade-offs and also to identify creative solutions and promote participation of all the stakeholders (Fusco Girard & Nijkamp, 1997; Fusco Girard, 1987) founded on trust. Trust depends on – *inter alia* – 'good', impartial, evaluations by public institutions, and not on formal ones.

Evaluation is able to make economical/social/environmental feasibility explicit by foreseeing, interpreting and comparing the quantitative and qualitative impacts of new actions. By means of evaluation approaches it is possible to develop priorities, assuming multiple, multi-dimensional and conflicting criteria/objectives. This is fundamental for decision-making in a time of crisis, with growingly scarce resources. *New developed solutions* are able to improve current choices (in design, in planning, in managing, in governing) and to increase the existing values.

Governance oriented towards city creativity collects data and information systematically, to improve knowledge for a critical judgment/assessment required in planning. Data, information and knowledge are to be structured in a systemic way, to allow comprehensive evaluations: to compare new actions in their implementation and in satisfying needs.

Thanks to 'good evaluation' processes creative city is able to better interpret its context and the alternative reactions to it: it is able to make comprehensive comparisons of future scenarios, programmes, plans, projects, in order activate reflection and learning mechanisms able to stimulate the identification of new solutions. So the evaluation process can become an *engine of creativity.* City creativity is confirmed by its results, continually assessed, in a process of learning through ex post evaluations (Landry, 2000, 2006). Ex post evaluation of best practices and worst practices gives the possibility to define adequate approaches for ex ante evaluation, able to support the new plan, design and management for the city.

17.3.2 The Evaluation of Intangibles

The evaluation of creative/innovative projects is linked to evaluation of intangibles (uniqueness in the organisational structure, brand, cooperative activity, etc.), that are becoming more and more important in value creation processes.

Evaluation of intangibles is related to all the processes of value creation: they are becoming more and more relevant not only in the creation of individual/social welfare but also in the development of enterprises value.

Many enterprises, even the small and micro ones, are creators of intangible values since they are able to keep specific knowhow and professional abilities along time.

Examples of intangibles are good will, brand, cooperation capacity, knowledge production ability, organisational capacity and so on. Intangibles are growing in importance because they are the foundations of economic development and social welfare.

A complex set of indicators (able to express vitality and identity of a place, the sense of community), and in particular subjective judgement expressed by perception indicators, is needed in order to monitor continuously the effectiveness of coordinated actions in the urban/metropolitan context on *soft*, intangible, immaterial values and their role in the local development (Fusco Girard & Nijkamp, 2009; Nijkamp, 1989). These indicators allow creative management of cultural/environmental heritage site plans.

Evaluation means *forecasting, interpreting and comparing* different actions in relation to specific goals.

We can distinguish different levels of the evaluation process: strategic processes, regulatory master plan and management level (Fusco Girard & Nijkamp, 1997). For example, at strategic level the crucial problem is evaluating the competitive capacity and the attractiveness of an area taking into account its position and characteristics:

the existing accessibility infrastructure, services provided, user service costs, opportunities opened to alternative use choices, the time taken to supply and carry out services, the existing logistic platforms, the performance of ICT infrastructure, the quality of the links, the urban knots and networks the area is connected with. This is the first step to improve competitiveness.

At the level of master plans evaluation mainly refers to land use choices, among various possible alternatives, such as the identification of the best combinations for commercial, industrial, residential uses, green areas, service production, tourism, areas for production, public spaces and private spaces.

Evaluation also refers to different foreseeable *hard* impacts (like economic impacts upon direct and indirect employment, etc.) of a new project to be compared with *soft* ones, related to the existing cultural landscape of areas. Impact assessment reveals whether the new man-made capital is compatible with the 'spirit of place' and whether this 'spirit of place' has been adequately interpreted as an *engine of creativity* able to contribute to local development.

Evaluation refers to urban-architectural choices in a memory/tradition and innovation/modernisation perspective. It highlights the *net* value of economic benefits, which are often over assessed. It is also fundamental for preserving and reconstructing the cultural landscape of urban areas, increasing its values and managing them.

Indeed, during the management phase of spatial planning processes, ex ante and ex post evaluation are means for stimulating and coordinating the choices of many actors on the basis of a comparison between received and lost benefits.

The improvement of governance – achieving a consensus among different actors – asks for a renewed ability to coordinate actions. The identification of a priority is possible even within diverse heterogeneous and conflicting options or criteria, using multi-criteria, multi-group, quantitative and qualitative methods of evaluation, that complement economic/financial/real estate evaluations.

These methods are also 'open' to participation and inter-subjective communication. They allow the laying out of a decision-support system – valid for the transformation of areas – useful in categorising priorities when faced with several alternatives.

Evaluation not only helps in comparing 'given' solutions, above all it is a stimulus for developing new design solutions/alternatives, starting from knowledge derived from good practices.

17.3.3 The Ex Post Evaluation of Good Practices

A significant role is played by ex post evaluations: good practices can be analysed and their relevant components can be learnt thus enabling the creation of new knowledge to promote and develop new experiences.

Ex post evaluation is considered a 're-interpretation' through which new strategies can be elaborated founded on the interaction between cultural resilience, urban metabolism and land sustainable use. When correctly interpreted and evaluated, good practices are examples of processes able to create and develop new values in multi-dimensional (economic, social, ecologic, etc.) spaces.

Good practices are considered technical/specialist knowledge heritage useful to improve urban governance and design/planning processes, and also to promote new diverse modes to live (in) the cities.

Good practices also provide not technical, not bureaucratic, not rigid knowledge; that knowledge is rather flexible, connected to real life and to daily experience, considering the citizens' perception of urban phenomena. If it becomes the reference point for the reduction of the gap between status-quo and the desired reality and enables the heritage of intangible resources to become 'tangible', this knowledge can help to modify the vision of the reality and also to intervene on the current lifestyle and culture.

The transformation of specific knowledge into generalised one introduces a new wider problem: how can new knowledge be produced to be used for change strategies oriented to human sustainable development? In other terms: how can knowledge be produced being the result of trans-disciplinary integration between scientific excellence, social relevance and operational capacity? How can knowledge be produced being really usable by different subjects involved in the promotion of human sustainable development?

The theory/practice gap of sustainability is very often due to the lack of knowledge able to 'connect' science/technology with culture/behaviour.

Knowledge derived from good practices reveals the strict relation between people and territory. Good practices represent special kinds of 'good ending stories' originated in streets, squares, schools and so on. They show how specific problems have been resolved (in this sense they are good ending stories. When they are appropriately evaluated at the institutional or scientific level, they can be communicated to everybody and not only to the owners of scientific knowledge. They should be object for a 'public interpretation', of an evaluation process accessible to all the citizens in order to supply citizen themselves with concrete tools to make critical choices based on simple indicators based on common knowledge.

17.4 Evaluations in Strategic Planning: Towards an Integrated Methodological Approach

17.4.1 Qualitative and Quantitative Evaluations

The relevance of relations in promoting human sustainable development asks for opportune evaluations of the relations themselves. Relations produce use value that is able to produce other values. In particular, inter-subject relations produce economic plus value in goods/services production; at the same time they produce a cultural and social plus value.

The value of these relations can be derived from the 'difference' between the value related to the relations existence and that one related to the lack of relations; this difference being the reference of the complementary value.

In the production sector, the relations enabling cooperation are sources of costs and time reduction and of better products and results: they are sources of value added. This notion can be extended to the multi-dimensional space shaped by inter-dependency relations.

In the social sector, relations oriented to cooperation improve the quality of life, the sense of community membership, the perception of private and public *happiness*.

In the policy sector relations stimulate anti-government self-organization processes.

The evaluation protocol is anyway a common protocol. It consists of comparing the *value* of situations which are analogue, although different, with regard to specific aspects (brand, innovation, etc.). Such a comparison refers not only to the economic, market, use value but also to the non-use value 'computed' trough participation processes.

In the above-mentioned perspective and through specific projects strategic sustainable planning:

- valorises pre-existent resources, connecting them in a relationship of complementarity/synergy;
- promotes the commercialisation of some services/products which are typical of the site;
- stimulates the production of new services, making the fruition of cultural resources richer;
- stimulates the production of new experiences which integrate just the services, linking the fruition with new knowledge (and not only information), emotions, remembrances, suggestions;
- recognises existing values, and produces new ones, both economic and non-economic values; indeed, the attention is posed on the creation of *new values* after the investment and not only on the conservation of existing ones; in particular, it stimulates the coordination of actions of different actors, thus also producing relational/cooperative values;
- it is founded on immaterial capital; the strength/efficacy of this plan is its capacity of eradicating in community culture, in civil culture, in the civic power of existing cooperative civil/social networks.

Strategic plan always contain the identification of all the actions/activities to be undergone, through the deduction of their 'combined' order of priority, on the basis of a careful evaluation of cultural values together with economic values.

Actually, evaluations should 'go beyond' economic values, which overestimate benefits and underestimate social/cultural and environmental costs.

Economic values cannot express the ecological or the social truth; that is the economic value of benefits arising from investments does not reflect either the ecological value or the social/cultural value of lost benefits. Net benefits of development projects should be assessed.

The purely economic evaluation may lead to project choices which are fragile under other points of view: it can lead to the ecological and to the social/cultural un-sustainability.

It is necessary an 'integrated' evaluation, that is a 'complex' evaluation (Fusco Girard, 1987) where the achievement of specific interests, is combined with the achievement of the public interest (that is of the common good).

We want to remind some of the numerous evaluation issues related to strategic plans.

- The evaluation of alternative conservation/development projects (representing possible implementation actions of a strategic plan), in order to develop a priority list of different 'poles' of the envisioned polycentric territorial asset; this evaluation should consider multi-plying effects (direct, indirect and induced impacts upon economic activities, etc.) consequent to different investments (for instance, a tourist harbour becomes a driver for the local economy and 'sustains' it).
- The evaluation of attractiveness capacity of a site consequent to its values/characteristics: this is the evaluation of the 'quality of its space', its vocation potential value to become the catalyst for development.
- The evaluation of 'places'. The challenge, in this case, is to assess the relationships among different components producing a frame of values. Through the evaluation of such elusive, immaterial, intangible aspects, it is possible to show that investing in cultural heritage produces 'revenues' which are economic besides extra-economic (immaterial, symbolic, spiritual, etc.). It is also possible in the evaluation of compatibility between landscape conservation and development, in the construction of new landscapes (for instance, in the case of regeneration of harbour areas, new infrastructures are needed which determine impacts on landscape values, etc.).

Multi-criteria evaluation methods seem to be more appropriate to support the exploration of alternatives in the long-term perspective. They are capable of dealing with multiple dimensions, *soft* data, while supporting interactive strategies and trying to give larger attention to conflicts rising among various stakeholders involved in the decision-making process (Fusco Girard & Nijkamp, 1997, 2004; Nijkamp, Rietveld, & Voogd, 1990).

Qualitative evaluations characterise this step, in which critical thinking is fundamental. Multi-criteria evaluation methods (CIE, Regime, AHP, Electre, etc.) represent useful decision-support tools in the strategic phase due to the lack of information for the decision-makers, to the uncertainty linked to the future perspective and to the diversity of the involved subjects.

The CIE approach (Lichfield, 1996), for example, can combine qualitative and quantitative objects within the general framework of integrated evaluation: an economic, financial, environmental, social and cultural analysis, making the distribution of net benefits among different involved groups explicit, allowing the detection of the socio-economic and socio-environmental effectiveness through a multi-dimensional impact analysis, resolving the problem of the identification of priority *vis a vis* multiple, heterogeneous and conflicting objectives.

The evaluation requires, at strategic level, the construction of a set of qualitative/quantitative indicators for assessing the impacts of each scenario on the existing context (Nijkamp & Giaoutzi, 2006).

In this chapter, for example, we will focus our attention on the assessment of the potential attraction capacity of a place. This could be a useful information for both private investors and public found allocation. Moreover, the evaluation of indicators could represent a tool to manage the results obtained during the process as well as a communication tool.

Various monetary and non-monetary methods can be introduced to assess the complex value of the various configured scenarios/alternatives.

17.4.2 Evaluation Criteria at Strategic Level

Net benefits have to be considered in relation to economic, social and environmental criteria.

For example, Strategic Environmental Assessment should consider not only economic but also environmental and social criteria. At strategic level, they can be summarised as in the following selection (Tables 17.1, 17.2 and 17.3). The indicators include per capita planted surface, the use of renewable sources in the total amount of energy use, new jobs in the year, and also indicators expressing intangible elements, the self-organising capacity of a community through cooperative networks, the 'glue' or inherent value of the site and so on.

Table 17.1 A selection of economic criteria

Economic criteria
Improvement of economic attractiveness and competitiveness
Regeneration capacity of economic activities
Attractiveness capacity for green industrials activities
Attractiveness capacity for creative people
Multi-functional and efficient use of harbour areas
Diversification of the existing economic activity and rise in production
Establishment of new activities (micro-businesses, small companies, medium-sized enterprises)
Localisation of essential specialised services to enterprises
Localisation of "clean" industrial production activities
Development of a flourishing tourist industry
Localisation of services for tourism, culture and leisure
Increase of the attractiveness of harbour areas for financial reinvestment
Increase of market values of areas/spaces
Localisation of innovative research activities
Cooperation networks among enterprises, public institutions and research centres
Improved interconnections of underground, railway and airport networks

Table 17.2 A selection of environmental criteria

Environmental criteria
Conservation, management and increase of green areas (planting and maintenance)
Promotion of green roof and green façade technology
Preservation of biodiversity
Implementation of cycle paths and pedestrian network
Conservation and improvement of landscape quality
Conservation and enhancement of existing cultural heritage
Minimisation of the need to shift from one place to another
Reduced car travel demand
Reduction of motorised traffic generators
Soil decontamination
Air pollution reduction
Water pollution reduction
Noise pollution reduction
Reduction of CO_2 emissions
Recovery of recyclable waste material (plastic, glass, cans, paper, etc.)
Water recycle (rain water recovery, etc.)
Waste reduction
Self-organised waste management
Use of renewable sources
New electric power plants localisation, based on energy innovation (wind energy, photovoltaic, geothermal, etc.)
Cogeneration

Table 17.3 A selection of social criteria

Social criteria
Employment development
Availability of residential areas (at convenient prices)
Availability of commercial areas (at convenient prices)
Availability of tertiary areas (at convenient prices)
Perception of belonging to a specific community
Perception of specific motivation of people/users
Promotion of social security
Upgrading of existing public spaces
Implementation of public spaces
Conservation of elements expressing the area's cultural identity
Integration between workplaces and leisure places
Community infrastructure uses (school, culture, sport, etc.)
Promotion and coordination of public/private spaces
Involvement of the III sector in specific programmes/projects/activities
Density of cooperative and partnership networks
Protection of the "spirit of the place"

Table 17.4 A selection of indicators

Indicators
Square metres of pedestrian and cycle surface/total road surface
Square metres of commercial surface in the area/total square metres
Square metres of contaminated land/Inhabitants
Number of innovative activities in the area
Variations across the time of the number of innovative activities localised in the area
Number of modern eco-compatible buildings/total number of buildings
Number of cooperatives enterprises/total number of enterprises
Number of micro-businesses/total number of enterprises
Number of illegal actions in the area (such as the ones connected to social corruption)/total number of illegal undertakings
Capacity for coordination of the various operators within specific plans/projects
Density of networks among public authorities, enterprises, research centres and university
Level of interpersonal trust
Number of festivals, events, festivities, ceremonies in the year, as expression of the spirit of place, of collective/social memory
Number of connections for each node of the neural network

It is, in fact, necessary to use criteria and indicators able to make objective technical issues together with perceptive and subjective aspects. In particular, we can propose the following list as example (Table 17.4):

It is clear that the economic approach is necessary but not sufficient in evaluation processes at strategic level.

17.4.3 Evaluation Criteria of a 'Place'

Strategic city plan is implemented starting from 'excellence areas'. Some of these areas are 'places'. Places are the entrance points to build a city polycentric model and to define the multi-dimensional profile of a site.

The competition among tourism sites depends on their attractiveness, which can be defined as the capacity to attract new investments, activities and tourists within an area and also the capacity to maintain the pre-existing ones (Coccossis & Nijkamp, 1995). It depends on:

- accessibility (road, railway, harbours, airports, etc.);
- the level of tourism infrastructures/equipments (hotels with different characteristics, etc.) and on their maintenance conditions;
- the intensity of existing cultural/artistic/historical/environmental values;
- integrative services related to pollution reduction (water, air, soil, landscape, etc.) and to the preservation of climate stability (heating and cooling avoiding CO_2 production, etc.);
- the availability of specific integrative services, such as multi-media services for the fruition of artworks, of history (recreations in space and time), able to determine emotional involvement of users; specific integrative services improving the

quality of other services currently supplied; services related to lasting in time characteristics, despite the more and more accelerated change, and determining the site identity;
- the level/quality of other immaterial services: on the civil/social/human quality, which the site identity itself depends on, reflecting into the spirit of places: its autopoietic capacity, its organisational structure; it depends on culture, traditions, know-how, knowledge, social/human quality and on the density of existing social/civil networks.

The specific concentration of such components and their particular combination determine the 'profile' of a site, its quality of life and therefore its attractiveness capacity, with the specific competitive vantage of attractiveness with respect to other sites.

Intervening on some (or all) the above components means improving the attractiveness capacity and therefore the competitiveness of a site. In general, only the first three items are considered in assessing attracting capacity. In this perspective it is relevant to consider the intangible cultural heritage and the attractiveness capacity of a place.

Strategic city plans try to improve the positioning of an area with respect to other sites. They are generally characterised by a key role of culture, conceived as a strength feeding and promoting economic development, both directly (creativity, innovation, etc.) and indirectly, for its capacity to link different social subjects.

We want to stress here the role of intangible heritage in determining the attractiveness of a site.

The convention on the Safeguarding of the Intangible Cultural Heritage (UNESCO, 2003) considers the role of the intangible cultural heritage as a 'glue' factor in bringing human beings closer together and ensuring exchange and understanding among them. This intangible cultural heritage, transmitted from generation to generation, is constantly recreated by communities and groups in response to their environment, their interaction with nature and their history, and provides them with a sense of identity and continuity.

The intangible cultural heritage, manifested in different domains (oral traditions and expressions, including language as a vehicle of intangible cultural heritage; performing arts; social practices, rituals and festive events; knowledge and practices concerning nature and the universe; traditional craftsmanship), has a strong role in determining the specific identity of a site/place: its *spirit*.

The identity can be evaluated through different criteria: territorial identity (the brand of a territory); environmental identity (the sense of places); cultural identity (the spirit of a community; the relationship between people and its physic environment); historic identity (the roots of a community); social identity (the sense of belonging); civic identity (the sense of citizenship).

The spirit of a place is very often the result of an age-old creative process. In a globalised economy, it distinguishes itself as an element of identity, authenticity and uniqueness.

The spirit of the place expresses the structure of interdependency of multi-dimensional elements linked to each other at different scales in a latent order that represents the intrinsic value. It is what survives over time not with standing continuous changes of urban assets: it is the element of continuance in the increasingly accelerated dynamics of the city/territory.

This intrinsic value depends on the relationship between physical elements and people's lifestyle. It is an intangible value capable of determining specific choices, behaviour and actions, because it expresses the link existing between space and people, between the past and the present.

17.5 The Evaluation of the 'Spirit of Place'

The 'spirit of place' can become the engine of evolution, the driving force to build future and also the engine for a new local development: the real matrix of human sustainable development of the city.

The spirit of places is an elusive and ambiguous notion. There is not a rigorous definition of spirit of places. But we can consider some characteristics of these particular spaces. Spirit of places reflects the structure of interdependences among multiple components at different scales, linked in a latent order that should be interpreted with hermeneutic processes.

This order survives to continuous changes of physical assets, as a permanence element in urban dynamics.

Hard values are tangible ones. They reflect specific relationships, correlations, in the space among multiple elements.

Soft values are intangible values. They include inherent value, glue value, etc. Intangible values have shaped our physical environment: they come before tangible ones. They produce added values of a site/space.

Mapping tangible and intangible values is important to make the spirit of places explicit and to better design transformations.

Places express the relationships between past and present, history and today life. Assessing the spirit of places means to be able to evaluate the relationships activating meanings and sense. The spirit of place expresses the significance that certain material factors acquire in shaping people's lives; the relationship between the material and the immaterial. Therefore the spirit of place is the 'inherent' value of a certain area. It is the 'glue' value that stimulates cultural resilience.

A project of transformation can strengthen or contradict the spirit of place. It can take the shape of a legible sign in a certain space; it can be meaningful for the recollection and the sense of direction. It can enhance the relationship between public and private spaces, encouraging a multiple use of the land, in order to respond to real local needs; it can foster a sense of liveability, vitality, the continuity of tradition, a sense of community, the perception of connections/relationships.

A new project may be successful from the economic point of view as producer of positive external effects on businesses and the real-estate market. It may also be in conflict with the spirit of place: it might be generator of new congestion, pollution and so on. It can determine a sense of estrangement, of diversity, of remoteness or refusal. It may happen that the spirit of place is exalted by a new project, but it may be that such a project is unproductive from the economic point of view.

The spirit of place cannot be evaluated in a single dimension but in many ones. It cannot be carried out only on the basis of the willingness to pay, but it is necessary to use also non-monetary indicators and ordinal or nominal rating scales.

The assessment of the spirit of places requires qualitative evaluations of hidden, latent, inherent, intangible values.

It can be assessed through the complex social value approach (Fusco Girard & Nijkamp, 1997; Fusco Girard, 1987).

In places market values are present together with other value signals: use values and use independent values.

They constitute the total economic value (TEV). But it is necessary to underline the limits of the TEV. The TEV of a place is compounded by tangible and intangible values.

Tangible values are the following:

– direct-use value for people using/enjoying directly the place (visitors, tourists, inhabitants, residents, owners, developers, etc.);
– use value for indirect users concerning people living near or far from the place that receive impacts from existing place;
– use value for potential users concerns people living "somewhere", far from the place, but that can use or visit it in the future.

Intangible values are the following:

– non-use existence value, that concerns the place as a resource for future generation;
– non-use intrinsic value: socio-cultural value, symbolic and so on.

Non-use (*soft*) values are the fundamental or the prerequisite for economic (*hard*) values. The more intensive non-use values are, the more use values increase. It can happen that when use values become too intensive, non-use values can decrease.

A rich literature discusses examples of dynamics of cultural heritage's use value (Mason, 2002).

Non-use values (I) can depend on the values/meanings assigned by inhabitants to the heritage of places and new procedures are required to assess people perceptions, preferences and so on.

The comprehensive complex value model that ties *hard* and *soft* values is

$$\text{Social complex value} = (\text{TEV}, \text{I})$$

TEV can be assessed on the basis of three different approaches: market approach, implicit market approach and simulated market approach.

'I' can be assessed through multiple-criteria and multi-group evaluation methods. It is a social construction based on common knowledge that includes hidden or inherent values.

If evaluation means forecasting, interpreting and comparing, the assessment of 'I' stresses the role of meanings, values, sense interpretation by people.

This bottom-up participated evaluation is a process of critical thinking, connected to the ability of perceiving values, even latent. It is a process of value creation, of new values production using inhabitants' knowledge. Inhabitants transform individual memory into collective memory, through dialogue and communication.

The evaluation of 'I' as interpretation takes into account the common feeling and emotions of users/people, so that all senses can be involved. Interpretation enables the addition of new hidden values, making people more and more aware of spaces' attributes.

Interpretation is a *creative process of values production.* It allows the understanding of the relationships between spaces and people (reflected in the way of life, in urban rhythms, social rituals, celebrations, traditions, arts, specialised skills, etc.), and helps to improve them.

It allows the recognition and the development of new priorities and hierarchies.

The set of indicators described below (Table 17.5) gives an operational validity to the complex values. It helps to capture the particular atmosphere of a place/site, its character. They remind traditions, memories and stories to contemporary generations so that they are not dissolved in the current collective amnesia. They promote cultural resilience, because they fix and transmit over time the cultural memory of a site/space/place that becomes the strength to build future.

Table 17.5 Intangible values of cultural heritage

Identity pattern	Category	Basic indicator
Territorial identity	High-quality contemporary architectures	Number/total buildings
The brand of landscape	Listed monuments	Number/cultural heritage
	Contemporary art production	Pro capita investment for budget/year
	Quality typical wine-and-food products	Number of products with mark guaranteeing the quality
	Quality typical handicrafts products	Number
Environmental identity	Cultural landscape	Percentage of city territory bound under landscape plan
The sense of places	Integration between built heritage and urban green spaces	Public green spaces square meters per capita
Cultural identity	Exhibitions	Number/year
The spirit of a community		
	Religious events	Number/year
	Folks events	Number/year
	Theatrical events	Number/year
	Cultural events for young people	Number/total events

Table 17.5 (continued)

Identity pattern	Category	Basic indicator
	Sporting events	Number/year
	Festivals	Number/year
	Local dialects	Yes-no
	Musical traditions	Yes-no
	Popular tradition and customs	Yes-no
	Rituals and ceremonies	Yes-no
	Investments for culture and for arts	Public per-capita investments
	Existing creativity index	Florida index
	Care to the future (long-term strategic plan)	Yes-no
	City rhythms	Holiday days/year
	Scientific research	Number of Patents/year
Historic identity *The roots of a community*	Historic events	Number of celebration/year
	Existing symbols recognised by people as historic reference	Yes-no
Social identity *The sense of belonging*	Local community	Number of associations/10,000 inhabitants
	Local micro-young communities	Number/total associations
	Streets closed to traffic and opened to meeting people	Square meters traffic-free zones/inhabitant
	Fair economy shops	Shops/1,000 inhabitants
	Urban security indicators	Offences against heritage/1,000 inhabitants
		Time bank
Civic identity *The sense of citizenship*	Civil history events	Number/year
	NGO	Number/total associations
	Voluntary association	Number/total associations
	Third sector	Number of employees in third sector/total employees
	Sport associations	Number/total associations
	Participation to building of local collective decisions	Average percentage of voting population
		Participants to Local Agenda 21, Forums, Participatory Budget

Evaluations of intangible values help people to manage, to internalise and metabolise *soft* values. Cultural values internalised and metabolised become more resistant in comparison to economic values.

The importance of evaluation as interpretation is founded on the strengthening of *soft* values and thus improving city resilience by avoiding *soft* values from remaining abstract and making them able to shape concrete experiences activating non-instrumental behaviours.

A new local endogenous development can start only if widespread people culture is able to overcome instrumental rationality. Coordination among actors, players, institutions and people, self-organisation, self-development can be implemented if self-centred economic culture is really overcome. Shared cultural symbolic values become educational and creative engines.

Evaluation as interpretation, involving people and not only experts (Copeland & Delmaire, 2004; Herbert, Prentice, & Thomas, 1989; Prentice & Cunnell, 1997), can contribute to face educational challenge (Fusco Girard, 2007).

Many dialogic and collaborative tools are required for the assessment of 'I', involving resident/occupiers on site and off site, developers, managers, conservation specialists, local workers, visitors, tourists, urban services operators, city government representatives and national institutions.

Each participant is asked to assign one or more historic, social, civic, cultural and so on, value to different places.

The number of preferences assigned to each site, expressed in percentage, reflects the perceived complex values of places (Vv.Aa, 2009). For example, for the three urban places A, B and C, the number of time that a value has been recognised

Table 17.6 Multi-dimensional profile of three urban places

	Identity						
Places	Territorial	Environmental	Cultural	Historic	Social	Civic	Percentage
A	10	18	19	8	24	21	100
B	12	0	48	12	28	0	100
C	28	30	18	0	0	24	100

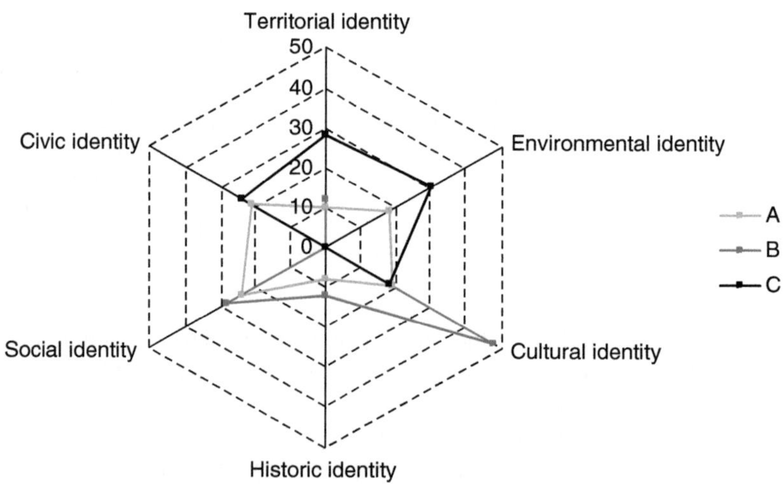

Fig. 17.3 Multi-dimensional profile representation of three urban places

to each site allows the identification of this multi-dimensional profile (Table 17.6, Fig. 17.3):

These multi-dimensional profiles increase if connection, relationships and complementarities are stressed. They give a first assessment of the spirit of places, if they are integrated with other approaches (Nijkamp & Reggiani, 2009).

17.6 Conclusions

The human dimension of development concerns tangible and intangible, material and immaterial, *hard* and *soft* needs and values. It deals with human aspiration to achieve *happiness* that is referable not only to jobs, income or green environments but also (and first of all) to relationships and bonds, i.e. to community construction.

The research about indicators of intangible in a technical/positivistic and constructivist perspective is fundamental for improving evaluations that include subjective experiences and perceptions. The strength of intangibles is on the consensus that they are able to stimulate.

Strategic evaluation depends on tangible and intangible capitals.

Creative evaluations are able to manage *hard* and *soft* values, economic and economic goals/objectives, quantitative and qualitative attributes in a 'creative' combination.

The future of cities will depend on the valorisation of their many differences in the growing standardised general context. Cities will be more and more spaces where creativity is practised.

Creativity depends on interactions/relationships.

Places are spaces of creativity and differences, where tangible and intangible values are strictly intertwined. In particular, priority should be assigned to intangible values.

Evaluation stimulates the production of new solutions aimed at improving the original hypotheses. Through evaluation it is possible to identify a ranking of various alternatives: priorities under multiple, multi-dimensional and conflicting criteria.

The evaluation of the best (and the worst) practices is fundamental for our legacy of knowledge and experience. It is essential for elaborating a vital project which aims at transforming cities into new 'places', and conservation of the spirit of places is a key element for the city cultural resilience.

Through evaluative processes it is possible to interpret the complex scenario of cities and to react to it by new plans, projects or management programmes that are able to transform the spirit of place in a local sustainable development engine.

Through evaluations as interpretations of heritage it is possible to create new values that can improve urban cultural resilience. A city is resilient not only when it has an action plan for mitigation/adaptation to change, but also when each inhabitant is able to transform data into information, information into interpretation, interpretation into critical knowledge and knowledge into everyday actions (Zeleny, 2005).

Creativity and resilience are, most of all, required for the production of *humanity* in the more and more fragmented, uncertain and instable city to promote a *human* sustainable development.

Continuous evaluation and monitoring processes become more and more necessary to select really creative choices and actions for regenerating urban economy and socio/ecological system, to concentrate scarce resources on innovative initiatives and to identify their priorities. Systematic identification of existing best practices in urban economic regeneration, governance, planning and design, conservation of cultural heritage, access to housing and services helps city in understanding its position compared to other experiences, and its strengths and its weakness. Creative and resilient city needs and demands sound evaluations, to develop principles, approaches and effective choices.

References

Ache, P. (2000). Vision and creativity. Challenge for city regions. *Futures*, *32*(5), 435–449.

Albrechts, L. (2005). Creativity as a drive for change. *Planning Theory*, *4*(3), 247–269.

Coccossis, H., & Nijkamp, P. (Eds.). (1995). *Sustainable tourism development*. Aldershot: Ashgate.

Copeland, T., & Delmaire, Y. (2004). *Heritage interpretation in the framework of the European heritage days*. Report on the training course 2003 for European heritage days co-ordinators. Council of Europe, Strasbourg.

de Botton, A. (2008). *Architettura e Felicità*. Milano: Guanda.

Frey, B., & Stutzer, A. (2002). *Happiness and economics: How the economy and institutions affect well-being*. Princeton, NJ: Princeton University Press.

Fusco Girard, L. (1987). *Risorse Architettoniche e Culturali: Valutazioni e Strategie di Conservazione*. Milano: FrancoAngeli.

Fusco Girard, L. (2007). Urban system and strategic planning: Towards a wisdom shaped management. In Y. Shi, D. L. Olson, & A. Stam (Eds.), *Knowledge and wisdom: Advances in multiple criteria decision-making and human systems management* (pp. 316–340). Amsterdam: IOS Press.

Fusco Girard, L., Cerreta, M., De Toro, P., & Forte, F. (2007). The human sustainable city: Values, approaches and evaluative tools. In M. Deakin, G. Mitchell, P. Nijkamp, & R. Vreeker (Eds.), *Sustainable urban development. Volume 2. The environmental assessment methods* (pp. 65–93). Abingdon: Routledge.

Fusco Girard, L., & Nijkamp, P. (1997). *Le Valutazioni Integrate per lo Sviluppo Sostenibile della Città e del Territorio*. Milano: FrancoAngeli.

Fusco Girard, L., & Nijkamp, P. (2004). *Energia, Bellezza e Partecipazione: la Sfida della Sostenibilità. Valutazioni Integrate tra Conservazione e Sviluppo*. Milano: FrancoAngeli.

Fusco Girard L., & Nijkamp P. (Eds.). (2009). *Cultural tourism and sustainable local development*. Aldershot: Ashgate.

Hall, P. (2004). Creative, culture, knowledge and the city. *Built Environment*, *30*(3), 256–258.

Healey, P. (2002). On creating the "city" as a collective resource. *Urban Studies*, *39*(10), 1777–1792.

Herbert D. T., Prentice R. P., & Thomas C. J. (Eds.). (1989). *Heritage sites: Strategies for marketing and development*. Aldershot: Avebury.

Kunzmann, K. R. (2004). Culture, creativity and spatial planning. *Town Planning Review*, *75*(4), 383–404.

Kunzmann, K. R. (2005). Creativity in planning: A fuzzy concept? *disP*, *162*(2), 5–13.

Landry, C. (2000). *The creative city: A toolkit for urban innovators*. London: Earthscan.

Landry, C. (2006). *The art of city making*. London: Earthscan.
Layard, R. (2005). *Happiness: Lessons from a new science*. London: Allen Lane.
Lichfield, N. (1996). *Community impact evaluation*. London: UCL Press.
Mason, R. (2002). Assessing values in conservation planning: Methodological issues and choices. In M. De La Torre (Ed.), *Assessing values of cultural heritage*. Los Angeles, CA: GCI.
McHarg, I. L. (1969). *Design with nature*. Garden City: The Natural History Press.
Nijkamp, P. (1989) *Quantity and quality: Evaluation indicators for our cultural-architectural heritage*. Serie Research Memoranda 0046, VU University Amsterdam, Faculty of Economics, Business Administration and Econometrics, pp. 1–41.
Nijkamp P., Button K., Baycan Levent T., & Batey P. (Eds.). (2008). *Urban planning*. Cheltenham Glos: Edward Elgar Publishing.
Nijkamp, P., & Giaoutzi, M. (2006). *Tourism and regional development. New pathways*. Aldershot: Ashgate.
Nijkamp, P., & Reggiani, A. (2009). *Complexity and spatial networks: In search of simplicity*. Berlin: Springer.
Nijkamp, P., Rietveld, P., & Voogd, H. (1990). *Multi-criteria evaluation for physical planning*. Amsterdam: Elsevier Science.
Norberg Schulz, C. (1979). *Genius Loci. Paesaggio Ambiente Architettura*. Milano: Mondadori Electa.
Prentice, R., & Cunnell, D. (1997). Response to interpretative media as a basis of multivariate market segmentation for museums and heritage centres: The case example of the people's story, Edinburgh. *Museum Management and Curatorship*, *16*(3), 233–256.
Ruskin, J. (1849). *The seven lamps of architecture*. Italian translation *Le Sette Lampade dell'Architettura* (1982). Milano: Jaca Book.
Sartorio, F. S. (2005). Strategic spatial planning: A historical review of approaches, its recent revival, and an overview of the state of the art. *disP*, *162*(3), 26–40.
Sen, A. (2002). *Rationality and freedom*. Cambridge, MA: Harvard Belknap Press.
Soleri, P. (2006). *Paesaggi Tridimensionali*. Venezia: Marsilio.
Stein, U. (2005). Planning with all your senses. Learning to cooperate on a regional scale. *disP*, *162*(3), 62–69.
UNESCO (United Nations Educational, Scientific and Cultural Organization). (2003, October 17). *Convention for the safeguarding of the intangible cultural heritage*. Paris: UNESCO.
Vv.Aa. (2009) *ISAAC. Integrated e-services for advanced access to heritage in cultural tourist destinations*. Six Framework Programme, Commission of the European Communities, Information Society and Media Directorate-General, http://www.isaac-project.eu .
Zeleny, M. (2005). *Human systems management: HSM, integrating knowledge, management and systems*. London: World Scientific Publishing Company.

Chapter 18
Economic Evaluation: The Contemporary Debate

Giuseppe Munda

18.1 The Rationale for the Use of Money Values in Public Policy Analysis

My main assumption here is the impossibility to deal with the concept of *economic value* (and related economic policy instruments) as an objective, value free category. Indeed, the key question is: *value for what and for whom*? Economic development implies the creation of new assets in terms of physical, social and economic structures. Within a process of *creative destruction* traditional environmental, social and cultural assets derived from a society's common heritage may disappear. For example, if the objective is to reduce the tourist pressure on Venice, one may think of limiting the number of visitors by introducing an entry ticket. The collected money could be used to maintain the city's cultural heritage. However, one could argue that due to the *relative scarcity* of a particular economic good, in this case Venice, people will be willing to pay the price of the ticket anyway. Thus, the economic instrument *entry ticket* will be useful for collecting money, but not for reducing the tourist pressure (by the way, in Venice the problem of overcrowding is still unresolved).

Secondly, can we use money values as a social decision tool for policy evaluation? If the answer is positive, a measurement of social costs and benefits should be made on the basis of the so-called *compensation principle*, usually associated with the names of Hicks (1939) and Kaldor (1939). According to this principle, the social cost of a given event is defined as the sum of money paid as compensation to those who have suffered damage. The level of utility that the damaged had before the event took place should determine the amount of compensation to be paid.

Sometimes, social policies based on principles of compensation and substitution may work, but one should be very cautious in applying such principles as a general guideline. There are allocations without any possibility of transactions in actual or

G. Munda (✉)
Department of Economics and Economic History, University of Barcelona,
Bellaterra (Barcelona) 08193, Spain
e-mail: giuseppe.munda@uab.cat

M. Cerreta et al. (eds.), *Making Strategies in Spatial Planning*,
Urban and Landscape Perspectives 9, DOI 10.1007/978-90-481-3106-8_18,

fictitious markets. Who would be willing to accept compensation for the destruction of the Sagrada Familia, the Statue of Liberty or the Coliseum?

We could argue that the presence of irreversibility and uncertainty urges us to abandon the compensation principle in favour of the *precautionary principle* (as more prudent social conservationist approach). The application of the precautionary principle surely introduces some additional costs but how much would the non-application cost?

The burden could be enormous, as admitted by the European Environment Agency (EEA, 2001). In 2002 *The Economist* (distant from radical environmentalism) suggested, as a possibly *positive* consequence of the accident of *Prestige* (a ship which heavily contaminated the coasts of Galicia in northwest Spain), stiffening of the European legislation on the subject of maritime transports.[1]

There is no doubt that from the viewpoint of society it is ecologically and economically more convenient to apply the precaution principle to prevent disastrous accidents. Of course, this principle implies that the majority of the society (mainly the non-experts), outside the economic system (i.e., outside the market mechanisms), would decide on the *amount* of, for example, cultural or natural capital to be protected. Thus, in the Venice example, the *maximum number of visitors* allowed per day should be clarified, and this can only be done on heuristic grounds, since tourist-carrying capacity can hardly be computed precisely.

In this context, from an economic point of view, the only instrument left is *Cost-Effectiveness*, that is, given a certain *physical* target (e.g., the amount of cultural heritage to be preserved or the amount of contamination to be accepted), it is rational to try to get it by means of the lowest possible use of resources (i.e., at the minimum social cost).

In general two approaches are possible:

1. according to the lowest cost;
2. according to the physical target (e.g., the more monuments preserved, the better).

A discussion on most appropriate approach could lead to the conclusion that improvement of a physical target is worth extra economic cost; or perhaps the opposite conclusion would be reached. In both cases we would have an ordinal ranking of alternatives and *Cost-Effectiveness* would 'fall down' into *Multi-Criteria Evaluation*, that is, two criteria and two different rankings must be explicitly dealt with.

From the above discussion the following conclusion can be drawn: to attach prices to non-market assets (such as most of environmental and cultural ones) gives a positive signal to society and *may* contribute to a more rational use increasing the chances for a better conservation. When one wishes to preserve a monument or a natural area, a fundamental question is: is there any resource which society is willing to assign to meet this objective? When dealing with such questions/issues, the concept of 'Total Economic Value' becomes immediately relevant. Attributing monetary values to, for example, historical heritage implies capturing user (actual, option and bequest) and non-user values (existential, symbolic, etc.). Of course, to compute total economic values has nothing to do with the 'true' or 'correct' value. All monetary valuation attempts will suffer deep technical uncertainties such as:

– which monetary valuation technique has to be used?
– what time horizon has to be considered?
– what social discount rate has to be utilized?

Moreover, one should remember that the market alone may be successful in efficient allocation of resources, but does not give any guarantee for preservation of the cultural or natural heritage at all. Once something is on the market, it can be bought or sold and so the willingness to accept the compensation principle may easily cause the destruction of any asset.

As a first conclusion, we could state that monetary compensation is, without any doubt, the only possible tool when an irreparable and irreversible damage has already occurred. This way, if an accident involving serious contamination occurs – as in the case of Seveso in Italy (1976), of Bhopal in India (1984), of Exxon Valdez in Alaska (1989) or, more recently, of the oil-tanker *Prestige* offshore the coasts of Galicia (2002) – it seems justified and opportune to indemnify the victims of such contamination. But it stays to verify if, in the long run, compensation is an effective tool to prevent enormous social costs, given that it doesn't guarantee the preservation of natural or cultural goods and services. The economic value is different from the environmental or artistic-cultural value. If we had to decide whether to save the Galapagos Islands or the inside sea in Holland, which value should we use? The economic one would favour the inside sea, which, since totally eutrophised, offers an important economic service receiving all the nutrients coming from human activity. The ecological one would obviously favour the Galapagos Islands. Is the choice of the values to be considered as socially predominant a scientific or a socio-political issue?

18.2 The Distributional Issue and the Existence of Multiple Social Values

In many real-world applications it is necessary to place monetary values on non-market goods. Several methodologies have been developed to cope with such estimation requirements. The principal ones are Contingent Valuation, the Travel Cost Method, Hedonic Pricing and the Shadow Project Approach. Among these only Contingent Valuation is universally applicable. The aim of Contingent Valuation is to elicit valuations (or 'bids'), which are close to those that would be revealed if an actual market existed. Respondents say that they would be willing to pay or willing to accept compensation if a market existed for the good in question. In order to determine the value of intangible goods and services, economists try to identify how much people would be willing to pay (willingness to pay, WTP) for these goods in artificial markets. Alternatively, the respondents could be asked to express their willingness to accept (WTA) compensation.

The quality of results in this method depends on how well informed people are. However, the problem with this method is that respondents may answer 'strategically'. For example, if they think their response may increase the probability of implementing a project they desire, they may state a higher value than their true value (*free rider problem*). In order to avoid free rider behaviour people should really pay the amount of money they indicate; unfortunately in this case, WTP depends upon the *ability to pay*, and thus projects which benefit higher income groups might be considered to be the best. Furthermore, society as a whole may have values that deviate from aggregated individual values. Society has a much longer life expectancy than individuals; thus the value society attaches to, for example, natural resources, is likely to deviate from individual values. Hence the simple summation of individual preferences may imply the extinction of species and ecosystems. This implies that public policy *cannot be merely based upon the aggregation of individual values, and estimation of willingness to pay at any particular point of time*. Thus, it is worth remembering that economic values depend on inter-generational and intra-generational inequalities in the distribution of the burdens of social costs (e.g., pollution) and in the access to useful resources.

Externalities can then be seen as 'cost-shifting'. In general, if the affected people are poor (or even not yet born), the cost of the internalisation of the externality will be low. This explains why a lot of multinationals locate particularly dangerous production plants in the developing countries: in case of accidents they have to pay monetary compensations which are much lower than in the western countries. The accident at the chemical plant of the Union Carbide in Bhopal, India, in 1984, is a sad example. Obviously, the institutional and juridical context is fundamental. In the case of oil contamination caused by Texaco in Ecuador (which seriously affected human health), the fundamental issue was deciding on whether the trial should be held in the United States or in Ecuador. Texaco insisted on Ecuador.

Accepting low values for a negative externality that provokes an impact on poor community is a 'political decision', far from being ethically neutral. Some years ago, an internal document of the World Bank, subsequently made public, suggested that toxic waste should be located in Africa, since the cost of compensation was extremely low and therefore such solution has to be considered as the most efficient one. One should note that the issue of *value free* science is a *key* issue for real-world policy and not just a philosophical debate. For example, David Pearce claimed that his work for the Intergovernmental Panel on Climate Change (IPCC), where lives of people in rich countries are valued up to 15 times higher than those in poor countries, is a matter of scientific correctness versus political correctness.[2] Is it really a matter of value free scientific correctness to use valuations based on assessments of a community's *willingness and ability to pay* to avoid risks of death?

One has to note that the issue is not maintaining that a human life has infinite value; for example, a reduction in road accidents can be secured at some cost, but society is unlikely to devote the whole of the national income to this end. The point is that often this valuation is made *implicitly* and presented as a technical issue, when in fact it is a political one.

Monetary valuation methods are based on phenomena such as consumer's surpluses, market failures and demand curves, which are just a partial point of view, because they are connected with one institution only: *the market*. From a social point of view, issues connected with actions outside of markets and behaviour of people different from the class of consumers should also be taken into account. The EU *White Paper on Governance* (CEC, 2001), where principles such as *transparency*, *participation* and *accountability* are emphasised, goes in this direction.

It has to be reiterated that the point is not to be against giving economic value to natural resources, to environmental sinks, to natural spaces or to cultural heritage. A location may be valuable for its biodiversity (measured in richness of species or genetic variety), and also as a landscape, and also have economic value (measured by differential rent, and also by the Travel Cost Method, or Contingent Valuation). These are different types of value. *The point is that it is misleading to take social decisions based on only one type of value.*

18.3 Implications for Planning

The world is characterised by deep *complexity*. This obvious observation has important implications on the manner in which policy problems are represented and decision-making is framed. Each representation of a complex system reflects only a sub-set of the possible representations of it.

A consequence of these deep indeterminacies is that in any policy problem, one has to choose an operational definition of 'value' in spite of the fact that social actors with different interests, cultural identities and goals have different definitions of value. That is, to reach a ranking of policy options, it is necessary to decide *what is important* for different social actors as well as *what is relevant* for the representation of the real-world entity described in the model. It may well be that in the process of assessing the cost of cross-boundary transactions in securities, and the possible benefit of regulating them, the relevant actors may accept that an increased volume of transactions and an associated medium-term GDP increase constitute 'value'. What constitutes 'value' and who the stakeholders are in the case of Venice are more difficult to chart. In particular the assessors should consider to what extent the proposed values correspond to the relevant constituency and try to avoid omission of relevant values because this may lead to polarisation of the debate.

One should note that the representation of a real-world system depends on very strong assumptions about (1) the *purpose* of this construction, for example, to evaluate the sustainability of a given city, (2) the *scale* of analysis, for example, a block inside a city, the administrative unit constituting a municipality or the whole metropolitan area and (3) the set of *dimensions*, *objectives* and *criteria* used for the evaluation process. A reductionist approach for building a descriptive model can be defined as the use of just *one measurable indicator* (e.g., the monetary city product per person), *one dimension* (e.g., economic), *one scale of analysis* (e.g., the commune), *one objective* (e.g., the maximisation of economic efficiency) and *one time horizon*. Thus, instead of focusing on 'missing markets' as causes of allocative disgraces, or trying to explain economic values by means of energy or other common rod measures (clearly a nonsense from an economic point of view), we should focus on the creative power that missing markets have, because they push us away from commensurability (i.e., a reductionist approach), towards a Social Multi-Criteria Evaluation (SMCE) of evolving realities.[3]

Public policy analysis should deal not only with the merely measurable and contrastable dimensions of the simple parts of the system under study, but also with the higher dimensions of the system, i.e. those dimensions in which power relations, hidden interests, social participation, cultural constraints and other 'soft' values become relevant and unavoidable variables that heavily, but not deterministically, affect the possible outcomes of the strategies to be adopted.

Any mathematical model, although legitimate in its own terms, cannot be sufficient for a complete analysis of the reflexive properties of a real-world problem. These reflexive properties include the human dimensions of, for example, the ecological change and the transformations of human perceptions along the way. The *learning process* that takes place while analysing the issue and defining policies will itself influence perceptions and alter significantly the decisional space in which alternative strategies are chosen. At the other end, *institutional and cultural representations* of the same system, also legitimate, are on their own insufficient for specifying what should be done in practice in any particular case.

The various dimensions are not totally disjointed; thus the institutional perspective can be a basis for the study of the social relations of the scientific processes. To take any particular dimension as the true, real or total picture amounts to *reductionism*, whether physical or sociological.

As a consequence, any attempt to fit the real world in a closed model leads to simplification, which violates reality. In most cases the marginalised dimensions are the reflexive properties of the systems. These characterise the problem in a fundamental way but are hardly identifiable and measurable.

In general, these concerns have not been considered very relevant by scientific research in the past when time was considered an infinite resource. On the other hand, the new nature of the problems faced in this third millennium (e.g., mad cow or genetically modified organisms) implies that very often when deciding on problems that may have long-term consequences we are confronting situations where facts are uncertain, values in dispute, stakes high and decisions urgent (Funtowicz & Ravetz, 1991, 1994). In this case, scientists cannot provide any useful input without interacting with the rest of the society, and the rest of the society cannot perform any sound decision-making without interacting with the scientists. That is, the question of 'how to improve the quality of a policy process' must be put, quite quickly, on the agenda of 'scientists', 'decision-makers' and, indeed, the whole society. This extension of the 'peer community' is essential for maintaining the quality of the process of decision-making when dealing with reflexive complex systems.

18.4 Conclusion

Various authors claim that modern public economic policy needs to expand its empirical relevance by introducing more and more realistic (and of course more complex) assumptions in its models. According to complexity theory at least, three different types of uncertainty exist: epistemological, scientific and technical (Giampietro, 2003; Munda, 2004).

To sum up, a system is complex when the relevant aspects of a particular problem cannot be captured by using a single perspective. To make things more difficult, human systems are *reflexive* complex systems. Reflexive systems have two peculiar properties: *awareness* and *purpose*, which imply an additional 'jump' in describing complexity. In fact, the presence of self-consciousness and purposes (*reflexivity*) means that these systems can continuously add new relevant qualities/attributes that should be considered when explaining and describing their behaviour (i.e., human systems are learning systems). One important feature of reflexivity is that the human representation of a given *policy problem* necessarily reflects perceptions, values and interests of those structuring the problem. Since in this case the source of uncertainty is mainly social in nature, we can call it *epistemological uncertainty*. Monetary valuation methods are based on phenomena such as consumer's surpluses, market failures and demand curves, which are just a partial point of view, because they are connected with only one category of institutions: *markets*. From a social point of view, issues connected with actions outside of markets and behaviour of people different from the class of consumers should also be taken into account. In this context, one of the most interesting research directions in contemporary economics is the attempt of taking into account political constraints, interest groups and collusion effects explicitly (Laffont, 2000, 2002; van Winden, 1999); as a consequence, transparency becomes an essential feature of public policies (Stiglitz, 2002).

The existence of *different levels and scales* at which a hierarchical system can be analysed implies the unavoidable existence of non-equivalent descriptions of it. As discussed by Giampietro (2003), even a simple 'objective' description of a geographical orientation is impossible without taking an arbitrary subjective decision on the system scale considered relevant. In fact, the same geographical place, for example, in the United States, may be considered to be in the north, south, east or west according to the scale chosen as a reference point (the whole United States, a single state, etc.).[4] Since in this case the source of uncertainty is 'more objective' in nature, we can call it *scientific uncertainty*. A well-known approach for dealing with this type of uncertainty in policy-making is the precautionary principle (see, e.g., Gollier & Treich, 2003).

A consequence of these deep uncertainties and indeterminacies is that in any policy problem, one has to choose an operational definition of 'value' in spite of the fact that social actors with different interests, cultural identities and goals have different definitions of 'value'. In empirical evaluations of public projects and public-provided goods, Multi-Criteria Decision theory seems to be an adequate policy tool since it allows taking into account a wide variety of evaluation criteria (e.g., environmental impact, distributional equity, etc.), and not simply profit maximisation, as most private economic agents would do. This implies that to reach a ranking of policy options, there is a previous need for deciding about *what is important* for different social actors as well as *what is relevant* for the representation of the real-world entity described in the model.

Social Multi-Criteria Evaluation has been explicitly developed for tackling such epistemological and scientific uncertainties (Munda, 2004, 2008). SMCE puts emphasis on the transparency issue, the main idea being that results of an evaluation exercise depend on the way a given policy problem is structured. Thus the assumptions used, the ethical positions taken and the interests and values considered have to be made clear. In this framework, mathematical models still play a very important role: the one of guaranteeing consistency between assumptions used and

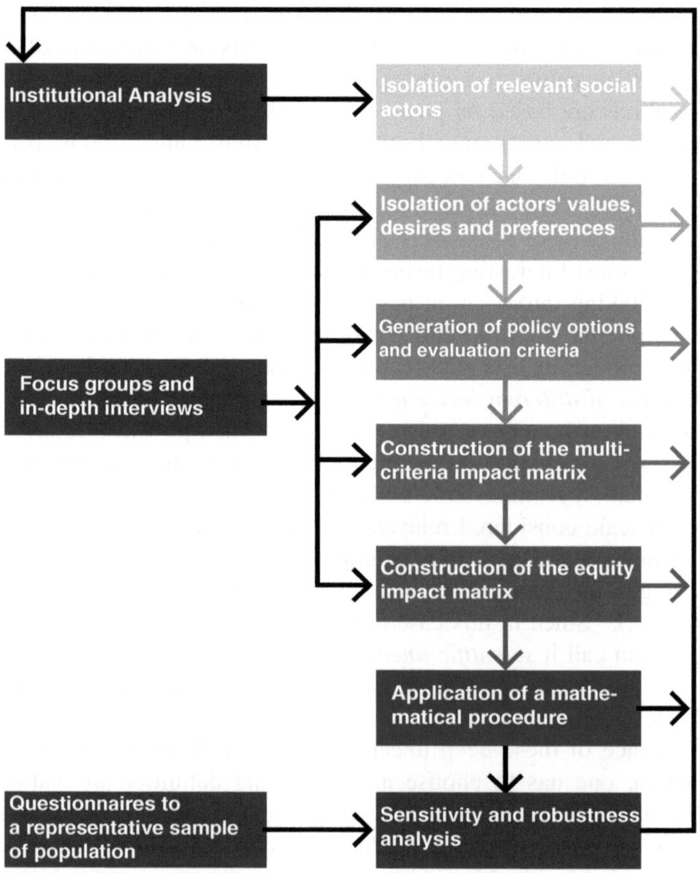

Fig. 18.1 The ideal problem structuring in Social Multi-Criteria Evaluation (Munda, 2005, p. 975)

results obtained. This implies taking into account *technical uncertainties* properly (i.e., those ones that can be simulated by means of mathematical tools such as probabilities, fuzzy sets and sensitivity analysis) (Dubois, Prade, & Sabbadin, 2001; Markowitz, 1989; Saltelli, Tarantola, Campolongo, & Ratto, 2004).

In empirical applications, one of SMCE's main objectives is to avoid the pitfalls of the technocratic approach by applying different methods of sociological research. For example, 'Institutional Analysis', performed mainly on historical, legislative and administrative documents, can produce a map of the relevant social actors. By means of focus groups it is possible to form an idea of people's desires and then to develop a set of policy options and evaluation criteria. The main limitations of focus group techniques are that they are not supposed to be a representative sample of the population and that sometimes people are not willing to participate or to state publicly what they really think (above all in small towns and villages). For this reason anonymous questionnaires and personal interviews are an essential part of the participatory process.

One should note that policy evaluation is not a one-shot activity.

On the contrary, it evolves as a *learning process* which is usually highly dynamic, so that judgements regarding the political relevance of elements, alternatives or impacts may undergo sudden changes. Hence a policy analysis must be flexible and adaptive in nature. This is why evaluation processes have a *cyclical nature*.

By this is meant the possible adaptation and modification of elements of the evaluation process due to continuous feedback loops among the various steps and consultations among the actors involved (Fig. 18.1).

Of course, the steps of the process are not rigidly set out. On the contrary, flexibility in real-world situations is one of the main advantages of SMCE.[5]

Notes

1. See *The Economist*, 23–29 November, 2002, p. 79.
2. See *New Scientist*, 19 August, 1995.
3. "There is great pressure for research into techniques to make larger ranges of social value commensurable. Some of the effort should rather be devoted to learning – or learning again, perhaps – how to think intelligently about conflicts of value which are incommensurable" (Williams, 1972, p. 103). A call for dealing explicitly with incommensurability can also be found in Arrow (1997) and in Martinez-Alier, Munda and O'Neill (1998).
4. These multiple-identity/multiple-scale systems can be defined as 'Learning Holarchies'. A 'holon' is a whole made of smaller parts (e.g., a human being made of organs, tissues, cells and atoms), and at the same time it forms a part of a larger whole (an individual human being is a part of a household, a community, a country and the global economy) (Koestler, 1969).
5. See, e.g., Vargas-Isaza (2004), for an application of SMCE in Colombia, where there was an extreme situation involving social actors belonging to various informal armies (the so-called actor armado); Martí (2005), who conducted a study with indigenous communities in Peru; or Sittaro (in this book), who applies SMCE in the context of indigenous communities in the Amazonian region of Ecuador.

Acknowledgements I wish to thank Andrea Saltelli for his precious comments on previous drafts of this chapter. The usual disclaimer applies.

References

Arrow, K. (1997). Invaluable goods. *Journal of Economic Literature, 35*(2), 757–763.

CEC (Commission of the European Communities). (2001). *European governance: A white paper*. Bruxelles: European Commission.

Dubois, D., Prade, H., & Sabbadin, R. (2001). Decision-theoretic foundations of qualitative possibility theory. *European Journal of Operational Research, 128*(3), 459–478.

EEA (European Environmental Agency) (2001). *Late lessons from early warnings: The precautionary principle 1896–2000*. Environmental Issue Report 22. Copenhagen: European Environmental Agency.

Funtowicz, S. O., & Ravetz, J. R. (1991). A new scientific methodology for global environmental issues. In Costanza R. (Ed.), *Ecological economics* (pp. 137–152). New York: Columbia University Press.

Funtowicz, S. O., & Ravetz, J. R. (1994). The worth of a songbird: Ecological economics as a post-normal science. *Ecological Economics, 10*(3), 197–207.

Giampietro, M. (2003). *Multi-scale integrated analysis of agroecosystems*. New York: CRC Press.

Gollier, C., & Treich, N. (2003). Decision-making under scientific uncertainty: The economics of the precautionary principle. *The Journal of Risk and Uncertainty, 27*(1), 77–103.

Hicks, J. (1939). The foundations of welfare economics. *Economic Journal, 49*(196), 696–712.

Kaldor, N. (1939). Welfare propositions in economics and interpersonal comparisons of utility. *Economic Journal, 49*(195), 549–552.

Koestler, A. (1969). Beyond atomism and holism: The concept of the holon. In A. Koestler & J. R. Smythies (Eds.), *Beyond reductionism* (pp. 192–232). London: Hutchinson.

Laffont, J. J. (2000). *Incentives and political economy*. Oxford: Oxford University Press.

Laffont, J. J. (2002). Public economics yesterday, today and tomorrow. *Journal of Public Economics, 86*(3), 327–334.

Markowitz, H. M. (1989). *Mean-variance analysis in Portfolio choice and capital markets*. Oxford: Basil-Blackwell.

Martí, N. (2005). *La Multidimensionalidad de los Sistemas Locales de Alimentación en los Andes Peruanos: los Chalayplasa del Valle de Lares (Cusco)*. Ph.D. Thesis, Doctoral Programme in Environmental Sciences. Barcelona: Universitat Autonoma de Barcelona.

Martinez-Alier, J., Munda, G., & O'Neill, J. (1998). Weak comparability of values as a foundation for ecological economics. *Ecological Economics, 26*(3), 277–286.

Munda, G. (2004). Social multi-criteria evaluation (SMCE): Methodological foundations and operational consequences. *European Journal of Operational Research, 158*(3), 662–677.

Munda, G. (2005). Multi-criteria decision analysis and sustainable development. In J. Figueira, S. Greco, & M. Ehrgott (Eds.), *Multiple-criteria decision analysis: State of the art surveys* (pp. 953–986). New York: Springer.

Munda, G. (2008). *Social multi-criteria evaluation for a sustainable economy*. New York: Springer.

Saltelli, A., Tarantola, S., Campolongo, F., & Ratto, M. (2004). *Sensitivity analysis in practice: A guide to assessing scientific models*. New York: Wiley.

Stiglitz, J. E. (2002). New perspectives on public finance: Recent achievements and future challenges. *Journal of Public Economics, 86*(3), 341–360.

van Winden, F. (1999). On the economic theory of interest groups: Towards a group frame of reference in political economics. *Public Choice, 100*(1–2), 1–29.

Vargas Isaza, O. L. (2004). *La Evaluación Multicriterio Social y su Potencial en la Gestión Forestal de Colombia*. Ph.D. Thesis. Doctoral Programme in Environmental Sciences. Barcelona: Universitat Autonoma de Barcelona.

Williams, B. (1972). *Morality*. Cambridge: Cambridge University Press.

Chapter 19
Social Multi-Criteria Evaluation Applied: A Community Planning Experience

Federico Sittaro

19.1 Introduction

The necessity of the international NGO UCODEP[1] to re-define its role in the communities settled inside the *Reserva de Producción Faunistica Cuyabeno*, Ecuador, offered the opportunity to face a quite exotic planning exercise, within the framework of a research project granted by the European Union.

The NGO had a certain amount of financial resource available, and no predefined field of intervention, which made it, was different from other similar situations. Usually a budget is assigned/allocated to projects, that is, to health, sanitation, education or other specific fields.

How to allocate such resources? What kind of interventions to promote? How to select interventions that, at the same time, could fulfill community's aspirations, the protected area status and the NGO's mission? For example, is installation of latrines more (or less) appropriate than a workshop on organic pesticide?

It was finally agreed that the Multi-Criteria Analysis (MCA) approach could be used by means of its most socio-political interpretation: the Social Multi-Criteria Evaluation (SMCE) (Munda, 2004).

The challenge was extremely stimulating, above all because it concerned a region that was, and still is, facing a dramatic clash between environmental conservation and economic expansion. This part of the Amazonian forest is both a world's mega-diverse hot spot and the location of the nation's greatest petroleum reserves.

This chapter is divided into two parts. The first part describes the case study: the context, the specific problem and the steps of the process used to solve it. Field activities and the results obtained are also presented in this part. The second part explores interactions between methodological choices and their impacts on a real-world case. Because of its dramatic value, the case offers general reflections easily applicable to other planning experiences.

F. Sittaro (✉)
Médecins Sans Frontières, Brussels 1090, Belgium
e-mail: federico.sittaro@gmail.com

M. Cerreta et al. (eds.), *Making Strategies in Spatial Planning*,
Urban and Landscape Perspectives 9, DOI 10.1007/978-90-481-3106-8_19,

19.2 The Cuyabeno Reserve Case Study

In July 1979,[2] the Ecuadorian Ministry of Breeding and Agriculture (later, Ministry of Environment) declared the Cuyabeno Reserve, a total area of over 600,000 hectares, to be part of the national system of protected areas. Located in North-Eastern Ecuador, it is one of the most ecologically diverse places in the world, more than 470 species of trees, and a habitat of 514 species of birds, 117 species of mammals and at least 176 species of amphibians and reptiles (see *Plan De Manejo De La Reserva De Produccion Faunistica Cuyabeno*, 1993).

Despite its exceptional ecological value, recent investigations show that deforestation, soil erosion and ecological fragmentation are concrete threats. Satellite images taken between 1986 and 2000 revealed a loss of 20,000 hectares of primary forest (Tapia-García, 2006), constituting 19% of the original asset of the Reserve's land cover.

In the late 1960s, this part of the Amazonian region started to gain attention due to the discovery of crude oil reserves. The surging petroleum industry attracted massive migration to the region (Araya & Peters, 1999).[3] Today national and international extraction companies are covering almost the entire Ecuadorian Amazonian basin.

This development imposes dramatic changes to the livelihood of indigenous populations, the traditional inhabitants of that area. The low-income populations are experiencing, on the one hand, a better access to the labour market, and thus to different opportunities to improve their existence, and, on the other hand, the heaviest environmental impact.[4]

The national government[5] allocated the ancestral territories to the indigenous communities settled within the reserve's borders thus giving a primary role to indigenous settlements. For example, in the area of the Cuyabeno Reserve a form of co-management between the Ministry of Environment and the communities has been established. This innovative land governance involves ecological zoning, reciprocal monitoring and surveillance between ministry rangers and the community's members, controlled resource exploitation and other norms.[6]

The actual population within the borders of the Cuyabeno Reserve is now almost 800 people, shaping seven communities, characterised by different histories, cultures, languages and, in general, livelihoods.[7]

The communities' sites are extremely isolated: from 5 h up to 2 days by the river. This is a major constraint to the establishment of any productive activity.

Furthermore, basic services provided by governmental institutions, such as education, health care and infrastructure, are much weaker and rarer than the country average (Bustamante & Jarrín, 2005).

In recent years these communities have been dealing with a great amount of requests: permissions for petroleum investigation surveys, land and services concessions for tourism agencies, endorsements for academic or media works and many proposals of collaboration with NGOs and foundations.

As mentioned, the legal status of the land owners gives to the reserve's indigenous communities the role of no-excludable players in the negotiation tables created for the definitions of such activities and services. However, the unequal distribution of power, which distinguishes these negotiations, does not guarantee an equitable planning of the communities' own development.

This case study refers to the period when UCODEP, an international NGO leading an integrated project promoting institutional assistance in the area, worked on revising its activities with the Cuyabeno Reserve's communities.

The budget available allowed only a single intervention in each community. The discussion point was: which intervention and how to select the most appropriate? The Cuyabeno Reserve case study represents an attempt to answer to this question.

The following was agreed on at the very beginning:

1. the assessment must cover a full range of services and activities to be developed in synergy by the NGO, the communities, the Ministry of Environment officers and the other institutions involved (i.e., the Ministry of Health in case of a sanitary activity); furthermore, activities should be in compliance with the protected area status;
2. decisions should be based not only on a diffuse knowledge of the communities' perspectives of their own development but also on realistic analysis of the available resources attempting to ensure the highest degree of intervention's sustainability;
3. the assessment must actively involve multiple social actors; the decision-making process should be transparent and traceable, even to actors not directly involved in the assessment.

19.3 The Process Logic

The inclusion of different voices speaking at the different hierarchical levels of the system and thus the integration of each party's knowledge into the strategic assessment were immediately recognised as crucial aspects of the planning process.

Participation was identified as a critical precondition for the accomplishment of the overall goal. A compromise among theoretical requirements, logistics and contextual conditions was requested.

Moreover, the case study represents an arena for the clash of contradictory values, belonging to the different actors involved: formal experts (biologists, farming technical assistants, tourism experts, etc.), indigenous people, governmental functionaries, local rangers, tourist operators and other NGO agents.

The SMCE approach was considered the most appropriate methodological framework (Munda, 2004).[8] Following SMCE process structure, the strategy outlined below was adopted (Fig. 19.1).

The first phase is called Institutional Analysis (IA); it defines the social actors within the problem's boundaries. Their role, interactions and, above all, perceptions of development are depicted here. IA is focused on the definition of the decision-making dynamics, the legal, political and administrative structures and processes

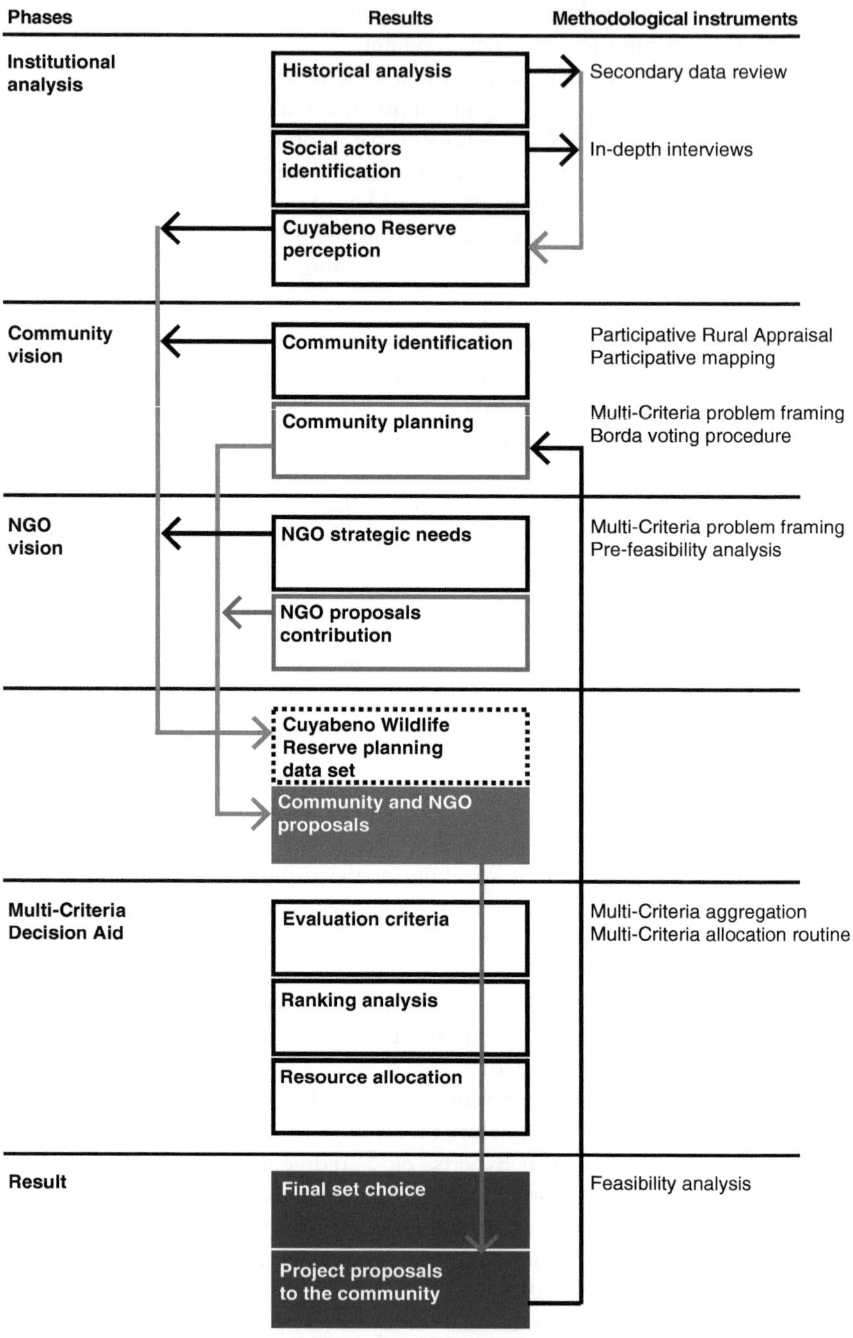

Fig. 19.1 Process description: phases, results and methodological instruments

through which decisions are made with respect to public policy (Funtowicz, De Marchi, Lo Cascio, & Munda, 1998).

The second phase is finalised to put up two pieces of information: a data set of the Cuyabeno Reserve with a special focus on the community's interaction with their own territories and a proposal set to be submitted to the following evaluative phase.

Throughout the third phase the process moves forward to the evaluation of alternatives and the exploration of results. This task is to be considered in a cyclic way. The feedback from the social actors involved during the previous phases is included; it makes it possible to evaluate all decision-making process in real time.

19.3.1 Institutional Analysis

The Section 19.2, sketching the history of the Cuyabeno Reserve, has already presented some of the results obtained during the IA phase.

Semi-structured interviews, focus group and secondary data review permitted to collect such information. Due to their prominent role, a major effort was required to better define the two mayor players involved: the seven communities and the NGO structure.

This peculiar effort and the way it has been included into the wider framework of SMCE is now presented.

19.3.2 The Community Vision

It was proposed to adopt Participative Rural Appraisal techniques (PRA)[9] (Chambers, 1992, 1994, 2002) as a leverage to locally conduct the participative process. These procedures focus on local societal structure and livelihoods, not imposing, but asking, waiting for an answer, giving everybody time and chance to talk, explicitly seeking minority positions and, above all, not advising.

Such approach was than translated into two main phases:

- the 'community identification', where the planning knowledge is constructed and exchanged;
- the 'planning session', where a set of alternatives is assembled to represent the community vision of its own development.

Creating a detailed picture of the interaction between the community and its territory was the main target of the 'community identification' phase. This phase allowed inclusion of as many people as possible into the process.[10] It was agreed that participants would work in groups divided by gender and distributed in time over the first 2 days.[11]

An usual criterion of competence was used: men were questioned about hunting practices, grazing, farming and common land management practices, while women were involved in the reconstruction of the community livelihood and activities (Fig. 19.2).

Participative mapping	Socio-economic appraisal
Community land sketch	Annual and daily calendar
Identification of access ways to/from the community	Institution diagramming
Identification of hunting trails and fishing spots	Disease assessment
Identification of petroleum extraction site and geo-seismic trekking path	Socio-economic assessment: community sketch, family units description and categorisation of belongings
Identification of tourism lodges inside the community land	Transect walks
Tourist trekking paths	Hunting trails: monitoring of animal occurrence
	Farming plot walks: identification of type and state of farming activities
	Hunting trails: identification of potential tourism attraction and weaknesses. Monitoring of animal occurrence

Fig. 19.2 Activities performed in the communities

For male activities the cartographic approach proved to be very effective. Maps, sketches, superimposition of transparent sheets to maps or aerial photos became very powerful icebreakers and a valuable 'blackboard' to record a wide range of issues.

If mapping was naturally focused on the relation with the territory, female groups were centred on the family unit approach (Moser, 1993). All activities were constituted by both 'desk' work and the direct observation of relevant spots in the surroundings.

The more authentic planning moment came in the 'planning session' phase conducted in the communities. Expected results included:

- a detailed list of proposals;
- a set of criteria deemed relevant to define communities' 'own development vision'.

Two instruments were used in this step, that is, the Multi-Criteria Problem Structuring and the Borda Matrixes.[12] All community members were requested to take part in plenary meetings.

The meetings were often held at their homes and food was provided for all the families. Typically, there were three meetings. They were intended to touch on all major dimensions of community's life, for example, tourism services and infrastructure, health and sanitation, productive activity or educational issues. During the first meeting the framework was often similar to brainstorming sessions.

The participants were committed to produce a list of at least 3–4 items that reflected the participants' position on issues discussed. Every issue was further broken down into the specific activities needed to realise it. The second meeting was, usually, focused on deeper insights into the proposals. Issues that emerged earlier (i.e., income generation, benefits distributions inside the community, costs and commitments) were further discussed.

These provided then the evaluation criteria for the further analysis. It was deemed crucial that these covered both positive and negative aspects (i.e., potential incomes but also intervention costs, or time needed to take care of tourists, and, thus, consequently time no longer available for farming). Preliminary conclusions were developed during the last meeting.

All previous information was summed up. The Borda voting framework was used to provide a collective prioritisation of the different issues emerged. Each participant, considered as a single voter, classified the activities distributing a fixed amount of seeds according to his/her personal preferences. The ordering, according to each specific dimension/criterion, is thus the overall aggregation of the assembly's opinion. This led to a rank that could immediately be shown to the assembly and collectively discussed (Fig. 19.3).

19.3.3 NGO's Vision

A special care was reserved to the NGO vision (regarding the activities to implement) as main driver of the entire process.

The NGO vision was developed locally by a continuous brainstorming with the local co-ordinator of the project and his staff and, remotely, with the International board of the UCODEP NGO. The NGO staff, holding both a specific knowledge and an experience in technical assistance to such remote areas, generated suggestions for additional activities.

These suggestions were then formalised into proposals through a pre-feasibility analysis: time and budget assessment was acquired and a survey to identify potential constraints and/or supports was prescribed, when proposals involved other institutions.

Furthermore, the following considerations were formulated:

- to consider each community as a single case (seven impact matrixes and seven independent evaluations);

Options \ Criteria	Working time to put the proposals into action	Residence time of the tourists in the community	Material resource present in the community	Benefits distribution between women and men	Benefits distribution between all the members	Amount of revenue generated	Necessary management capabilities	Lack of basic services such as water and sanitation	Final rank
1. Museum	0.26	0.40	0.36	0.44	0.48	0.44	0.45	0.36	3.19
2. Doing nothing	0.52	0.10	0.72	0.11	0.12	0.11	0.60	0.48	2.76
3. Trek	0.39	0.20	0.54	0.22	0.24	0.22	0.30	0.24	2.35
4. Cabaña	0.13	0.30	0.18	0.33	0.36	0.33	0.15	0.12	1.90

Fig. 19.3 Proposals and criteria emerged in Tarapuy

Fig. 19.4 Criteria evaluation list

Community vision
1. Community priorities
2. Endogenous resources of the community
3. Fair sharing of the benefits between community members
NGO's vision
4. Possibility to achieve external financing
5. Visibility of project activities
6. Availability of project internal resources
7. Technicians motivations
External view
8. Institutional complexity
9. Technical solution appropriateness
10. Change to survive to the end of the project (sustainability)

– to concentrate efforts in areas of greatest need, to promote efficient networking with the other social actors;
– to raise awareness of possible further improvements of intervention, thus intending every activity as a potential catalyst for new grants.

Finally, an overall list of criteria was worked out to be applied to the whole set of communities.

The list was split respectively into: the community vision, the NGO vision and the 'external' vision (Fig. 19.4).

19.3.4 Multi-Criteria Decision Aid

The Multi-Criteria Decision Aid phase was developed by Multi-Criteria aggregation and Multi-Criteria allocation routine. The Multi-Criteria aggregation was performed by means of the NAIADE aggregation procedure.[13] A sorting was obtained: acceptable, possibly acceptable or unacceptable interventions. Breakdowns and discussions within the NGO staff further clarified the possible choices to obtain a set of

Fig. 19.5 Final activity selection

Community	Proposed activity
Teikua	Short cycle farming
Charap	Short cycle farming
Playas	Rainwater harvesting
Zabalo	Minor animals breeding changed to rainwater collection
Sancudo	Management plan
Puerto Bolivar	Ethno-botanic path
Tarapuy	Ethnological museum

three options for each community. These were further processed so that one activity was selected for each community. These seven proposals, as a whole, were selected as making full use of the budget and time resources and, singularly, representing the best compromise among the different communities' visions (Fig. 19.5).

The final proposals were submitted for further feasibility analysis and for the review of the Reserve's officers and, finally, presented to the respective communities. Community assembly had the chance to accept or refuse the proposal. Only one proposal was contested and the choice moved to the second best.

19.4 Methodological Reflections

Despite uniqueness, I believe, the quite peculiar experience of the Cuyabeno Reserve can apply to other settings.

The clash between social and environmental issues adds an extensive paradigmatic value to the case. The relatively confined system of analysis makes of it an analytically manageable realm in which interactions between methodologies and results can be traced.

19.4.1 Context and Methodological Choices: A Dynamic Interaction

The first point to examine is the mutual influences between the context and the planning process. It is argued that both sociological and very hands-on facts deserve to be carefully considered, because both leave a remarkable print into the overall process.

Logistical constrains are evident. Reaching the sites was not easy. Even providing food for the families, as an obvious compensation for the time spent, represented a not trivial task (the food alone overloaded the carrying capacity of the pirogue).

Moreover, people had to be personally convinced of the worthiness of contributing to this effort and they initially revealed a quite strong hostility towards the collaborative scheme of planning: participation could not be assumed; it had to be earned.

The attitude of hostility, definitely legitimate in absolute terms,[14] was also strongly locally motivated. Many past experiences were reported: communities were first involved, or consulted for project proposals or investigations and then left; not a simple report was presented to them, not even with the information they had provided. All these experiences obviously generated a strong mistrust towards any external actors.

At the same time, as mentioned, there were some special actors who were capitalising on the attention of the communities: the so-called indigenous relations offices of the petroleum companies. These powerful players have repeatedly contributed to the spread of a very compliant behaviour towards the indigenous people (Fontaine, 2003, 2005).

In many cases, such as the case of the NamPaz foundation of the Encana Petroleum Company, long-term interventions are looking for sustainable outcomes. Nevertheless, the very opposite tendency of resolving 'urgent' disputes versus the communities with ready-to-go donations can be observed: in some cases offshore engine are promised in order to make an exploring mission be accepted and allowed; similarly the charming construction of a 'community house' is presented as a compensation for one-year massive geophysical explorations in the community's land.

It is understandable, from a communitarian point of view, that the involvement in a time-consuming participatory process does not appear so appealing as the presentation of a wishes' list to a plethora of actors who are just waiting for the chance to access to the Reserve asset.

However, from my point of view, the high involvement of people into the process was a non-negotiable requirement. It was considered crucial to make every possible effort to create a frank dialogue and trust.[15]

At the very end the situation can be resumed as such: there were a decision and an assessment to make. A trustful environment was then required. It was decided to avoid any gaps of information (Munda, 2000) in the overall process and, at the same time, to choose a 'language of evaluation' (Martinez-Alier, 2002) allowing the knowledge created to be shared and the preferences statements to be clear. The way the planning knowledge is gathered and, even more so, the way the preferences regarding proposals and priorities are stated have great practical implications. This may represent a turning point of the whole methodological strategy. As reported, other tools were used within the general framework proposed by MCA: that is, PRA techniques, in particular, Participative Mapping techniques (Chambers, 1992, 1994, 2002).

In fact, sketches or superimposed transparencies provided a sort of blackboard where information was collected and exchanged. The methodologies chosen had to be functional to the construction of a common effort.

The main theoretical question to answer was: is there an overall coherency among methodological tools adopted for this planning process? Or more specifically, is the MCA framework capable to harmonise the overall process?

19.4.2 A Tool for Complex Governance

It is necessary to make a step back and examine the reasons why such an approach was selected.[16] There is a main assumption: the approach has the capacity to handle both multiple dimensions and multiple scales.

19.4.2.1 Dealing with Multiple Dimensions

In the Cuyabeno Reserve case, the multi-dimensional context was clearly exemplified by the concurrent occurrence of environmental (biodiversity conservation, pollution), social (heath, education, gender emancipation and land dispute) and productive (tourism, farming, grazing and logging) concerns. It was crucial to adopt an assessment tool capable of keeping the intrinsic diversity of all these issues and reducing the risk for reductionism.

The Multi-Criteria approach is indeed naturally ascribed to solve the technical incommensurability issue (Munda, 1997) exactly because it deals with multiple dimensions and keeps the intrinsic units of measurement unaltered. In this way the assessment load is highly simplified, because there is no need to translate all the pertinent variables into a unique unit of measurement (Martinez-Alier, Munda, & O'Neill, 1998).

Moreover, if social conflicts, and especially environmental conflicts, are defined as a clash between radically diverse sets of values or livelihoods (Martinez-Alier, 2002), the proper assessment tool is not the one providing a single answer but, on the contrary, the one prone to supporting negotiation processes.

To better clarify this point, consider again the hostility towards participation in the process. An attitude that, as mentioned, may have local explanations and at the same time a general trend can be traced. For example, it can be related to a lack of legitimacy of the process's leading agent (Sittaro, 2007), or it may be connected to cultural causes (see the case of Uwas people in Colombia, Fontaine, 2003, 2005).

I do not believe that compromise solution is always reachable; I rather argue that negotiation processes may reveal the conflict's roots in the information data set generated.

In addition, I would like to point out that the principle that social conflicts should be always, necessarily, 'harmonised' is not supported here. A concept, applied to environmental conflicts, has a hidden ideological premise: natural resources are 'raw materials' available on the 'free market'. As the anthropologist Laura Nader explains, the areas of high biodiversity are also those with high linguistic diversity. The loss of native languages means loss of knowledge and replacement by new language as well as a new ecological frame, new resource economies and new discourses of 'ecological modernization' that delegitimise conflict-based response in favour of coercive harmony (Nader, 1996).

Methodological choices should reflect such premises. The conviction that mathematical aggregations are neutral is refused and it is advised that a preference model holding respectively the non-compensability and veto features (Munda, 2004) needs to be adopted.

These latter two represent very fine technicalities but, as I will try to show, they have very discriminative implications. Under the compensability property it is always allowed to trade-off among different dimensions. An increase in the infrastructure level, for example, can compensate for a loss in environment conservation or vice versa.[17] Consequences are then obvious: compensation might help the negotiation and the formulation of a compromise; at the same time, such a property is not able to protect minority positions and therefore it is inappropriate when power patterns among actors are unevenly distributed. Moreover, if the option of a veto is given to each of the parties involved, it is mathematically consequent that minority positions do not disappear behind majority. When minorities express a strong opposition, this might represent a crucial and not excludable outlook of the system (Funtowicz & Ravetz, 1993): such opposition should emerge and even stop, that is, veto, the evaluation of a certain alternative.

19.4.2.2 Dealing with Multiple Scales

The multi-scale approach can be basically referred to (Giampietro, 2003) can be basically referred to the very definition of complex social systems: if manifold equivalent descriptions are needed, then a system is complex (Ramos-Martin, Giampietro, & Mayumi, 2007). This implies that the involved decisional structures lay on different hierarchical levels or even on different holarchies.[18]

Taking into account the multiple scales issue implies the following: very different actors with disparate roles and binding power are included; the knowledge generated during the decision-making has to be filtered (how?) during the ascending from one level to another (this issue has been scarcely explored in literature). In other words, how to deal with different decisional levels where punctual and localised information is progressively synthesised?[19]

To further explore this point, the dichotomy between actors involved and the amount of information handled has to be mentioned. This is an intrinsic phenomenon of any decision-making process (Simon, 1976; Steelman & Ascher, 1997), a node where the greatest reductionist effect can concentrate. It raises two risks: to oversimplify the plurality of views and to loose relevant descriptions of the problem.

Chambers (2002) recently explored this issue in the PRA literature. After almost a couple of decades of PRA applications, it has become evident that despite the great efficacy in gathering a rich and detailed local pictures, PRA techniques have shown limits to be transposed to higher levels, exactly because of a too rich and inclusive data set.[20]

Coherently with the above considerations, and due to the great versatility of MCA-based procedures, it was possible to structure the process flow in different sections connected by 'filtering nodes'. In this case study, intermediate results were presented at each stage (thus activating a filtration) with 2-fold positive effects: (1) the evaluation outputs could be rapidly restituted to participants; and (2) gathered information could be rapidly synthesized, validated and transferred to the higher level of the evaluation process.

19.4.2.3 The Management of the Preference's Sets

The 'filtering nodes' have another crucial feature: the transmission of each actor's priorities. Following the Cuyabeno Reserve case the first preference set pertains to each indigenous settlement and it describes the perception of what is urgent and how it could be solved. It is collected during the community workshops, filling up the Borda Matrixes. A second wider preference set is generated during the NGO workshops when a further viewpoint was aggregated with the previous. The third stage corresponds to the submission of the final proposals to the officers of the Cuyabeno Reserve for review and finally to communities' decisional assembly for endorsement (or refusal).

All these levels are supposed to be synthesised in the final list of criteria and then made available to any external review. The management of this procedural node is responsible for the inner quality of the entire process.

The problem is: how to maintain the consistency between levels when earlier preferences and expectations are shifted to the next phase without the chance to directly monitor the handling of those.[21] Main actors (first the communities, then the NGO staff and, finally, external institutions) are requested to produce an analysis of the problem and a ranked set of solutions, but no assurance is given that the final choices will be exactly those they proposed.

It is assumed that the most appropriate solution is obtained by the integration of all the voices involved, none having a prominent role, but all having a specific strategic competence to provide. This represents a plurality that the final judgement criteria set should fully witness.

In the Cuyabeno Reserve case, this issue clearly presents a contentious point: it was deemed to fully include communities into the decision-making process but it was not possible to physically involve them into the very final stage of the decision assessment, when the final set of alternatives was selected.

A structural solution was sought: the process is looped in its shape and the final endorsement is given to the first party involved, that is, the communities.

19.4.2.4 The Role of Accountability

The concept of accountability can be formalised by the means by which individuals and organisations report to a recognised authority and are held responsible for their actions (Ebrahim, 2003).

Accountability is traditionally seen as a powerful vertical mechanism that enables superiors to impose performance standards and financial probity on subordinates. The accountability concept brings a sense of value-based practice: something standing half-way between a procedure and a moral obligation. It is then seen as a powerful means to improve efficiency, due to higher motivation, and fostering of consciousness by sharing relevant information at the different levels of management.

In the Cuyabeno Reserve case accountability emerges as key issue. In particular, the concept of 'downward accountability' (Edwards & Hulme, 1996) assumes great relevancy. The 'downward' adjective constitutes the real novelty, downward to the beneficiaries involved, and not only upward to the donors financing interventions.

It is claimed that the upward direction has a clear scope in allowing the evaluation of intervention, and thus measuring effectiveness in the use of funding. Such effectiveness can be pursued only if accountability is also implemented locally. The vicious circle, where I started my reasoning, is broken: the lack of trust towards external subjects, which communities feel strongly, can potentially be eliminated. The opportunistic approach (Michener, 1998) of beneficiaries of development interventions can be solved only if shadows of the agents promoting development interventions are unveiled.

Participation, seen not as a formal exercise, but as the closest process's 'watchdog', has the responsibility to ensure creative solutions, to pinpoint specific knowledge and thus to testify the inner quality of the planning itself.[22]

With regard to accountability, transparency also turns out to be a key issue: only evaluation methods that guarantee transparency can allow to boost creativity solutions, to collect robust knowledge and to guarantee a revisable process.

I also maintain/believe that transparency, and thus the disclosure of the rationality used to take decisions, is one of the few means to deal with the unbalanced pattern of power that always characterised such negotiations. It helps clarifying where 'vested interests' (Banville, Landry, Martel, & Boulaire, 1998) occur and, at the same time, it creates the only possible room for a deliberate trade-off between relevant dimensions and thus disclose the path for a durable compromise.

A method is referred to as transparent when it provides assumptions, criteria of choice, scale for measuring the impacts (Munda, 2006) and, therefore, presents minimum requirements to support trusted processes.

Operationally, transparency cannot be considered as an absolute concept; on the contrary, it should be grounded in the specific context of application, and particularly correlated with the different comprehension capacities.

The huge variance of educational levels involved in this case study makes very clear this concept.

In this case study an adaptive strategy, involving the adoption of a parallel set of tool, was applied. Starting from the highest decisional levels, I made full use of impact matrixes, reference tables utilised to assess performance's scores and final ranks plots. It emerged that these had allowed a fast recognition of both assessment ends and means.

The same conclusion is possible if the issue is analysed from the other way around: the final impact matrix makes it very clear what was used but also what was not included into the evaluation. For example, the Reserve's officers made objections regarding the lack of cost-opportunities considerations in the evaluation of some proposals. Regardless of being in accord or not with it, it was possible to quickly check out not only the results but also the assessment procedure.

Finally a clear lesson emerged from the field: the continuous adaptation to locally based features is an essential requirement. This means that almost every concept, or topic touched, should be appropriately translated to well-known images. In the Cuyabeno Reserve case this was constantly dealt with as a background priority. As explained, maps or land sketches were used as primary communication means. These provided local-referred terms of references translating all the concepts emerged during the meetings.

Many experiences already exist that have pushed forward this concept. They are referred as 'Theatre and Development' (Save the Children, 2001; McCarthy, Galvao, & Chambers, 2005). This discipline involves the use of theatre, including street theatre and puppet theatre, to promote communication campaign, that is, regarding education and health practices. The adaptation of a Multi-Criteria process into such dynamics was able to enforce both the effectiveness and the creativity of the planning process. Once more the idea can be easily translated into other realms where the same idea can be shared to ground the participative process in ad hoc visual communication: images, movies, video games or role play (Gallopin, Hammond, Raskin, & Swart, 1997; Guimaraes Pereira et al., 2001).

19.5 Conclusions

If applications of MCA approaches in the western world are widespread, is it then possible to make full use of its potentialities to depict complex systems in far contexts like the one of the Cuyabeno Reserve? The solution proposed and discussed in this chapter is based on the integration of MCA and PRA approaches: is such an integration appropriate?

I believe that this integration is consistent, because both MCA and PRA approaches share the same epistemological premises, such as values-pluralism and the rejection of objective 'reality' or 'truth', depicting a world instead, where there are multiple layers of 'realities' depending on positions in cultural and social settings.

In fact, it is during the 'planning session' phase that the influence of MCA way of thinking was strongest, put into practice by the use of Borda Matrixes. This allowed a continuous switch between advantages and drawbacks of each proposal. This also contributed to create a frank and maybe even conflicting environment, but above all, it was constructive and led to a clear explication of visions and preferences. This was made possible due to the transliteration of the assessment requirements by means of PRA techniques that have structured, and not imposed, an appropriate language of evaluation thus taking profit of the MCA way of structuring problems.

Notes

1. See: http://www.ucodep.org.
2. Acuerdo Interministerial No. 0322, 26/7/1979.
3. From national census data, INEC 2000, the inhabitants of the entire province of Sucumbios grew between 1974 and 2000 at a 9% rate, passing from 15 to 130 thousands.
4. Historically the Informe Yana Cutri had great relevance. It was led by Dr. Miguel San Sebastian, who conducted an epidemiologic survey in the households close to extraction sites and denounced a cancer rate 6–12 times higher than the rest of the country (Instituto de Epidemioogìa 'Manuel Amunàrriz', 2000). Actually many international campaigns are denouncing the huge environmental damage connected to the Amazonian petroleum extraction, (http://www.globalaware.org; http://www.texacorainforest.org; http://www.oilwatch.org.ec.)
5. Ministerio de Agricultura y Ganadería (1992), Convenio Entre El Ministerio De Agricultura Y Ganadería, La Asociación De Comunidades Indígenas De La Nacionalidad Cofán (Acoinco) Y La Comunidad Cofán De Zábalo, Para La Conservación, Usos Y Servicios De Los Recursos De La Reserva De Producción Faunística Cuyabeno En La Jurisdicción De La Comunidad De Zábalo. Quito, Ministerio de Agricultura y Ganadería.
6. Agreements were signed between indigenous nationality associations and the Ministry of Environments,; see Convenio Para Conservación Manejo Ecológico Y Aprovechamiento En Un Sector De La Reserva De Producción Faunística Cuyabeno (1995, 1996, 1999) of the Instituto Ecuatoriano Forestal y De Areas Naturales y Vida Silvestre (INEFAN), Quito.
7. Three indigenous nationalities are represented: Kitchua, Shuar and Cofan.
8. It is not the purpose of this chapter to recall theoretical premises regarding Multi-Criteria theory. For a general formulation of a Multi-Criteria problem and its application in public policy evaluation, see Roy, 1996; Munda, 2004.
9. Participative Rural Appraisal (PRA) is a family of approaches and methods finaliszed to allow rural people to share, amplify and analysze their own knowledge. An enthusiastic description of the philosophy of PRA is reported by the International Institute for Sustainable Development (IISD, 1995).
10. As a reference average participation was around 75–80% of the population. This features is due to the fact that the stay in each settlement was quite long, 7–10 days, meaning that it was feasible to directly communicate with each of the families (each community has an average population of 100–150 people, around 20 families).
11. Gender separation was immediately understood to be necessary, and it gave rise to a much richer picture of the communities. It also guaranteed a greater efficiency, allowing work to be carried out in parallel groups. Furthermore, group separation allowed to integrate specific data and information. The field work was shared by Jesus Placencia, community mediator,; Luis Tonato, biologist,; Susana Anda, gender anthropologist student,; Catalina Suquillo, biology student; and the author.
12. Jean Charles de Borda, in early 1770s, proposed a simple summing preferences to achieve a social ranking (see Young, 1988, 1995; McLean & Urken, 1995 for discussions on voting schemes). Under this simple voting method alternatives are ordered in terms of each criterion and then a score is assigned proportionally to each relative position. For example, the first, out of four alternatives, obtains 4 points, the second three3, the third two2 points and, finally, the fourth only one1 point. Points acquired, according to each criterion, are then summed up to provide an overall rank.
13. NAIADE (Novel Approach to Imprecise Assessment and Decision Environments) is a discrete Multi-Criteria method whose evaluation matrix may include either crisp, stochastic and fuzzy measurements of the performance of an alternative with respect to an evaluation criterion (Munda, 1995; JRC, 1996).
14. This attitude, in development practice, represents an increasing and documented trend (Michener, 1998; Talbot, 1995; Gough et al., 2003).

15. This is really a crucial point. The relaxed atmosphere if reached, it is a powerful tool to release emotional restraints and to include more people into the process. In the community of Puerto Bolivar, due to past friction between community leaders and the previous NGO's co-ordinator, we were totally refused, even to explain our willingness to change previous way of relationship. The setting up of a shared work scheduling was allowed only after a dramatically intense football match, NGO vs. Puerto (4–6), ended only when a swarm of gad-flies invaded the field.
16. I am here referring to the use of MCA to social problems. For detailed analyses, see Gampier and Turcanu (2006) and Munda (2006).
17. This is the same idea of summing up costs and benefits and then selecting the alternative with the most positive flow.
18. Holarchy is a combination of the Greek word 'holos' and the word 'hierarchy'. It is an organised structure of units or entities that are called 'Holons'. Each 'Holon' could be regarded as either a whole or as a part depending on how one looks at it.
19. For example, in the Cuyabeno Reserve case, the information pertaining to one community had to be first aggregated with the information belonging to all other six before it could be discussed with the Ministry of Environment, the formal Reserve authority. Furthermore, such information combined with Ministry evaluation (but also NGO staff evaluation) was presented to the departmental health authorities, or to other pertaining authorities, for further consultation. To make this point very explicit, the very first level of information was the exact position of the cabana of Don Jaime Yacelga and Doña Carmen Yumbo (and the knowledge that they have 6 sons, during the dry season they cultivate yucca in an half-hectare plot, using traditional seeds and no chemicals). At the highest level such information was transformed, for example, in the proposal to realise an itinerating workshop about self-production of organic pesticide in all Cuyabeno Wildlife Reserve (CWR) communities and in the surrounding buffer area (thus including Don Jaime and Doña Carmen too).
20. To better put together this theoretical point I refer also to Participative Geographic Information System (GIS) literature (Abbot et al., 1998; Harmsworth, 1998; Minang, 2003; Zurayk, 2003) that has perfectly caught this point of parallel skimming between scale of reference and information handled, advocating the use of appropriate GIS techniques as a powerful tool to solve this node.
21. The proper management of expectations is as very critical issue, in development aid. In this case, for example, the open agenda, communities decide on the issues to be discussed. It is fascinating to draw a relevant description of the community vision, open to any kind of suggestions, but if misunderstood it can create too many expectations about the project's actual means and then create a vision projected too strongly on expectations than on reachable goals.
22. For a more systematic ex post evaluation of the process or for its meta-evaluation (Wenstøp & Seip, 2001), I refer to a previous work (Sittaro, 2007) where a specific MCA- based routine is proposed to compare this or similar planning experiences. By means of pedigree matrixes, meta-criteria and amoeba plots (van der Sluijis, Kloprogge, Risbey, & Ravetz, 2003), the following features have been considered: methodological appropriateness, quality of information input, degree of systemic reductionism along the process and, finally, the process's dependency on local, unpredictable events.

References

Abbot, J., Chambers, R., Dunn, C., Harris, T., de Merode, E., Porter, G., et al. (1998). Participatory GIS: Opportunity or oxymoron? *PLA notes IIED*, *33*(5), 27–33.

Araya, I., & Peters, H. (1999). *La Colonización Espontánea Del Nororiente Ecuatoriano Y La Reserva De Producción Faunística Cuyabeno*. Quito: Curso para Guías Naturalistas de la Reserva de Producción Faunística Cuyabeno.

Banville, C., Landry, M., Martel, J. M., & Boulaire, C. (1998). A stakeholder approach to MCDA. *Systems Research*, *15*(11), 15–32.

Bustamante, T., & Jarrín, M. C. (2005). Impactos Sociales de la Actividad Petrolera en Ecuador: un Análisis de los Indicadores. *Íconos, 21*, 35–56.
Chambers, R. (1992). *Rural appraisal: Rapid, relaxed and participatory*. Report of Institute of Development Studies. Brighton: University of Sussex.
Chambers, R. (1994). The origins and practice of participatory rural appraisal. *World Development, 22*(7), 953–969.
Chambers, R. (2002). *Participatory numbers: Experience, questions and the future*. Report of Institute of Development Studies. Brighton: University of Sussex.
Ebrahim, A. (2003). Accountability in practice: Mechanisms for NGOs. *World Development, 31*(5), 813–829.
Edwards, M., & Hulme, D. (1996). Too close for comfort? The impact of official aid on non-governmental organizations. *World Development, 24*(6), 961–973.
Fontaine, G. (2003). *El Precio del Petróleo. Conflictos Socio-Ambientales y Gobernabilidad en la Región Amazónica*. Quito: FLACSO, Institut Français dÉtudes Andines.
Fontaine, G. (2005). Microconflictos Ambientales y Crisis de Gobernabilidad en la Amazonía Ecuatoriana. *Íconos (Dossier), Petróleo y Medio Ambiente en la Amazonía Andina, 21*, 35–46.
Funtowicz, S. O., De Marchi, B., Lo Cascio, S., & Munda, G. (1998). *The Troina water valuation case study*. Research Report prepared by ISIS. Ispra: European Commission Joint Research Centre.
Funtowicz, S., & Ravetz, J. R. (1993). Science for the post-normal age. *Futures, 25*(7), 739–755.
Gallopin, G., Hammond, A., Raskin, P., & Swart, R. (1997) *Branch points: Global scenarios and human choice*. Resource Paper of the Global Scenario Group, PoleStar Series [http://www.sei.se/publications; accessed 5 April 2009].
Gampier, C., & Turcanu, C. (2006, 15–19 December) *The governamental use of multi-criteria analysis too expensive to be required or too important to be abandoned?* Proceedings of the 9th Biennial Conference of ISEE, *Ecological Sustainability and Human Well-Being*, Delhi.
Giampietro, M. (2003). *Multi-scale integrated analysis of agrosystems*. Boca Raton, FL: CRC Press.
Gough, C., Darier, E., de Marchi, B., Funtowicz, S., Grove-White, R., Guimaraes Pereira, A., et al. (2003). Context of citizen participation. In B. Kasemir, J. Jäger, C. C. Jaeger, & M. T. Gardner (Eds.), *Public participation in sustainability science: A handbook* (pp. 37–61). Cambridge: Cambridge University Press.
Guimaraes Pereira, Â., Corral Quintana, S., Funtowicz, S., Gallopín, G., De Marchi, B., & Maltoni, B. (2001). *Visions – Adventures into the future*. The VISIONS project at the JRC. Ispra: European Commission.
Harmsworth, G. (1998) *Indigenous values and GIS: A method and a framework indigenous knowledge and development monitor*. Netherlands Organisation for International Cooperation in Higher Education (Nuffic), 6(3) [http://www.landcareresearch.co.nz/research/sustainablesoc/social/ikdmpap.asp; accessed 5 April 2009].
Instituto de Epidemioogìa 'Manuel Amunàrriz' (2000). *Informe Yana Curi*. Quito: Abya-Yala.
International Institute for Sustainable Development (IISD) (1995). *Guidebook for field projects on adaptive strategies: Adaptive strategies for sustainable livelihoods in arid and semi-arid lands*. International Institute for Sustainable Development, Winnipeg, Manitoba [http://www.iisd.org/; accessed 5 April 2009].
Joint Research Centre (JRC) (1996). *NAIADE manual*. Ispra: JRC.
Martinez-Alier, J. (2002). *The environmentalism of the poor: A study of ecological conflicts and valuation*. Cheltenham: Edward Elgar.
Martinez-Alier, J., Munda, G., & O'Neill, J. (1998). Weak comparability of values as a foundation for ecological economics. *Ecological Economics, 26*(3), 277–286.
McCarthy, J., Galvao, K., & Chambers, R. (2005). *Enacting participatory development: Theatre-based techniques*. London: Earthscan Publications.
McLean, I., & Urken, A. B. (1995). *Classics of social choice theory*. Ann Arbor, MI: University of Michigan Press.

Michener, V. J. (1998). The participatory approach: Contradiction and co-option in Burkina Faso. *World Development*, *26*(12), 2105–2118.
Minang, P. A. (2003) *Assessing participatory geographic information systems for community forestry planning in Cameroon: A local governance perspective*. Master Degree Thesis at the International Institute for Geo-information Science and Earth Observation, Enshede.
Moser, C. (1993). *Gender planning and development: Theory, practice and training*. London: Routledge.
Munda, G. (1995). *Fuzzy information on multi-criteria environmental models*. Heidelberg: Physika-Verlag.
Munda, G. (1997). Environmental economics, ecological economics and the concept of sustainable development. *Environmental Values*, *6*(2), 213–233.
Munda, G. (2000) *Conceptualising and responding to complexity*. Policy Research Brief, 2. Cambridge Research for the Environment, Cambridge [http://www.clivespash.org/eve/publ.html; accessed 5 April 2009].
Munda, G. (2004). Social multi-criteria evaluation (SMCE): Methodological foundations and operational consequences. *European Journal of Operational Research*, *158*(3), 662–677.
Munda, G. (2006). Social multi-criteria evaluation for urban sustainability policies. *Land Use Policy*, *23*(1), 86–94.
Nader, L. (1996). *Civilization and its negotiators: Harmony as a pacification technique*. Brasilia: XIX Reunião Brasileira de Antropologia, Associação Brasileira de Antropologia.
Ramos-Martin, J., Giampietro, M., & Mayumi, K. (2007). On China's exosomatic energy metabolism: An application of multi-scale integrated analysis of societal metabolism (MSIASM). *Ecological Economics*, *63*(1), 174–191.
Roy, B. (1996). *Multicriteria methodology for decision aiding*. Dordrecht: Kluwer.
Save the Children (2001). *Using Theatre for Development in child rights programming: TfD training manual*. Kathmandu: Office of the South and Central Asia Region.
Simon, H. A. (1976). From substantive to procedural rationality. In S. J. Latsis (Ed.), *Method and appraisal in economics*. Cambridge: Cambridge University Press.
Sittaro, F. (2007). Meta-evaluation of participatory community planning. In M. Balbo (Ed.), *International aid ideologies and policies in the urban sector*. 7th N-Aerus Conference. Venezia: Department of Planning, IUAV.
Steelman, T. A., & Ascher, T. A. W. (1997). Public involvement methods in natural resource policy-making: Advantages, disadvantages and trade-offs. *Policy Sciences*, *30*(2), 71–90.
Talbot, A. (1995). Transitional societies and participatory development: Reflections based on Malian realities. In H. Schneider & M. H. Libercier (Eds.), *Participatory development from advocacy to action* (pp. 141–152). Paris: Organization for Economic Cooperation and Development.
Tapia-García, C. (2006). *Assessment of land cover change in the Cuyabeno Wildlife Reserve and confining area using FAO's land cover classification system (LCCS)*. Firenze: Land Cover Department.
van der Sluijis, J., Kloprogge, P., Risbey, J., & Ravetz, J. (2003, 19–21 August) *Towards a synthesis of qualitative and quantitative uncertainty assessment: Applications of the numeral, unit, spread, assessment, pedigree (NUSAP) system*. International Workshop on Uncertainty, Sensitivity, and Parameter Estimation for Multimedia Environmental Modeling, Rockville, MD [www.nusap.net, accessed 5 April 2009].
Wenstøp, F., & Seip, K. (2001). Legitimacy and quality of multi-criteria environmental policy analysis: A meta-analysis of five MCE studies in Norway. *Journal of Multi-Criteria Decision Analysis*, *10*(2), 53–64.
Young, H. P. (1988). Individual contribution and just compensation. In A. E. Roth (Ed.), *The Shapley value* (pp. 267–278). Cambridge: Cambridge University Press.
Young, H. P. (1995). Optimal voting rules. *Journal of Economic Perspectives*, *9*(1), 51–56.
Zurayk, R. (2003) *Participatory gis-based natural resource management: Experiences from a country of the south*. Arid Land newsletter, 53 [http://ag.arizona.edu/OALS/ALN/aln53/zurayk.html; accessed 5 April 2009].

Chapter 20
Driving Forces and Spatial Impacts: An Integrated Approach for Small- and Medium-Sized Cities

Begüm Özkaynak

20.1 Introduction

Today, any discussion of urban development must address, in an integrative manner, issues of demography, housing, infrastructure, transportation, economic development, employment, poverty, health, social coherence, land use, pollution and environmental degradation. However, defining objectives and setting priorities at the local level become a difficult task, as the nature and direction of urbanisation, is increasingly influenced by global economic integration and the struggle of countries – and indeed of individual cities- to be competitive in the global marketplace (Cohen, 2004). In general, local strategies have to be formed and reformed according to the logic of macro-level factors regarding feasibility, and actors' responses and political judgements about the values and interests they wish to promote (Healey et al., 1995).

Acknowledging the multi-dimensionality of urban development and the difficult inter-relations of the economic, environmental and social spheres, on the one hand, and of local and global factors, on the other, makes research in the field decidedly problematic. Therefore, an interdisciplinary and integrative approach allowing for the communication of multiple codes is needed for framing urban research (Castells, 2000; Martinez-Alier, 2002; Munda, 2006). In this context, and with the increased understanding of the role of uncertainty in policy-making, interest in scenario analysis has grown in recent years. The scenario approach is now widely seen as a valuable analytical device and a key aid to decision-making processes as it provides a background against which to consider alternative strategic policy options. Today, it is widely acknowledged that scenario analysis is also relevant to discussing and coordinating urban policies, which are becoming increasingly complex and challenging.

However, as cities are not independent entities isolated from the larger economic and social forces that operate on them, one of the challenges in constructing

B. Özkaynak (✉)
Department of Economics, Boğaziçi University, Istanbul 34342, Turkey
e-mail: begum.ozkaynak@boun.edu.tr

M. Cerreta et al. (eds.), *Making Strategies in Spatial Planning*,
Urban and Landscape Perspectives 9, DOI 10.1007/978-90-481-3106-8_20,

scenarios to assess urban development is the need to bridge macro-structural influences, local driving forces and social actors. In this context, van Asselt et al. (1998) and European Environmental Agency (EEA) (2000) note that scenarios built at the local level, but taking notice of national and global developments, remain rare.

The project presented here was conducted in the province of Yalova, Turkey, and was aimed at making an integrated assessment of possible future states of the province in 2020, given the local driving forces (e.g., demographic trends; people's attitudes, perceptions and priorities regarding socio-economic and ecological issues; and the nature of governance) and the external factors (at the regional, national and global levels). It is believed that the questioning of local driving forces and external factors that today prevail in the province will help us understand the possible directions of change in the city. This analysis can also provide us with a background against which to explore and formulate the paths that need to be taken within a multi-layered system of governance to strengthen those forces that would favour more sustainable modes of urban development with reference to problem areas. Therefore, this chapter focuses on the multiple forces operating at various spatial scales impacting on Yalova.[1] It is hoped that the approach reported here will be of wider relevance to other regional and local authorities, considering the need for integrating external developments with local factors.

The structure of the chapter is as follows: the next section will very briefly introduce the scenario methodology and clarify what is meant here by driving forces for those who are not familiar with the approach; this will be followed by an introductory section to the Yalova case study. Subsequently, local and external factors important in shaping the future of the province are carefully identified and listed. The chapter concludes with a discussion of ways in which local, domestic and international contexts can interact and impact Yalova.

20.2 The Scenario Approach and Its Relevance for Strategic Planning

The definition of 'scenario', which appears in the United Nations Environmental Programme's *3rd Global Environmental Outlook* (GEO3), states the following: "[S]cenarios are descriptions of journeys to possible futures. They reflect different assumptions about how current trends will unfold, how critical uncertainties will play out and what new factors will come into play" (UNEP, 2002, p. 320). In other words, scenarios develop, in a structured way and with an internally consistent logical plot, a set of diverse and plausible stories about how the future may unfold based on 'if-then' propositions (Alcamo, 2001; Clayton et al., 2003; Gallopin et al., 1997; Raskin et al., 2002).

There is no single approach to scenario-making, but as Gallopin et al. (1997) and Rothmans et al. (2001) point out, all scenarios should be relevant, coherent, consistent, credible and transparent.[2] To those ends, it is often argued that some standard and useful steps to elaborating scenarios should be observed (Gallopin et al., 1997; Ghanadan, 2002; Schwartz, 1991). After identifying the focal issue (e.g., a sector, a region), and setting out spatial and temporal boundaries, the point of departure for all scenarios is the description of the current state of affairs – the historical context, key characteristics of the study area, events, main issues of concern and social actors – and the list of driving forces, or variables, with potential to motivate fundamental changes and deviations from current trends. The evaluation of these drivers according to their degrees of importance and uncertainty serves to identify which critical issues can significantly alter the course of events. Given this, scenarios possessing an internal logic that links all elements into a coherent plot can be formulated to address key variables. Of course, the development of scenario narratives may be followed by the quantification of items, if needed, and where possible. The final part of the scenario will consist of an image of the future situation, the so-called end-state.

In the following sub-section, an attempt will be made to present the key driving forces that prevail in the city of Yalova from a hierarchical scale perspective, within the local system first, and then subsequently, up through regional, national and global levels. However, before proceeding further, it is necessary to clarify what is meant here by driving forces – establishing the distinction with trends – and critical uncertainties.

Driving forces represent the key factors, variables or processes that influence a situation and dominate the dynamics of the whole system (Gallopin & Rijsberman, 2000; Gallopin et al., 1997). The actions of institutions, such as businesses, political parties, government agencies and international bodies, may also play a part in determining the key drivers. The driving forces considered in most of the scenario studies are as follows: demographic (e.g., age structure, population growth); economic (e.g., level of growth, sector distribution); social (e.g., distribution of wealth, poverty, needs and expectations); cultural (e.g. values, nationalist impulses); technological (e.g., efficiency, lock-in); political (e.g., stability, laws and action of political parties); and environmental (e.g., material and energy, pollution levels).

Commission of the European Communities (CEC) (1999) and Gallopin and Rijsberman (2000) rightly note that driving forces can differ in the degree of certainty that can be ascribed to each of them. It is argued that some drivers are predetermined and invariant, at least for the scale of analysis and over the planning term.[3] For instance, in the case of European scenarios, a predetermined element is ageing (see, e.g., *Visions Project*, Rothmans et al., 2001; *Scenarios Europe-2010*, CEC, 1999). The predetermined elements have implications for all future scenarios and are taken as givens. At the same time, there are some driving forces, but the ways in which they will evolve represent critical uncertainties, as, for example, major political decisions, and environmental and social tensions. It is pointed out that the ways in which these uncertainties are resolved determine the shape of particular scenarios.[4] Therefore, while the initial drivers are the same in all scenarios, the trajectory followed is different in each case due to the influence of key driving forces, which can lead to different futures.

Extending the discussion above, it is possible to clarify the differences between trends and driving forces. A trend is the direction that a particular driving force takes for a period of time. For instance, the supply of qualified labour can be important for the level of growth, and a shortage or abundance of qualified labour are trends that might possibly impact the course of events. In this context, the so-called conventional or business-as-usual scenarios adhere to present trends into the foreseeable future (van Notten et al., 2003). However, Gallopin and Rijsberman (2000, p. 2) note that "projections of [current] trends in human affairs may be legitimate over the short-term, but they become unreliable as time horizons expand from months and years to decades and generations. Fundamental uncertainty is introduced both by our limited understanding of human and ecological processes, and by the intrinsic indeterminism of complex dynamic systems."

Having reviewed some characteristics of the scenario approach, we will now move on to the Yalova case and explore the driving forces on the local, regional, national and global scales that influence the study area.

20.3 The Yalova Case Study

With a current population of nearly 170,000 (TUIK, 2002), Yalova is located in the Marmara Region, in North-Western Turkey, and situated in the middle of three large industrial cities, namely, Istanbul, Kocaeli and Bursa. The province is bordered by the Marmara Sea in the north and the west, making sea transportation between Istanbul and Yalova possible. In terms of the land area it covers some 850 km^2. Yalova is Turkey's smallest province and it has a population density of 199 inhabitants per km^2.

There are large-scale chemical plants and small-to-medium-scale industries on the eastern side of the city; agriculture and floristry are still important pursuits; and thermal spring facilities, albeit below their full potential, as well as summer vacationing spots make a contribution to the city's economy. Education levels in the city are on average higher than the rest of Turkey.

From an environmental point of view, Yalova is one of the greenest areas in the region with half of its land area consisting of forests. However, it has also witnessed heavy migration over the last two decades, and become an example of rapid and unplanned urbanisation, a trend prevalent in Turkey since the mid-twentieth century. Yalova also occupies a special position as a city severely affected by the Marmara Earthquake in August 1999. In fact, the earthquake and the economic crisis following it prepared the ground for rethinking the problems of the province in economic, social and environmental terms, their root causes and alternative solutions.

When the issue of 'what is to be done for the future' is raised, there are conflicting views. The question of whether the region needs to be industrialised further or not is a lively topic of discussion. Those who see industry as a source of employment and income are in favour of the establishment of a 'general organised industrial zone' in order to attract investment. Those who oppose further industrialisation are mainly concerned about the risk of unqualified migration to the city and unplanned urban growth. The municipality, in this context, focuses on the possibilities of generating employment outside the industrial sector, mostly in the services sector such as tourism, and IT.

At the local level, the project was presented under the name of *Sustainable City: The case of Yalova* as a 1-year joint project of Boğaziçi University and Autonomous University of Barcelona.[5] Universities in Turkey are among the institutions most trusted by the public. Boğaziçi University's good reputation and the involvement of a foreign university amplified the significance of the project and helped in conveying its legitimacy.

The overall case study conducted in the city of Yalova, with the complementary desktop research, made it clear that multiple actors and factors (sometimes conflicting) on local, regional, national and global scales are influencing the study area. Given the set of relevant issues which must be considered at different stages of the analysis, a listing of these driving forces and a closer examination of them in terms of givens and critical uncertainties will enable us to obtain a clearer understanding of the possible processes of change in the city. In the scenario development exercise, the question becomes: 'Which of these internal and external forces will put their stamp on the city?'

20.4 Local and External Driving Forces Prevailing in Yalova

Investigating local driving forces, an empirical study was conducted in the province of Yalova in 2003. The study consisted of 36 in-depth interviews, three focus groups, three workshops and a survey administered to a total of 1196 respondents representative of the urban and rural population. These qualitative and quantitative studies were used to decipher the position of multiple actors in the community vis-à-vis different alternative development paths – industry, agriculture-tourism and IT services-university – and to obtain insights into the external driving forces that prevail today in the study area. These insights into external driving forces were complemented by desktop research. For global driving forces, several documents related to EU enlargement and policy were reviewed, that is, *Great Transitions* (Raskin et al., 2002) of the Global Scenario Group; for possible developments in Europe, the *Scenarios Europe – 2010* project of the European Commission (CEC, 1999), the *Visions Project* (Rothmans et al., 2001) and the *Four Futures of Europe* by Centraal Planbureau (de Mooij & Tang, 2003).

Regarding the national and regional driving forces, a general literature review on Turkey was conducted with a focus on its potential accession to the European Union (EU), economic growth prospects and the environmental and social aspects of economic development and urbanisation.

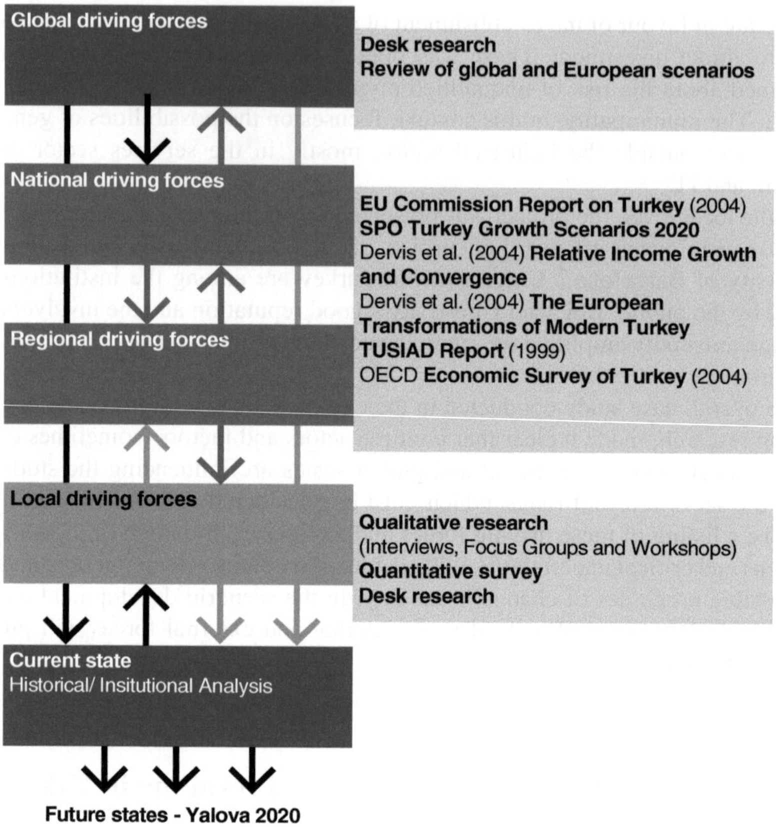

Fig. 20.1 The research methodology for the identification of driving forces prevailing in Yalova

Figure 20.1 summarises these ideas and briefly presents the research methodology used to identify the driving forces prevailing in Yalova, which will be explained here. While reading this section, it is important to keep in mind the fact that driving forces do not unfold in isolation from one another but interact on different scales. One familiar example is land use at local level, which results from local institutions and actions but is also shaped by national policy frameworks and global economic markets.

In fact, the existence of multiple forces operating at various spatial scales poses empirical and theoretical problems, and over the last decades numerous theorists have addressed the problem of their interaction (Beauregard, 1995; Gibson et al., 2000). Flyvbjerg (2001, p. 138), in this context, notes the following: "[a]s anyone who has tried it can testify, it is a demanding task to account simultaneously for the structural influences that shape the development of a given phenomenon and still craft a clear, penetrating narrative or micro-analysis of that phenomenon."

However, Flyvbjerg (2001) believes that it is still important to deliberately seek out information to answer questions about what structural factors influence individual actions, how those actions are constructed and their structural consequences.

20.4.1 Local Driving Forces in Yalova

20.4.1.1 Availability of Human Capital in Yalova

In Turkey, Yalova is among the top ten provinces in terms of rapid local population growth rates. The city grew annually by 22% in the period 1990–2000, primarily because of migration to Yalova from different regions of Turkey, the Balkans and Europe. Moreover, the most important characteristic of Yalova's population pyramid is that half of the population is younger than 30 years of age and around 14% are above 50 years of age (TUIK, 2000). This constitutes two different sides of Yalova's image: it is both a potentially dynamic place with a young and active workforce and a place of retirement. Moreover, 92% of the respondents in Yalova replied 'Yes' when asked, 'Do you think that you will continue to live in Yalova?' The acceptance of this fact leads to the conclusion that the proportion of the active population will continue to increase in Yalova due to its young demographics, at least in the short-medium term. Therefore, depending on the local employment opportunities, Yalova is likely to experience further migration pressures. In the absence of employment opportunities, there is always the risk that the active and young workforce will move from Yalova to the major cities in Turkey and to foreign countries if necessary.

In terms of education, the level of literacy and schooling, together with the number of high school and university graduates, Yalova is above the national average (TUIK, 2000). Hence, Yalova already has a well-educated and young workforce and offers opportunities to support a variety of investments in the region. In addition, in terms of labour force participation, considering the fact that educational attainments in Yalova are high, an increasing proportion of women in the younger cohorts is also expected to enter the labour market in the coming decades.

20.4.1.2 The Availability of Natural Capital in Yalova

Although the city centre is quite ugly today, Yalova is the greenest area in the Marmara Region with around 55% forest cover and 30% first-degree productive agricultural land. Moreover, it has beautiful mountains for ecotourism and perhaps weekend tourism; its thermal sources are within the priority list of thermal opportunities in Turkey.

In this context, the southern part of Yalova, constituting a large, forested area, has been included among areas in the Marmara Region for protection as part of Turkey's natural wealth and heritage by international agencies and NGOs (Yeşil Atlas, 2003).

However, the ways in which this natural wealth will be used depend to a great extent on the specific concerns about the environment and priorities at the local and national levels. Turkey's relations with the EU and developments within the Union in terms of environmental regulations will presumably be critical in shaping environmental policy in Yalova as well.

20.4.1.3 Two Motives Behind Migration to Yalova and the Local People's Perceptions and Priorities

When persons not originally from Yalova are questioned about what influenced their decision to come to the area, they usually cite two reasons, which are in some ways contradictory. The first is economic (e.g., finding better employment opportunities), the second concerns the smallness of the city, the climate and environmental qualities (e.g., for summer holidays or retirement), as might be expected since pensioners and summer residents constitute an important part of the city's population and identity. These two reasons for migration to Yalova constitute two different local tendencies that are likely to influence future policies in Yalova.

Moreover, the perception among the general public is that economic issues such as unemployment and economic stagnation are among the most important problems of the city, paralleling those of the country. In the quantitative survey, when the priorities in an urban development context were questioned, economic goals (particularly employment and job opportunities) were found to be considered relatively more important than social and environmental goals and also received relatively low satisfaction scores. As the quantitative and qualitative research findings show, apart from the seasonal residents, year-round residents, especially the young and low–middle income groups, expect greater employment opportunities and growth at the local level.

20.4.1.4 The Sense of Belonging to Yalova

In the quantitative survey, city residents in Yalova were found to have a strong sense of belonging to the city despite the fact that Yalova is a 'pulled together' city, and nearly two-thirds of the respondents were not born in Yalova. This degree of feeling that one belongs to Yalova is noteworthy. Although people in Yalova admit to feeling the disadvantages of living in a small city, it seems that they enjoy close personal relations, social interactions and trust. However, the case study also revealed that Yalova lacks mechanisms to encourage full integration of different groups within the urban structure. There are few venues offering opportunities to reconcile cultural and income-level differences within civil society. Those that exist include the following: cinemas, sports centres, festivals, concerts and restaurants. This is especially a concern among the young people. This problem of lack of cultural and social life, if not properly addressed, combined with migration pressures and unemployment, can lead to problems of integration, an elevated rate of crime or young people choosing to leave the city.

20.4.1.5 The Geographic Location of Yalova as Part of Istanbul Metropolitan Area and Its Hinterland

If Yalova – still as a small city – were located in another part of Turkey, developments within the country might have little significance or a delayed impact of lesser importance. But Yalova stands in the Marmara Region, geographically close to Europe and at the nexus of three large cities: Istanbul, Bursa and Izmir. Istanbul is a mega-city covering the land between Tekirdağ and Kocaeli and accounts for almost one-fifth of the nation's population. Bursa and Izmir are both large industrial cities located very close to Yalova. Surrounded by such developed areas, and given the economic geography of Turkey, it is very unlikely that Yalova becomes an area of depression.

20.4.1.6 The Economic Base of Yalova

Strong industry with a good local image. Since the 1970s a series of industrial investments in chemicals, textiles and energy investment have been made in Yalova by a large group, the Akkök Group[6] (composed of Aksa[7] and many other companies). Aksa is the world's largest acrylic producer under a single roof and, otherwise, with strong production margins and a robust financial structure, less dependent on the Turkish economy given its export-oriented nature. Legal cases have been filed against Aksa – some are still pending and there is one court decision against the company – claiming that the diffusion of chemicals in the area during the earthquake negatively affected human health and agricultural lands. Nonetheless, as revealed by a qualitative and quantitative survey, Aksa and the overall Akkök Group have a generally positive image in the local region thanks to the employment opportunities offered. It was noted that the activities of the Akkök Group in terms of social corporate responsibility are highly acknowledged by the public as well. Investment reports related to the company indicate that their sound performance is expected to be sustained in the near future. This means that the Akkök Group will continue to be an important player in the city with a view to expanding the industries in the region. Nevertheless, social and environmental regulations in Turkey in the next 15 years will be critical for this group and they will always be mindful of the possibility of moving their production base outside of Yalova.

Historical ties with tourism and looking ahead. From a historical perspective, tourism has long been important for the city: first between the 1930s and the 1970s, for its thermal sources, and then during the 1980s for sea tourism and summer homes. As such, Yalova, located in the middle of three large cities, still can play a role as a countryside retreat. It is argued that Yalova possesses the four vital attributes that are needed for health and thermal tourism: clean air, nature, thermal spas and good weather. For foreigners, these qualities can also be integrated with a wide variety of activities, such as historical site-seeing with daily trips to Istanbul and Bursa.

The future of local agriculture. In the last decades, an important tendency in the city has been the selling of agricultural land for construction as an easy way of earning money given problems in the agricultural sector in Turkey. Both official reports and the quantitative survey indicate that the share of workforce engaged in agriculture, which as a percentage of total employment today stands at 38%, will continue to decrease in Yalova.[8] While this transition will certainly happen, whether its pace is slow or fast will be critical in determining its impact on the labour market and on other local economic and social spheres. In Yalova, this transition certainly will not progress at a rate slower than the national average, which is itself difficult to predict. Presumably, while agriculture will be conducted in a more modern manner, its importance in the overall picture might well be maintained for some time, external factors permitting.

The IT services and tourism vision of the local government. Given Istanbul's tendency for gradual expansion as a metropolitan area, the local government in Yalova focuses on the possibilities of generating employment as part of Istanbul's hinterland in the services sector based on IT (e.g., call centres, process outsourcing), and on tourism (e.g., congress/meeting and weekend tourism and thermal/health tourism). It is argued that Yalova has the potential to offer businesses an adequate infrastructure and urban services at relatively lower costs than Istanbul has. Yalova has one of the best telecommunication infrastructures in Turkey and is the national pilot city for an e-government/e-municipality project. For congress/meeting tourism, it is argued that Yalova also offers a good location given its proximity to three large cities, on the one hand, and for its unspoiled countryside, on the other.

20.4.1.7 Public Support for Alternative Development Paths and Power Relations

During the quantitative survey, Yalova residents were presented with three different alternatives in relation to the future development of the city. These alternatives comprised a simplified version of different perspectives that have been on Yalova's agenda for some time and were offered by accentuating their positive aspects. The first alternative relies on industry as the engine of economic growth; the second involves the agricultural and tourism sectors; and the third foresees the development of the IT industry and relevant training services. In Yalova, the position of the social actors differed on these alternative development paths and two separate coalitions were formed around them. The first group, comprising the local government, those involved in tourism, farmers/florists, women, pensioners and summer residents, have all listed their preferences as:

Agriculture-Tourism > Information Technology-University > Industry

The second group, consisting of industrialists, merchants and young people, made their selection as:

Industry > Information Technology-University > Agriculture-Tourism

Given these differing views and tendencies among the local people, two issues will be critical in shaping local politics in this regard.

The first issue is the distribution of wealth and power among these groups and the second issue is the contexts within which strategic decisions are taken at the local level. It seems that the existing large-scale industry is a very powerful actor at the local level and much better organised as well. However, its position could significantly shift on the basis of developments within Turkey and the outside world.

20.4.1.8 Public Awareness of Environmental Problems and Environmental Values

To date, environmental problems have not been mentioned among the most important problems of the city and the environmental goals are not a priority for the city dwellers. The province has no record of a resistance movement or organised protests to protect the environment. Though the Yalova people had some complaints about the health hazards and environmental problems created by the acrylic fibre factory (Aksa), it is difficult to argue that they have reacted in large numbers. This can be related to the long-term suppression of civil society in Turkey by the state.

However, the quantitative survey results, when compared to responses obtained through other studies conducted across Turkey and in Istanbul, clearly show that Yalova residents have become more aware of environmental issues as a result of Earthquakes, sea pollution and fears of deforestation. Moreover, Yalova residents value the environment at a much higher level than is the case in the rest of Turkey. These results are promising and, combined with some other external factors, can be a shaping factor in future local politics.

20.4.1.9 The Role of the Earthquake

Yalova is a province located in a first-degree earthquake risk zone and was hit severely by the 1999 Marmara Earthquake. Looking ahead, this could have important implications for future investment decisions and types of settlements in Yalova. The actual law of industrial zones is against the establishment of an organised industrial zone in Yalova, mainly for three reasons: first, it is near the sea; second, it has first-degree agricultural land; and third, it is in an earthquake-prone area. However, the record of earthquakes in Turkey, with repeated devastation in the same regions, indicates that the country is very quick in forgetting and in returning to business as usual. Moreover, it is well known that while Turkey has no shortage of legal regulations, it lacks the will to enforce these regulations.

Of course, Yalova is not a city in isolation and is influenced by several external factors. It seems that Yalova residents are aware of this fact as well. Weighted responses provided to the question 'What is the most important external driving force that will be influential in Yalova? And what is the second?' reveal that: 35% think the economic circumstances in Turkey will come to the fore; 20% consider that political and economic developments in the world will determine what will happen to Yalova. Looking beyond the internal driving forces, we now make an attempt to delineate a series of factors on regional, national and global scales that will affect, either directly or indirectly, the future of Yalova.

20.4.2 Driving Forces at the Regional Scale

20.4.2.1 Marmara Region as a Pole of Economic Development and Migration Pressures

In Turkey, the share of the total population living in cities was 25% in the 1950s and showed a continuous increase, reaching 65% in 2000 (TUIK, 2000). Işık and Güvenç(1999) expect that this share will continue to rise, albeit at a slower pace, from the present 65% to 85% over the next 25 years. They also argue that the national development perspective that provided some degree of coherence between the cities and regions has eroded rapidly since 1980 and that regional development dynamics have undergone a radical change.

The economic and industrial development in the Marmara Region can be discussed in this connection. The region is home to approximately 26% of Turkey's population and has the highest annual population growth rate: 26.7 persons per thousand in the period 1990–2000 (TUIK, 2000). When the pattern of change of population and energy consumption shares in the 1983–1990 and 1990–1996 periods is examined, it is seen that the city of Istanbul spread to neighbouring provinces to form the infrastructure of a new pattern of settlement. As such, it is possible to argue that a new process of metropolitan development is likely to take place in Turkey, especially in Istanbul. This is also said to be consistent with the regional distribution of population movements in Turkey, as a shift from the northern and north-eastern to the western, southern and south-eastern regions is seen. Therefore, in the early years of the twenty-first century, the Marmara Region is likely to act as a pole of economic development and experience further migration pressures, particularly over the next 15 years, despite the marked decline in the population growth rate (Işık & Güvenç, 1999). The resulting migration pressures could be substantial unless effective policy measures to deal with inter-regional disparities are taken.

20.4.2.2 Environmental Policy-Making in the Marmara Region

In Turkey, the existing laws and regulations that would prevent pollution and protect the environment have not been put into effect in a serious or timely manner. The Marmara Region has suffered greatly from this situation. Sea pollution, deforestation, emissions of pollutants and loss of agricultural land are on top of the list of problems.[9] In the *2004 Regular Report on Turkey*, although progress is acknowledged with regard to the transposition of the *environmental acquis*, problems in the implementation and enforcement are also underlined (CEC, 2004). It is also noted that a framework law on natural conservation and protection and implementing legislation on birds and habitats need to be adopted. In this context, though the southern part of Yalova has been categorised as an area to be protected in the Marmara Region, the fear is that members of the cabinet and big business interests might sacrifice these natural endowments for the sake of industrial development and/or rent-seeking.[10] Therefore, the uncertainty remains with regard to how (or whether) Turkey will fill the gap in environmental regulation and implementation.

20.4.3 Driving Forces at the National Scale

20.4.3.1 The Role of the EU and Turkey in the Long Road to Membership

Turkey has long sought to become a member of the EU, and the approval of Turkey's candidacy for full membership has transformed Turkey's EU membership project into a concrete reality. Although the negotiations could take as long as a decade, if not more, the EU accession process is a very good anchor for Turkey and even the start of the negotiations is the engine for political and economic transformation.

Öniş (2003, p. 29) argues that "for countries like Turkey with significant legacies on client politics, a powerful external anchor is a necessary if not a sufficient condition for undertaking and consolidating major internal reforms." Therefore, while underlining the fact that it is difficult to make a decisive break with populist cycles and the state's transformation in the absence of a powerful EU anchor, "the evolution of Turkey-EU relations and the kinds of signals provided by the EU are likely to have a crucial bearing on Turkey's economic performance over the course of the next decade" (Öniş, 2003, p. 29).

Of course, Europeans have reservations about extending membership to such a new large country. Turkey is seen as a more complicated candidate given that it is culturally different, populous and poorer. In the recent evaluations of Turkey's prospects for membership, apart from some structural weaknesses, the main obstacles mentioned were political and related to democratisation and human rights (e.g., freedom of expression). Also, at a more general level, it is argued that many of the changes in legislation have not yet been implemented (Larrabee & Lesser, 2003; Öniş, 2003).

In short, Turkey today stands at an important crossroad. There are a number of uncertainties surrounding Turkey's chances of acquiring full membership in the Union and the road ahead towards EU membership looks at best bumpy. In this situation, the outcome will surely depend as much on Turkey's motivation and success in meeting membership requirements as on developments within the EU itself in terms of its approach to Turkey, integration and further enlargement.

Should the EU fail to offer Turkey a clear road to membership, it is generally claimed that the economic reform process could easily be reversed (Öniş, 2003). Moreover, as Kirişçi (2002) and Larrabee and Lesser (2003) argue, this would probably make the Turkish elite more nervous about the nation's future stability and security and would strengthen already potent nationalist forces. Accordingly, the result would be a more inward-looking and sovereignty-conscious Turkey, one characterised, at least in the short-to-medium term, by considerable instability and insecurity and less capable of addressing its economic and political problems in a productive manner.

20.4.3.2 Turkey's Changing Political and Economic Context

An interesting observation about Yalova is that political developments at the national level, and in turn, Turkey's economic conditions, are seen as the most important forces likely to influence Yalova's future. Without doubt, Turkey's future political and economic conditions will be leading determinants of Yalova's role on the national and international scene and the direction of local policy in the coming years.

Historically, Turkey has been, and is still, subject to political instabilities because of tensions between the secular establishment, some ruling governments and unstable patterns of growth. The resulting political instability creates difficulties with implementing structural reforms and further increases the vulnerability of the economy to possible crisis (e.g., the twin economic crises in November 2000 and February 2001). Indeed, Turkey is ambitious to achieve further economic growth, but it must generate the political and economic stability needed both to speed up the process of domestic investment and to attract direct foreign investment (Öniş, 2000a; Öniş, 2003). Although, as mentioned earlier, the role of the EU as an external anchor is crucial and constitutes a powerful driving force to accelerate the kind of changes needed in Turkey's domestic politics and economic structure, the national policy framework is still very important.

20.4.4 Driving Forces at the Global Scale

20.4.4.1 Turkey as Part of World Politics and Regional Security

During the Cold War, Turkey was a key part of the Western defence system against the expansion of Soviet influence into the Eastern Mediterranean and Middle East. Larrabee and Lesser (2003, p. 1) indicate that "with the end of Cold War, many in Turkey and the West assumed a much reduced role for Turkey as a regional actor and as an ally of the West. These assumptions, however, proved unfounded. Rather than declining, Turkey's strategic importance has increased." In fact, Turkey has emerged as an increasingly important regional actor with substantial military as well as diplomatic weight, at the nexus of the strategically and economically important Middle East, Central Asia and Caucasus (Larrabee & Lesser, 2003). The September 11 attacks, the Iraqi War and oil and gas pipeline projects underlined the fact that the way in which Turkey evolves is important, both to the United States and to Europe. A serious threat to Turkey would have important political and security consequences for both. Moreover, Kaygusuz and Arsel (2005) note that the continued needs of the EU for energy make Turkey an 'energy terminal' or an 'energy corridor'.

This also means, as Öniş and Keyman (2003) argue, that it is no longer possible to separate national interests from the international, and investments in Turkey, both domestic and foreign, can very easily increase or be postponed due to changes in world politics, or by a fall or increase of confidence in domestic markets. For instance, Turkey has been affected negatively by the US-led war in Iraq, and the quantitative survey revealed that the people in Yalova are aware of this fact. The impact of political and economic developments in the world was listed as the second most important external driving force, after domestic issues, that will be influential in the future of Yalova.

20.4.4.2 Globalisation, Liberalisation and the Future of Social and Environmental Policies

While acknowledging the absence of a single definition or theory of globalisation, in a political economy framework, three apparent realities of today's globalisation processes are put forward. The first is associated with the increasing integration of worldwide products and capital markets – given technological advances in the transportation, information and communication systems – with international trade liberalisation and greater mobility of capital for efficiency, global prosperity and welfare in the long run (Eder, 2002; Gallopin et al., 1997). The second is associated with the uneven process of development that tends to aggravate inequality within countries (e.g., the North–South Divide), as well as within themselves by favouring certain regions or certain social groups over others. Increasingly powerful multinational corporations shift the location of their activities to low-wage economies. The third is that of increasing flows of energy and materials in the world economy, the increasing amount of some forms of pollution (such as CO_2), now also coming from the newly industrialised countries (e.g., China, India), and the displacement of raw material extractions from the peripheries of the world.

It is often argued and generally accepted that such a market-centred globalisation has changed the conditions under which local and national economic development occurs as the state's role in economic development is becoming increasingly redundant, if not irrelevant. For instance, Öniş (2000b) notes that the capacity of the nation-state to redistribute wealth and its ability to provide welfare for the poor have been severely undermined, at a time when the demand for protection is greater. Keyder (2003), instead of thinking of globalisation as only the expansion of productive activities, brings forward its qualitative aspects. First, globalisation entails a progressively more difficult escape from the constraints of international law and norms and second, it contributes to the creation of the so-called international civil society (e.g., the post-Seattle movement), bringing together diverse groups concerned with issues of human rights, the impact of global capitalism on workers, communities and culture, as well as the environment and ecology, and on the interrelations between them.

Given the current challenges of globalisation, the recent literature (CEC, 1999, de Mooij & Tang, 2003; Eder, 2002; Keyder, 2003) indicates that in reality there are two competing models in the world – US and European as alternative ways of shaping and governing globalisation and market processes. On the one hand, the US neo-liberal model is characterised by individualist and uncontrolled free market values, reductions in all social protection systems (e.g., unemployment and health benefits) and lack of environmental policies, which is in fact an imperialistic and control-oriented approach. On the other hand, the European social market model, as a credible alternative, proposes a new mode of governing well beyond the national politics at a supra-national level that tries to promote the principles of decentralisation, openness, multi-culturalism and participation. The orientation is towards the provision of a decent quality of life for all citizens, the question of social justice, in terms of the distribution of wealth and income together with the recognition of cultural rights as the focal point, and an emphasis on international environmental regulations (for instance, the *Kyoto Protocol*).

In this context, there is a debate in the contemporary literature regarding the competitiveness of the European social market model compared to that of the US model of capitalism. In today's globalisation the conventional wisdom is that the US model, and in turn the American model, looms larger. It is also true that the European social protection systems are already under pressure given the problems associated with unemployment and an ageing population. However, CEC (1999, p. 63) argues that although the process of globalisation is likely to continue as technology overcomes existing barriers and new countries join the global market, the possibility of an anti-globalisation backlash based on trade blocs, in which the increasing influence of regional groupings shapes the international economic order and its related social conditions, should not be dismissed. Specific concerns about health and the environment, or cultural identity, may also provide *foci* for more concentrated forms of resistance to full global economic integration.

20.5 Discussion

As already mentioned, one of the challenges in constructing city-scale scenarios is that there is need to integrate not only the social, economic and environmental perspectives, but also the external developments and local driving forces. Van Asselt et al. (1998) and EEA (2000) note that there is no blueprint to link local, regional and global perspectives in a common and consistent framework. Noronha et al. (2002) argue that the main constraints on sustainable futures at the local level, and sometimes at the national level as well, seem to emerge from the 'room to manoeuvre' available in making choices. In general, it is argued that when larger forces create available choices on a smaller scale, local actors react: they resist, cooperate, form alliances, adapt and/or accept bargains. What actually happens is the result of the dialectic of structural change and actors' responses. City-scale scenarios and strategic plans have to take these interactions into account.

In this context, it may be important to organise the driving forces touched upon in the previous section into two major groups according to their degree of uncertainty, as depicted in Fig. 20.2.

The left-hand side in Fig. 20.2 denotes drivers that are predetermined or invariant, the so-called givens, within the time frame of the present analysis (until the year 2020), both at local and at external levels.

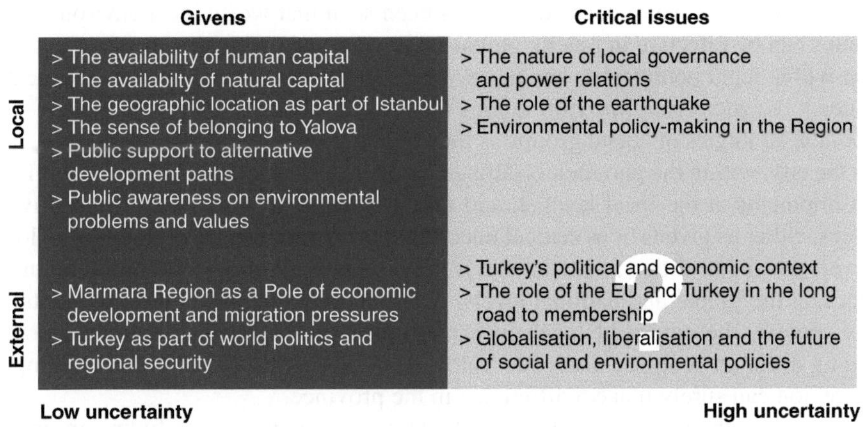

Fig. 20.2 The categorisation of driving forces as givens and critical uncertainties

The role these givens played and the way they affected various factors at local, regional and national levels were touched upon in the previous section. For instance, it was made clear that Yalova has a well-educated and young workforce, which encourages a variety of investments in the region. Similarly, the fact that Yalova is located in the Marmara Region, which is a pole of economic development, has profound impacts on its future and should be taken into account in determining the plausible combinations between city-scale and national-level scenarios.

Moreover, it should also be clear to the reader that Turkey's and Yalova's young and growing demography will put pressure on the domestic labour market in the short-medium term.

The right-hand side in Fig. 20.2 denotes local and external drivers with highest degree of uncertainty. The brief overview in section four made it clear that Turkey's quest for EU membership is no mere minor or technical issue, but something which lies at the heart of the country's present and future. As the accession negotiations began in October 2005, the already-considerable influence of the EU on Turkish economic, social and environmental policy-making is likely to intensify. Therefore, within the time frame of the current study, it is possible to merge two of the above-mentioned uncertainties, namely, Turkey's political and economic conditions and its relations with the EU. This being the case, the key external uncertainties remaining refer to globalisation, Europe and the future of social and environmental policies, on the one hand, and Turkey's relations with the EU, on the other. The resolution of these two critical uncertainties can strengthen a weak signal of change at the local level or become a driving force for changes in some other factors with a sequences of events linked in a logical and consistent manner.

Needless to say, while external forces can be particularly strong, cities are not influenced by these forces alone. Local driving forces and social actors and coalitions in the urban environment also play crucial roles in the future of the city.

For instance, on several occasions, it has been seen that world local environmental groups can be effective in raising global awareness and pressuring the state to function within legal boundaries, and to pay more attention to social and environmental issues.[11] Keyder (1999, p. 190) argues that "the evolving shape of the city results from the strategies of social groups as they negotiate an identity for themselves, and for the city, within the parameters defined by the forces of globalization." Therefore, developments at the local level should take particular account of the local driving forces, either as givens or as critical uncertainties. Without an understanding of local drivers and the coalition formations, it is not possible to combine the structural influences, at the global and national levels, with potential local-level developments. In this context, the nature of local governance and the future distribution of power among different coalition groups in Yalova and the environmental policy-making in the region can surely make a difference in the province.

Regarding the sector composition of Yalova's economy in the future, in principle, it can be argued that the chances of alternative development and infrastructure projects being implemented, for instance, those related to IT services, thermal tourism and congress tourism, will significantly increase with Turkey's EU membership. Supposedly, therefore, uncertainty with regard to accession to the EU would not only limit opportunities that Yalova might otherwise benefit from but also adversely affect its ability to establish the necessary business confidence and infrastructure for attracting capital. Therefore, the IT Services-University policy option can be considered more viable, if Turkey enters the EU.

Moreover, Yalova's strategic evolution can be considered as influenced by the chain of economic developments of which it will be part. To date, there is significant evidence that Yalova's large-scale industry is economically strong and enjoys a positive public image at the local level. They could expand their business in the region if they would get the opportunity. It is also worthy of consideration that the economic dimension and provision of employment opportunities are without any doubt primary concerns at the local level. In case there is no strong legal support, guidance or leadership from the European and national levels, it is difficult to think of any collective resistance or environmental movement in Yalova against further growth of the chemical industry. However, an EU which assumes the leading role in most global social and environmental concerns is likely to affect the legal and political structures in Turkey and to have a positive impact on the enforcement of environmental laws and regulations and the creation of a synergy between local actors and dynamics. One may also think that strong leadership and true commitment of some local NGOs, the governor or the mayor at local levels could also be crucial in giving priority to social and environmental issues and resisting external pressures. The reality for Yalova will, in the end, lie somewhere in between the interaction and combination of these external and local driving forces.

Given that the city itself is a unit that both influences and is influenced by the external environment, the establishment of cooperation and coordination efforts between cities, with a multi-layered system of governance from the local to the global, reveals itself as an important dimension. Such an undertaking is, in a sense, dependent on collaboration between the private sector, civil initiatives, local governments, the central government and international institutions. Obviously, the challenge is less a technical one than a profoundly political one. Governance, enhanced trust and the environmental consciousness of the residents at local level as well as the international agenda will play crucial roles in defining the terms of the debate and in pursuing more sustainable urban policies.

Notes

1. For details of the Yalova case study and Yalova Scenarios-2020, see Özkaynak (2005).
2. The aspect of coherency refers to the inclusion of all relevant dimensions and linkages between the various processes considered. Consistency implies that key assumptions are checked against each other.
3. Rothmans et al. (2001) note that 'givens' are relative to time and spatial scales. For instance, a given now may not be perceived to be a given in 2030 or 2050. Similarly, a given for Europe may not necessarily be a given for a region within Europe.
4. It is also important to note that an element can be predetermined and at the same time contain uncertain factors. Ravetz et al. (2000) indicates that while climate change, for instance, is widely accepted, the actual results and impacts are still very uncertain.
5. The Open Society Institute and the Boğaziçi University jointly financed the fieldwork of the project. The technical support of the Autonomous University of Barcelona and Begüm Özkaynak's IGSOC scholarship are also gratefully acknowledged.
6. See: http://www.akkok.com.tr.
7. See: http://www.aksa.com.
8. The share of agricultural labour force in Turkey declined constantly from 50% in 1988 to 33% in 2002 and will continue to decline (the pace of which will depend on the cut of the subsidies). Note that this share is 4% for EU-15 and that the average for accession countries is 15%.
9. According to the EEA (2004, p. 13), Turkey is a country "with an estimated 9,000 species, has the richest flora of any country in Europe, with an estimated 2,800 endemic species". Marmara Region, being a point of juncture between Europe and Asia, is part of this fauna, of course.
10. In Turkey, all forest areas (100%) are owned and managed by the state.
11. For a general account of few, but still important, environmental movements in Modern Turkey – mainly the Bergama movement and the oppositions to Akkuyu nuclear power plant and to the Ilısu Dam – see Kadirbeyoğlu (2005) and Arsel (2005).

Acknowledgements This chapter draws on Begüm Özkaynak's Ph.D. thesis, 2005, Universitat Autonoma de Barcelona. The support of Özkaynak's IGSOC scholarship is gratefully acknowledged.

References

Alcamo, J. (2001). *Scenarios as tools for international environmental assessments*. Environmental Issue Report 24. Copenhagen: European Environmental Agency.

Arsel, M. (2005). The Bergama Imbroglio. In F. Adaman & M. Arsel (Eds.), *Environmentalism in Turkey: Between democracy and development* (pp. 263–276). Aldershot: Ashgate.

Beauregard, R. A. (1995). Theorizing the global-local connection. In L. Knox & P. J. Taylor (Eds.), *World cities in a world system* (pp. 233–248). Cambridge: Cambridge University Press.

Castells, M. (2000). Urban sociology in the twenty-first century. In I. Susser (Ed.), *The Castells reader on cities and social theory* (pp. 390–406). Oxford: Blackwell.

Clayton, A., Wehrmeyer, W., & Ngubane, B. (2003). *Foresighting for development*. London: Earthscan.
Cohen, B. (2004). Urban growth in developing countries: A review of current trends and a caution regarding existing forecasts. *World Development, 32*(1), 23–51.
CEC (Commission of the European Communities). (1999). *Scenarios Europe-2010. Five possible futures for Europe*. Bruxelles: Forward Studies Unit of the European Commission.
CEC (Commission of European Communities). (2004). *Regular report on Turkey's progress toward accession*. Bruxelles: Commission of European Communities.
de Mooij, R., & Tang, P. (2003). *Four futures of Europe*. The Hague: Centraal Planbureau.
Eder, M. (2002). The challenge of globalization and Turkey's changing political economy. In B. Rubin & K. Kirişçi (Eds.), *Turkey in world politics: An emerging multi-regional power* (pp. 249–284). Istanbul: Boğaziçi University Press.
EEA (European Environmental Agency). (2000). *Cloudy crystal balls*. Environmental Issues Series 17. Copenhagen: European Environmental Agency.
EEA (European Environmental Agency). (2004). *Mapping the impacts of recent natural disasters and technological accidents*. Environmental Issue Report 35. Copenhagen: European Environmental Agency.
Flyvbjerg, B. (2001). *Making social science matter: Why social inquiry fails and how it can succeed again*. Cambridge: Cambridge University Press.
Gallopin, G. C., Hammond, A., Raskin, P., & Swart, R. (1997). *Branch points: Global scenarios and human choice*. Global Scenario Group. PoleStar Series Report 7. Stockholm: Stockholm Environment Institute.
Gallopin, G. C., & Rijsberman, F. (2000). Three global water scenarios. *International Journal of Water, 1*(1), 16–40.
Ghanadan, R. (2002). *Choices ahead: Three alternative development scenarios for California*. Berkeley, CA: The Nautilus Institute for Security and Sustainable Development [http://www.nautilus.org/archives/pub/ftp/napsnet/special_reports/RHGMAPaperNautilus062202.pdf; accessed 2 February 2007].Report of The Energy and Resources Group.
Gibson, C. C., Ostrom, E., & Ahn, T. K. (2000). The concept of scale and the human dimensions of global change: A survey. *Ecological Economics, 32*(2), 217–239.
Healey P., Cameron S., Davoudi S., Graham S., Madanipour A. (Eds.) (1995). *Managing cities: The new urban context*. London: Wiley.
Işık, O., & Güvenç, M. (1999) *The changing nature of Turkish urbanization and economic geography on eve of the 21st century: New challenges and new opportunities*. TUSİAD Report, pp. 52–83.
Kadirbeyoğlu, Z. (2005). Assessing the efficacy of transnational advocacy networks. In F. Adaman & M. Arsel (Eds.), *Environmentalism in Turkey: Between democracy and development* (pp. 101–116). Aldershot: Ashgate.
Kaygusuz, K., & Arsel, M. (2005). Energy politics and policy. In F. Adaman & M. Arsel (Eds.), *Environmentalism in Turkey: Between democracy and development* (pp. 149–166). Aldershot: Ashgate.
Keyder, Ç. (1999). *İstanbul: Between the global and the local*. Lanham, MD: Rowman and Littlefield.
Keyder, Ç. (2003). *Memalik-i Osmaniye'den Avrupa Birliği'ne*. (*From the Ottoman Empire to the European Union*). Istanbul: İletişim Publications.
Kirişçi, K. (2002). US-Turkish relations: New uncertainties in a renewed partnership. In B. Rubin & K. Kirişçi (Eds.), *Turkey in world politics: An emerging multi-regional power* (pp. 122–150). Istanbul: Boğaziçi University Press.
Larrabee, F. S., & Lesser, I. O. (2003). *Turkish foreign policy in an age of uncertainty*. Arlington: RAND Publications.
Martinez-Alier, J. (2002). *Environmentalism of the poor: A study of ecological conflicts and valuation*. Cheltenham: Edward Elgar.

Munda, G. (2006). Social multi-criteria evaluation for urban sustainability policies. *Land Use Policy, 23*(1), 86–94.

Noronha, L., Lourenco, N., Lobo-Ferreira, J., & Lleopart, A. (2002). *Coastal tourism, environment and sustainable development*. New Delhi: TERI Publications.

Öniş, Z. (2000a). The Turkish economy at the turn of the new century: Critical and comparative perspectives. In M. Abramowitz (Ed.), *Turkey's transformation and American policy*. Washington, DC: The Century Foundation Press.

Öniş, Z. (2000b). Neoliberal globalization and the democracy paradox: Interpreting the Turkish general elections of 1999. *Journal of International Affairs, 54*(2), 23–40.

Öniş, Z. (2003). Domestic politics, international norms and challenges to the state: Turkey-EU relations in the Post-Helsinki era. In A. Çarkoğlu & B. Rubin (Eds.), *Turkey and the European union* (pp. 9–34). London: Frank Cass.

Öniş, Z., & Keyman, F. (2003). *Helsinki, Copenhagen and beyond: Challenges to the New Europe and the Turkish state*. Paper presented at the conference on Cyprus' European accession and the Greek–Turkish rivalry, Yale University, New Haven.

Özkaynak, B. (2005). *Indicators and scenarios for urban development and sustainability*. Ph.D. Thesis. Spain: Autonomous University of Barcelona.

Raskin, P., Banuri, T., Gallopin, G., Gutman, P., Hammond, A., Kates, R., et al. (2002). *Great transition: The promise and lure of the times ahead*. Stockholm: Stockholm Environment Institute [http://www.tellus.org/seib/publications/Great_Transitions.pdf; accessed 2 February 2007].SEI PoleStar Series Report 10, Global Scenario Group.

Ravetz, J., Gough, C., & Shackley, S. (2000). *Urban development 2050. Draft scenario report from the new futures workshop*. Manchester: Manchester University.

Rothmans, J., van Asselt, M., Anastasi, C., Rothman, D., Greeuw, S., & van den Bers, C. (2001). *Integrated visions for a sustainable Europe. Final report*. Maastricht: ICIS.

Schwartz, P. (1991). *The art of the long view: Planning for the future in an uncertain world*. New York: Currency Doubleday.

TUIK (Turkish Institute of Statistics). (2000). *Census of population: Social and economic characteristics of population (by province)*. Ankara: Turkish Institute of Statistics.

TUIK (Turkish Institute of Statistics). (2002). *Population statistics of Turkey*. Ankara: Turkish Institute of Statistics.

UNEP (United Nations Development Programme). (2002). *Global environment outlook* (Vol. 3). London: Earthscan.

van Asselt, M. B. A., Storms, C. A. M. H., Rijkens-Klomp, N., & Rothmans, J. (1998). *Towards visions for a sustainable Europe: An overview and assessment of the last decade in European scenario studies*. Maastricht: ICIS.

van Notten, P. W. F., Rotmans, J., van Asselt, M. B. A., & Rothman, D. S. (2003). An updated scenario typology. *Futures, 35*(5), 423–443.

Yeşil Atlas. (2003). *Turkey's natural heritage: Zero extinction* (Vol. 6). Istanbul: Doğan Yayıncılık.

Chapter 21
Thinking Through Complex Values

Maria Cerreta

21.1 Introduction

Cities and territories are called upon to face strategic challenges of sustainable human development, based on the complexity of the interacting perspectives, interests and preferences of decision-makers and stakeholders, taking into account the existing resources and different forms of capital (human, social, economic, environmental, man-made, cultural, etc.) and their links and mutual relations (Fusco Girard, Forte, Cerreta, De Toro, & Forte, 2003; Kirdar, 2003).

In this perspective, integrated approaches to decision support for strategic planning can help to generate more efficient and effective results than sectoral approaches and, at the same time, are able to work in a multi-dimensional and cross-sectoral (inter-/trans-/multi-/sectoral) decision space (Wiek & Walter, 2009).

Indeed, integration is a complex concept, characterized by different dimensions that need to be defined and explored; it involves vertical and horizontal processes, which can be diffuse, fluid and multi-directional rather than rigid, hierarchical and unilinear; and it is related to different forms of spatial development activity (Allmendinger & Tewdwr-Jones, 2006). Integration in evaluation approaches means considering the dynamic interactions between different contextual dimensions, able to combine *existing relationships* and explore the potential to build *new relationships*.

The particularities and specificities of the context, in turn, suggest that the most appropriate integrated approach will depend on the *nature* of the decision-making situation in question (Leknes, 2001; Mayer, van Daalen, Els, & Bots, 2004) and on the manner of addressing it. Therefore, any given situated decision problem must be identified according to a multi-dimensional perspective.

Indeed, a decision-making situation can be considered an *opportunity* rather than a *problem* (Keeney, 1992), in which strategic thinking may creatively suggest further alternatives, starting from the awareness of existing values. Values not only guide

M. Cerreta (✉)
Department of Conservation of Architectural and Environmental Heritage, University of Naples Federico II, 80132 Naples, Italy
e-mail: cerreta@unina.it

M. Cerreta et al. (eds.), *Making Strategies in Spatial Planning*,
Urban and Landscape Perspectives 9, DOI 10.1007/978-90-481-3106-8_21,

the creation of suitable alternatives but also support the identification of decision situations.

"Value-focused thinking addresses the large void between unstructured creative thinking without bounds and very structured approaches to decision problems. It is the structuring of thinking to address decision opportunities and problems in creative ways" (Keeney, 1992, p. 8).

Value-focused thinking suggests a different paradigm for addressing decisions from the standard alternative-focused-thinking paradigm. This approach includes the following steps: the allocation of significant effort to articulating values, the articulation of values before other activities in decision situations and the use of the articulated values to identify decision opportunities and create alternatives.

In-depth and thorough understanding of the values inherent in a decision situation can provide important and sensitive insights into all aspects of decision-making and help to improve the decision process with synergic effects on the identification of opportunities and creation of alternatives. The recognition of existing values is closely linked to the identification of decision opportunities and guides the strategic thinking process.

Recognising the role of values means acknowledging the imperative need to move beyond the instrumental aspects of practice, starting from questions such as *what values? whose values? values for whom? values based on what point of view? values based on what kind of priority?*

The value-focused thinking approach considers as an essential assumption a multi-dimensional vision of value, a complex perspective, according to which it is possible to *integrate* values belonging to different and multiple dimensions.

In an integrated decision-making approach, *thinking through complex values* implies the inclusion of a multi-dimensional perspective, taking into account tangible and intangible values, *hard* and *soft* values, objective and subjective values, use values, non-use values and intrinsic values (Fusco Girard & Nijkamp, 1997 and Chapter 17, this book) and their synergic and complementary relationships.

Thinking through complex values means *thinking across boundaries*, considering soft spaces and fuzzy boundaries, overcoming different kinds of limits, and having plural 'insights' in order to formulate a 'situated strategy' (Liew & Sundaram, 2009) addressing a 'situated decision problem'.

This chapter explores how *thinking through complex values* can support the structuring of integrated decision-making by orienting it towards the elaboration of strategic goals and actions able to create new values from the plurality of knowledge and the specificity of the context.

With its normative, spatial, temporal, cultural, social and cognitive features, the context becomes the frame in which planning responses and behaviours[1] can be shaped.

In its first part, this chapter explores the connection between values, knowledge and strategies, focussing on their interdependencies. Values make explicit the relations between different knowledge forms; conversely, the interaction of knowledge makes it possible to recognize values. At the same time, knowledge orients value and value represents the *measure* of knowledge (Zeleny, 2006 and Chapter 15, this book).

The second part of the chapter discusses the role of evaluation within an integrated perspective, which is seen as an 'opportunity' to elaborate strategies and 'organize hopes' (Forester, 1999; Sandercock, 2003) in spatial planning.

The integrated perspective considers evaluation as an activity embedded in the planning process and supporting many other activities in that process, each time playing a different role (Alexander, 2006). Within an integrated perspective evaluation underpins the dialogue between knowledge and values in order to translate such dialogue into the planning of strategic objectives and actions (Friedmann, 1987); it enables the identification of relevant values and related meanings, the exploration of opportunities and the creation of alternatives; it measures possible impacts and effects while managing complex and multiple priority systems.

The third part of the chapter focuses on three case studies, in which the evaluation process was structured in an integrated perspective guided by complex value-focused thinking and based on a 'combinatorial philosophy'. The use of combinatorial assessment methodologies is becoming a widespread practice (Deakin, Mitchell, Nijkamp, & Vreeker, 2007; Fusco Girard, Cerreta, & De Toro, 2005; Krönert, Steinhardt, & Volk, 2001; Medda & Nijkamp, 2003; Miller & Patassini, 2005; Munda, 2008). They are seen as flexible tools able to overcome the limits of each single method, accommodate a multi-dimensional and plural perspective and improve the quality of the decision-making process.

These three cases represent different attempts to identify complex values as premises for the process at hand and to exploit the plurality and diversity of knowledge in order to identify situated strategies.

Finally, this chapter reflects the strengths and weaknesses of integrated approaches and highlights the need to view evaluation and planning as reciprocally embedded, mutually shaping activities. It may well be argued that within the field of integrative approaches, the recognition of value (economic, non-economic and intrinsic) assumes a fundamental role and is closely linked to different forms of knowledge. Through their interaction strategic objectives and evaluation criteria are identified, scenarios constructed, decisional rules deduced and sectoral assessments implemented in order to create and prioritise alternative options. The use of a combination of techniques penetrates and includes informal, 'soft spaces' of decision, able to complement the more formal process, combining flexible and functional approaches with formal development plan strategies (Allmendinger & Haughton, 2009), and considering decision support *versus* discussion support (Rinner & Bird, 2009).

21.2 The Interplay of Knowledge and Values

Thinking through complex values in spatial strategy-making takes on an 'exploratory' meaning. The way in which we deal with questions of value in planning was examined by Campbell (2002), who analysed how planners can make situated ethical judgements, based on a critical understanding of a given decision context. According to Campbell, in a world where knowledge can only be partial and transitory, we must rely on judgement, and that fundamental to the process of judging between better and worse, is the *question of value*.

In fact, actions cannot be value-free. Thus, explicit consideration needs to be given to the nature of the values that our decision processes and outcomes are seeking to promote. That is why planning situations require evaluation methods based on complex value-focused thinking: this helps to articulate values, identify decision opportunities and create alternatives.

Values and cognitive perceptions are two of the most significant influences on decision-making which affect how decision-makers, decision-takers and stakeholders will interpret and respond to particular stimuli and sets of conditions.

In complex, uncertain and conflict-ridden planning contexts, different categories of values can be identified: direct-use values, indirect-use, non-use and intrinsic values. The explicit recognition of the existence of multiple interdependent values establishes both the conceptual and empirical foundations for understanding just how these value categories may be applied to the planning context.

This means becoming aware of the 'complex social value' of a context and its resources (Fusco Girard, 1987 and Chapter 17, this book). Thus, the explicit recognition of the existence of multiple interdependent values makes it possible to include instrumental and intrinsic values in evaluation. Further, by prioritising values we can distinguish between them, highlight different perspectives and take into account various kinds of conflicts.

Intrinsic value allows us to move beyond the private sphere and reflect on collective benefits and externalities, explicating a clear ethical dimension. It expresses the 'glue value', the system of immaterial relations, its specific character and its particular identity (Fusco Girard & Nijkamp, 1997, 2004). It is a proactive value, capable of constructing integration, reducing marginalisation, overcoming fragmentation and stimulating vitality: a 'catalyst' of material and immaterial energies, able to blend various value dimensions, helping to capture its deep unity (Fusco Girard, Cerreta, De Toro, & Forte, 2007).

Intrinsic value is consistent with the concept of *value complex* formulated by Zeleny (1998, 2005) and conceived as a 'meta-criterion', anchored and integrated in fundamental values that are broadly accepted and not subject to choice. The value complex is the expression of a cognitive equilibrium, characterised by candour and trust, based on principles, ethics and rules, mostly qualitative and expressible only in imprecise and fuzzy language, but rooted in specific contexts.

Recognising all the diverse categories of values implies the recognition of the multiplicity and diversity of knowledge. Some questions arise when dealing with the interplay of knowledge and values: does knowledge affect values, and if so, how? How can values be acknowledged, managed and assessed? What knowledge is necessary to do this?

Any representation of a complex system reflects only one subset of its possible representations (Giampietro, Allen, & Mayumi, 2006; Munda, 2008). A consequence of these deep subjectivities is that, in any normative exercise connected to a public-decision problem, one has to choose an operational definition of 'value'. This in spite of the fact that social players with different interests, cultural identities and goals may have different definitions of value. Consequently, to formulate a ranking of policy options it is first necessary to decide what is important for different social players and what is relevant for the representation of the real-world entity described in the model (Funtowicz, Martinez-Alier, Munda, & Ravetz, 2002; Munda, 2004).

Multiple values correspond to multiple forms of knowledge. Relations and dynamics between knowledge and values are not linear and reflect different value interpretations within a self-feeding process.

According to Zeleny (2006 and Chapter 15, this book), knowledge, in its multiple forms, is measured by the *added value* that coordination of effort, action and process adds to the inputs of material, technology, energy, services, information, time and so on, necessary to produce knowledge. Therefore values represent the measure of knowledge and must be socially recognised and accepted.

In a decision context and in activity such as spatial strategy-making, fostering encounters between diverse forms of knowledge means understanding the existence of different values. Exploring the landscape of knowledge forms (*KnowledgeScapes* according to Matthiesen, 2005) means exploring the landscape of values too, in a space of coexistence and dynamic interaction. Simon (1983) highlighted the importance of 'social decision-making' and articulated the ways in which values, alternative knowledge and preferences derive from interactions with the social environment. The context plays a crucial role and can be considered as the decision-making environment, which is spatial and scalar in nature (Larner & Le Heron, 2002).

Thinking through complex values focuses on the structuring of the decision problem (Mingers & Rosenhead, 2001), in order to address complex problem situations, that is, situations that are 'ambiguous', 'ill-structured', 'wicked', 'messy', 'intractable' or difficult to manage (Cats-Baril & Huber, 1987; Rittel & Webber, 1973; Rosenhead, 1989; Schön & Rein, 1994), characterised by the existence of multiple players, multiple perspectives, incommensurable and/or conflicting interests, important intangibles and key uncertainties. According to Mingers and Rosenhead (2001), problems of this kind are more 'strategic', in the sense that they set the 'givens' of well-structured problems (Ackoff, 1979; Checkland, 1985; Rittel & Webber, 1973; Schön, 1987). This perspective aims at considering different aspects or dimensions of a problem situation, rather than different types of problem (Mingers & Brocklesby, 1997).

Consistently with the complex-values approach, problem situations are closely related to the decision-making environment, which is strongly dependent on the interaction between knowledge and values.

Complex values are, in fact, linked to the context and to the decision frame (Strauss, 2008) and emerge from the cognitive frame shaping the physical, environmental, social and economic environment.

Complex values should be the driving force for any decision-making process: they help to explore the decision context and structure the problem, by guiding information collection, uncovering hidden objectives, improving communication, facilitating involvement in multiple-stakeholder decisions and interconnecting decisions. Complex values are 'strategic values' able to guide strategic thinking, 'discover' decision opportunities and create alternatives. They are embedded in a given problem context.

Exploring a broader decision context can open up decision opportunities (Keeney, 1996), and the decision situation is tailor-made for complex value-focused thinking. Values are, at one and the same time, the values of context resources and those of the stakeholders involved in the decision-making process. The creation of alternatives involves 'interdependent values' that guide 'interdependent and interconnected decisions'. Strategic objectives can be considered as being the representation of values, which also specifies what is relevant for the strategy and the final decision.

The relationship between multiple knowledge, multi-dimensional values and possible strategies is fluid, dynamic and incremental and requires continuous interaction among/with the local stakeholders and decision-makers. This relationship develops progressively through continuous feedback thus activating and maintaining learning mechanisms.

Full awareness of relevant values in a decision context depends on the different kinds of information and knowledge, *hard* data and *soft* data that characterise the decision opportunity. It is necessary to move beyond multiple and complex values, towards the realisation that value judgements need to be made in the face of multiple and often conflicting ways of valuing (Richardson, 2005). Values are critical in determining how evaluation is carried out, from its inception, through the process at every stage: they are a key part of the decision-making process.

21.3 Towards Complex Multi-Method Evaluation Systems

Evaluation can assume different meanings and roles within decision-making processes, especially if it is related to spatial planning.

E. R. Alexander (2006), after highlighting the fact that evaluation is intrinsic to all types of decision-making, focuses in particular on 'evaluation in planning'. Indeed, the idea of 'evaluation *within* planning' (Fig. 21.1) seems to better interpret the concept of planning-evaluation proposed by Lichfield (1996) where the binomial makes explicit the close interaction and reciprocal framing of evaluation and planning: evaluation is conceived as deeply embedded in planning, affecting planning and evolving with it.

The evolution of evaluation methods reflects their evolving relationship with the planning process and also the way in which they interact with the diversity and multiplicity of knowledge and values.

In *Evaluation in planning* (2006), Alexander traces the history of the evolution in planning-evaluation and identifies four main generations of methods, "that represent progress from empirical positivism to post-positivist interaction" (Guba & Lincoln, 1989, quoted in Alexander, 2006, p. 11).[2] However, a direct match between planning-evaluation methods, planning models and form of rationality is not so obvious, although the diffusion of new paradigms and the identification of new

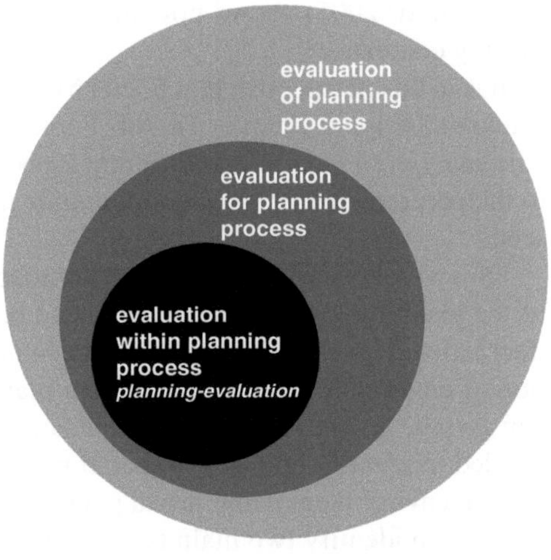

Fig. 21.1 Evaluation in relation to the planning process

rationalities has activated the development of new approaches and methods towards complex multi-method evaluation systems (Alexander, 2006; Deakin et al., 2007; Miller & Patassini, 2005).

These approaches move away from 'traditional evaluation' to embrace the integrated evaluation process, promoting wider communicative interaction with stakeholders in a dialectic and mutual learning process. Cognitive limitations, behavioural biases, ambiguity and variability of preferences and norms influence collective choices, where facts are uncertain, values in dispute, stakes high and decisions urgent (Funtowicz & Ravetz, 1991); this requires the adoption of evaluation tools that are scientifically sound, transparent with regard to the decision-making process, but also of the participatory type. One response to these questions has been the development of more integrated forms of assessment, variously called Integrated Assessment or Integrated Appraisal (IA), Integrated Impact Assessment (IIA) (Bond, Curran, Kirkpatrick, & Lee, 2001), Sustainability Assessment or Appraisal (SA) (Pope, Annandale, & Morrison-Saunders, 2004), Environmental Impact Assessment (EIA), Strategic Impact Assessment (SIA) (Partidario, 2000) and Strategic Environmental Assessment (SEA) (Fischer, 2007).

Moreover, focusing on the impact prediction does not guarantee the integration of multi-dimensional values in the decision-making process, nor does it take into account the many and diverse phases of spatial transformation. Strategic Environmental Assessment can be considered as a meaningful methodological environment for testing the applicability of the 'planning-evaluation' concept, moving beyond the impact assessment mindset (Bina, 2007) and integrating different and complementary approaches and methods oriented to strategic planning. It opens up possibilities for more inclusive and ongoing engagement processes, which are transdisciplinary, committed to methodological pluralism, participatory and context situated (Carlsson-Kanyama, Dreborg, Moll, & Padovan, 2007).

Similarly, the application of different methods in combination with Spatial Decision Support Systems (SDSS) and Problem Structuring Methods (PSM)

(Rosenhead, 2005) shows how multiple perspectives can be included in decision-making contexts.

It is in this perspective that Social Multi-Criteria Evaluation (SMCE) proposed by Munda (2004 and Chapter 18, this book), which extends the field of social benefit-cost analysis to incorporate different aspects referring both to impact evaluation and to the participation of local communities in the decision-making process should be seen.

On several occasions evaluation approaches that are functional to the application of the planning-evaluation concept have been considered to be 'planning tools'. At international level, many tools have been developed in order to facilitate the integration of environmental values in urban planning (de Roo & Visser, 2004; Rotmans, van Asselt, & Vellinga, 2000; Runhaar, Driessen, & Soer, 2009). These tools are applied to identify and assess spatial functions and process aspects and aim at integrating environmental and urban planning. According to Runhaar et al. (2009), it is possible to identify two main types of planning tool:

1. *Substance-oriented tools*, which take the form of knowledge on the state of the urban environment through indicators, Geographic Information System (GIS) and so on, and can be used to produce knowledge. This category includes computer-based Planning Support Systems (PSSs), covering a wide range of geo-information technologies, which are used to visualise environmental conditions and explore the effects of spatial developments.
2. *Process-oriented tools*, which facilitate dialogue, consensus-building and negotiation, and stimulate the search for and development of creative solutions, based on an interactive planning approach (Amler et al., 1999; Susskind, McKearnan, & Thomas-Larmer, 1999; Valentin & Spangenberg, 2000).

Hybrid planning tools, able to combine substance-oriented tools and process-oriented tools, aim to facilitate integration of environmental and urban planning, flexible modelling and support of interrelated decisions (Liew & Sundaram, 2009).[3]

Evaluation methods developed in planning, despite their widespread application, have been unable to bridge the gap between theory and practice. According to Alexander (2006), the challenge is to create an evaluation framework that is "responsive to complexity, transparent for communication, and enable effective interaction" (p. 274), seen as an *arena* for debating and resolving conflicting claims.

21.4 Integrated Evaluation Approaches: Some Situated Experimentations

Integrated evaluation approaches may enable the interpretation of material and immaterial relations characterising a context, the acknowledgement of existing tangible and intangible values, and the creation of strategies aimed at the production of new values and at the sustainable development of many local resources in a multi-dimensional perspective.

There is a need for developing useful models of contingent situated application, facilitating the development and institutionalisation of complex multi-method evaluation systems, which take into account specific purposes and are linked to the specific context.

In order to understand just how integrated evaluation approaches can be translated into complex multi-method evaluation systems and become planning tools, three Italian case studies are presented. They focus mainly on the evaluation/planning relation within the design of local development strategies: the first concerns the integrated plan for the Altilia-Saepinum archaeological site (in the Molise region); the second, the strategic plan for the Buccino historic centre (in the Campania region); and the third, the role of evaluation in the design of the Cava De' Tirreni master plan (also in the Campania region).

In spite of the very different scales and geographic specificities of these three cases, the same methodological framework was applied, adapted to each case by using the communication, analysis and evaluation methods and techniques best suited to each planning context. Each of them explores the relation among knowledge mobilisation, values identification and the construction of spatial strategies.

An internal path developing in small steps, able to 'integrate' and 'keep together' diverse elements is the leitmotif that runs through these three cases; it has an incremental rather than a cascade nature thanks to continuous feedbacks. This leitmotif underpins the methodological framework common to the three cases and, consistently with the Keeney's 'value-focused thinking' approach (1992, 1996), is conceptualised as *thinking through complex value* and is divided into four phases:

1. *recognising a decision problem/opportunity* – this phase is related to the definition of the problem situation and aims at highlighting the multiple dimensions of the context (spatial, geographic, economic, social, environmental, anthropologic and cultural) by analysing *soft* and *hard* data and activating various forms of knowledge (explicit, systematized, experiential/practical-contextual, implicit, etc.) (Healey, 2008) with respect to specific needs; the ultimate goal of this phase is to identify spaces for action representing opportunities for local development and not only solutions to a specific problem;
2. *specifying values* – this phase aims at identifying the values embedded in the activated knowledge; special attention is paid to capturing and exploring local complex social values, to explaining the situated frame and to disclosing existing strategic objectives and any existing conflicts;
3. *creating alternatives* – the action space is explored in terms of decision opportunities within a broader decision-making context able to deal with shared knowledge and values and fostering plural micro-decisions in the light of local potential and criticalities;

4. *identifying strategic actions* – this phase responds to the need for defining 'preferable choices' arising from dynamic and flexible learning processes and expressing integration and complementarity between procedural strategies and transformative strategies (Bina, 2008).

The three experiences are described by analysing main issues, methods adopted and outputs.

21.4.1 Altilia-Saepinum: From Archaeological Site to Territorial Catalyst

The archaeological site of Altilia-Saepinum[4] is an interesting Roman city, founded in 293 BC, situated on the Matese mountainside. It belongs to the municipal territory of Sepino (CB), a farming village in Molise and a well-known cultural and thermal centre. Sepino is also the centre of many tourist itineraries, combining 'the pathway of transhumance' and 'the Samnite civilization'; moreover, the Altilia-Saepinum site is situated at the intersection of two roads of age-old significance. Despite its important historical heritage and its relations with the social, economical and environmental context of the Campobasso Province, the site's potential is still largely untapped.

However, enhancement of this site is recognised by local institutions and inhabitants as key to creating a competitive territorial system leveraging on sustainable tourism, involving a larger geographical area, including other municipalities of the Campobasso province and comprising a total of about 18,500 inhabitants. These are small- and medium-sized communities with strong ties to their land, where the prevailing farming and craft economy limits changes, but guarantees the persistence of strong links between the natural and built environment, constituting an important *complex value* for this territory.

In 2007, the Sepino municipality and the regional archaeological heritage authority launched a strategy-making process with the aim of developing a program for submission for EU funding. The original stimulus for this initiative came from the cultural association 'Friends of Sepino' which was concerned about the future of the site and wished to promote its potential. Thus a decision-making process was launched to identify a strategy for the enhancement of the Altilia-Saepinum site, following the steps illustrated in Fig. 21.2.

The first step comprised in-depth analysis of *hard* and *soft* data. The main outputs of this phase were the identification of: (1) connections between Sepino and other key archaeological sites and urban centres in the region and in the whole of southern Italy; (2) plans and programs responsible for the current transformation of the area and at times causing stagnation rather than development. This binomial dispelled the perception of Sepino as an isolated resource and suggested that this site should be seen as a resource able to kick-start the stagnant economy through the development of tourist-related activities. The site's location and strong links

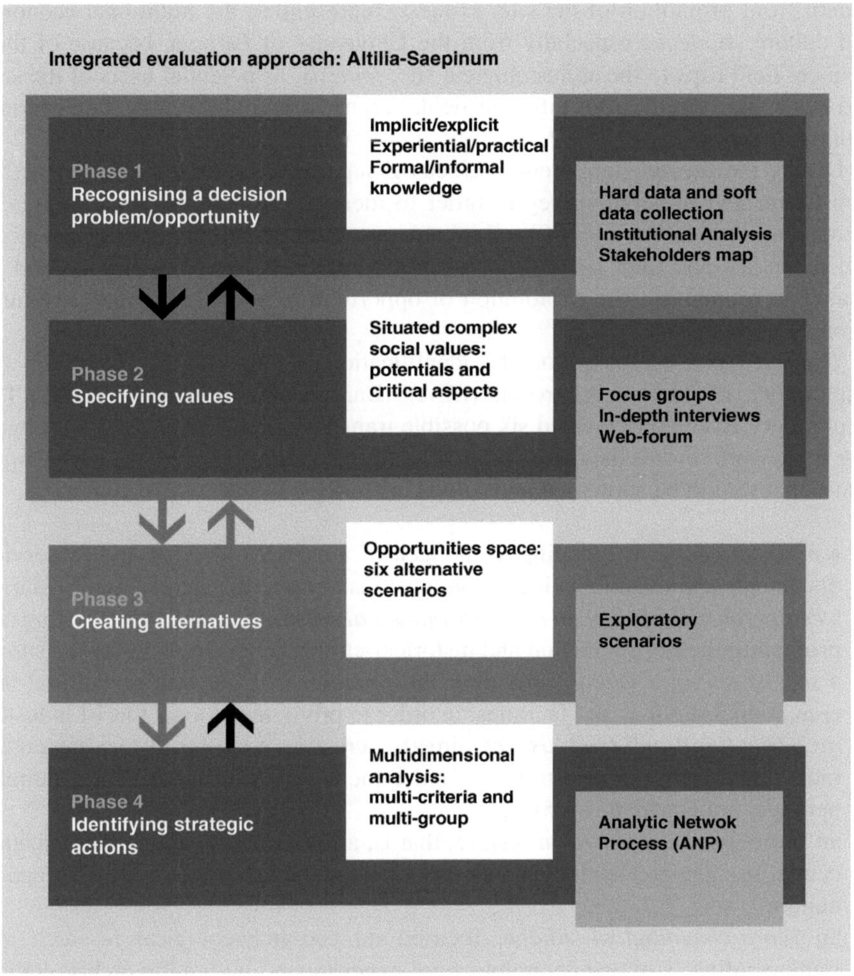

Fig. 21.2 Methodological approach: steps and tools

with other historical and cultural heritage sites made this opportunity feasible and suggested the possibility of reframing the site within a more complex and diversified archaeological system.

An Institutional Analysis (IA) (De Marchi, Funtowicz, Lo Cascio, & Munda, 2000) was carried out to transform this opportunity into reality. To this end, the first phase concluded with the identification of both relevant decision-makers and stakeholders. In particular, decision-makers were selected among those institutions able to follow the transformation process up to the implementation and management stage; stakeholder categories acknowledged as relevant included some *environmentalist groups*; *sports and cultural associations* directly involved in activities concerning this site; local *restaurateurs*, *retailers* and *small businesses*, who would benefit from promotion of the site; *farmers*, representing the traditional economy and culture; *students,* especially from the University of Perugia, because of their frequent field trips to the archaeological site; *tourists*, as potential users of the site; and academic experts from different fields (archaeology, philosophy, engineering, economics, etc.).

During the second step, thematic focus groups were organised together with a web-forum and in-depth survey in order to identify the cultural, social, environmental and economic views of local inhabitants, their perception of critical aspects and potentials and their preferences for the complex of Altilia-Saepinum and its related surroundings. The exploration of opportunities started here and continued throughout this phase.

The third phase focused on the identification of opportunities based on the Exploratory Evaluation Approach (EEA) (Barbanente & Khakee, 2005). The Exploratory Evaluation yielded six possible transformation scenarios. Their cognitive framework reveals the main *preferences* of the community and, the problematic issues that should be addressed and solved in strategic terms:

1. a *museum centre*, responding to the lack of a centre collecting and preserving documents and material evidence on local history and the archaeological site;
2. a *centre for archaeological and environmental research*, aimed at integrating and promoting the environmental and historical-cultural characteristics of the site;
3. a *tourist services centre*, providing information on local tourist, cultural and recreational services and facilities, in order to promote the local tourist industry;
4. *an agricultural and food district*, aimed at creating a cluster of local resources including: quality of the territory and typical agri-food productions, cultural heritage, crafts and art tourism;
5. an '*introduction to the town' centre*, that is, an information and communication centre near the archaeological site, also acting as a laboratory for the spread of culture;
6. an *eco-archaeological village*, focused on eco-archaeological research and implementing 'sustainable architectural experiments' within the archaeological site.

These scenarios clarified what the archaeological site of Altilia-Saepinum may become in the future, what is relevant for those who live in the territory of Sepino and also for those who visit it. At the end of this phase the existing system of tangible and intangible relationships and values that characterise the Altilia-Saepinum context finally became clear.

During the fourth phase, multi-criteria and multi-group evaluation of the six alternatives was carried out in order to identify the preferable, shared solution. The six alternatives were compared through the application of Analytic Network Process (ANP) (Saaty, 1996, 2001, 2004). This method was used since it provides a 'creative approach' to practical ways of thinking and solving decision-making problems. It allows representation of each alternative scenario within a reticular framework of goals, objectives, criteria and solutions and the managing of preferences by propagating them across the reticular structure consistently with the system of goals, criteria and solution interdependencies.

ANP models have two parts: the first is a control hierarchy or network of objectives and criteria that verifies the interactions in the system under study; the second includes the many sub-networks of influences among the elements and clusters of the problem, one for each control criterion.

In the Altilia-Saepinum decision context, the hierarchical structure of ANP helped to define the weights to be assigned to each criterion. For example, landscape was recognised as being among the principal resources of the territory, and job creation and partnerships are considered crucial for the development of the area. After all the comparisons were completed, the results showed that the preferable scenario that best combines environmental, social and economic criteria is the *centre for archaeological and environmental research*, followed by the *eco-archaeological village*, the '*introduction to the town' centre*, the *tourist services centre*, the *museum centre* and the *agricultural and food district*.

The decision-making process, designed so as to identify suitable alternatives through dialogue between scientific, technical and common knowledge, was able to implement a transparent and shared decision-making strategy for the enhancement of the Altilia-Saepinum archaeological site, recognising the main components of the opportunity context and including their mutual implications, incorporating multiple and heterogeneous dimensions and plural values.

21.4.2 Buccino: Building a Strategy for an Integrated Valorisation

Buccino,[5] a small town with about 5,000 inhabitants, stands in a favourable location in the basin of the River Sele and the Gulf of Paestum.

The discovery of archaeological findings witnessing human settlements in Ancient Neolithic times and the unearthing of an ancient Roman town under Buccino's historic centre gave birth to an archaeological park that is frequented by local inhabitants, but whose potential is still largely untapped.

The main goal of this study, conducted within a broader integrated program sponsored by the regional and provincial administrations, was to design transformation strategies consistent with a sustainable development perspective and with the resources of the context, expected demands and the local relational and institutional fabric.

In this case too the process followed the four phases illustrated at the beginning in Section 21.4 of this chapter (Fig. 21.3).

In phase one, an interpretative analysis of *hard* and *soft* data and information revealed the image of an area challenged by crisis and decline: economic activities are mainly traditional and unable to stand up to the challenge of competitiveness; this is causing a drain of young people, leading to an ageing population and preventing sustainable local development. On the strengths and opportunities side, Buccino lies in the heart of an area rich in high-quality agri-food products and

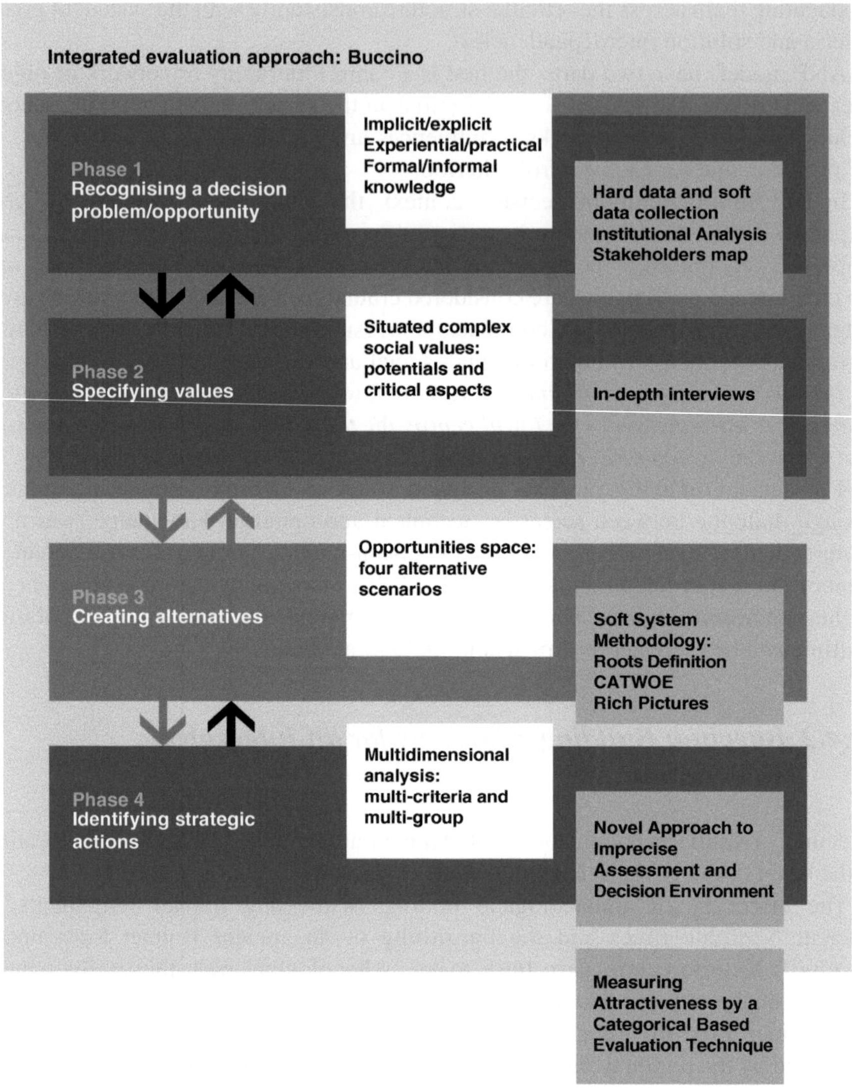

Fig. 21.3 Methodological approach: steps and tools

cultural resources and is part of a cluster of small towns able to develop significant interrelations and move in the direction of change.

The Institutional Analysis (IA) was carried out in this perspective hence looking for categories of stakeholders able to translate this vision of opportunities into concrete actions. Three categories were identified: (1) promoters, (2) users and (3) operators.

Promoters include local and regional institutions (the local administration and the regional authority for archaeological heritage) and education institutions. *Users* are the citizens of Buccino, and its surrounding area, who use the local services and are set to benefit from the area's development. *Operators* are divided into four main groups: (1) typical product producers (farmers and processors), (2) building enterprises, (3) tourism operators and (4) social operators.

All the stakeholders were interviewed; the interviews were structured combining *learning dynamic approaches* defined within the framework of Soft System Methodology (SSM; see Checkland, 1981, 1999, 2001) which aims at replacing positivistic epistemology with a phenomenological view in which actors are seen as subjects attributing meaning to a perceived reality. SSM provides a systemic framework to those involved in a problematic situation and fosters debate on ways of improving the problem situation. The interviews, structured according to the Roots Definition (RD) model and analysed using the CATWOE (*Customers, Actors, Transformation process, World view, Owners, Environmental constraints*) procedure, were represented by the Rich Pictures (RPs) technique (Checkland, 1981) useful for structuring the acquired data, information and knowledge and also for developing a graphic report to be used for further communication and interaction.

The RPs reveal that some elements are recurring, albeit in different combinations with other components; certain spatial characteristics, problematic issues and future perspectives are widely shared by different categories of stakeholders at different levels.

One RP was synthesised for each stakeholder category, and together they represent the main contents of the scenarios developed during the third phase. Each scenario interprets main strategic objectives:

1. *Buccino 'like Salerno'* – This is mainly an economic development scenario, based on a tourism development model supported by territorial marketing activities similar to the one adopted in the city of Salerno; this scenario is mainly based on the potentials of local resources (archaeological heritage, landscape and typical products) and on the possibility of making Buccino a node of a network with the neighbouring villages;
2. *Buccino productive city* – This scenario focuses on rehabilitation of the industrial sector. It foresees reorganisation of industrial activities, the replacement of abandoned ones and the use of alternative energy sources to meet industrial demand;
3. *Buccino 'nature city'* – This scenario leverages on environmental and agricultural resources as resources for the future; it focuses on incentives for greening the production sector and encouraging young people to improve the competitiveness of traditional activities though certification and traceability;
4. *Buccino 'slow city'* – This vision identifies sustainable tourism as the key local development strategy enhancing tangible and intangible resources; it includes the promotion of archaeology, exhibition venues and services, local production and the landscape, and it foresees improvement of accommodation services. The goal is not only to attract quality tourism but also to guarantee a better quality of life for the inhabitants of Buccino.

Finally, in the fourth phase multi-criteria and multi-group evaluation of the four alternative scenarios was carried out to identify the preferable and shared master plan.

The scenarios were compared by applying the Novel Approach to Imprecise Assessment and Decision Environments (NAIADE) (see Chapter 18 by Munda, this book), through which two evaluation matrixes are structured and implemented: the multi-criteria assessment matrix and the multi-group assessment matrix (interdependent in the Social Multi-Criteria Evaluation, SMCE; Munda, 2004) thus emphasising the social dimension of the decision at stake and guaranteeing an effective rigorousness in multi- and inter-disciplinary interaction.

Buccino 'like Salerno' and *Buccino 'slow* city' obtained the highest preference in the discussion and were assigned the role of guiding the master plan work in a 2-fold sense: providing a vision for the future and leveraging on the sustainable use of local resources for development. The two scenarios were incorporated in several elements of the master plan: the general goals – the main one being to guarantee a good quality of life – and also the many interventions[6] proposed to drive the city towards the envisioned change. An additional multi-criteria assessment was carried out to identify priority actions able to set in motion development and change. This latter assessment was based on the method outlined in Measuring Attractiveness by a Categorical Based Evaluation Technique (M-MACBETH) (Bana e Costa & Vansnick, 1999; Bana e Costa, de Corte, & Vansnick, 2005) for two main reasons: this approach enables the use of qualitative judgements, crucial when judgements are collected in multi-disciplinary and multi-expertise environments; and its outputs are organised as preferable groups of actions in the light of the different emphasis assigned to groups of criteria and thus supporting a strategic composition of actions.

The integrated assessment described in this case is not a technique but rather an interdisciplinary and participatory process of combination, interpretation and sharing of knowledge and values among the various scientific disciplines to promote the understanding and management of complex problems, and the identification of shared enhancement and sustainability strategies.

21.4.3 Integrated Spatial Assessment in the Cava de' Tirreni Experimentation

Cava de' Tirreni[7] is a municipality that acts as a 'junction' in the area of the Amalfi coast and the province of Salerno. An SEA was carried out to support the development of a master plan: this offered a practical opportunity to test the Integrated Spatial Assessment (ISA) approach (Fusco Girard, Cerreta, & De Toro, 2008).

This approach was developed to integrate multi-dimensional aspects within a complex development of strategies and choices in planning, acknowledging the importance of the environmental, social and economic effects of a decision-making process focused on the creation of alternative transformative options.

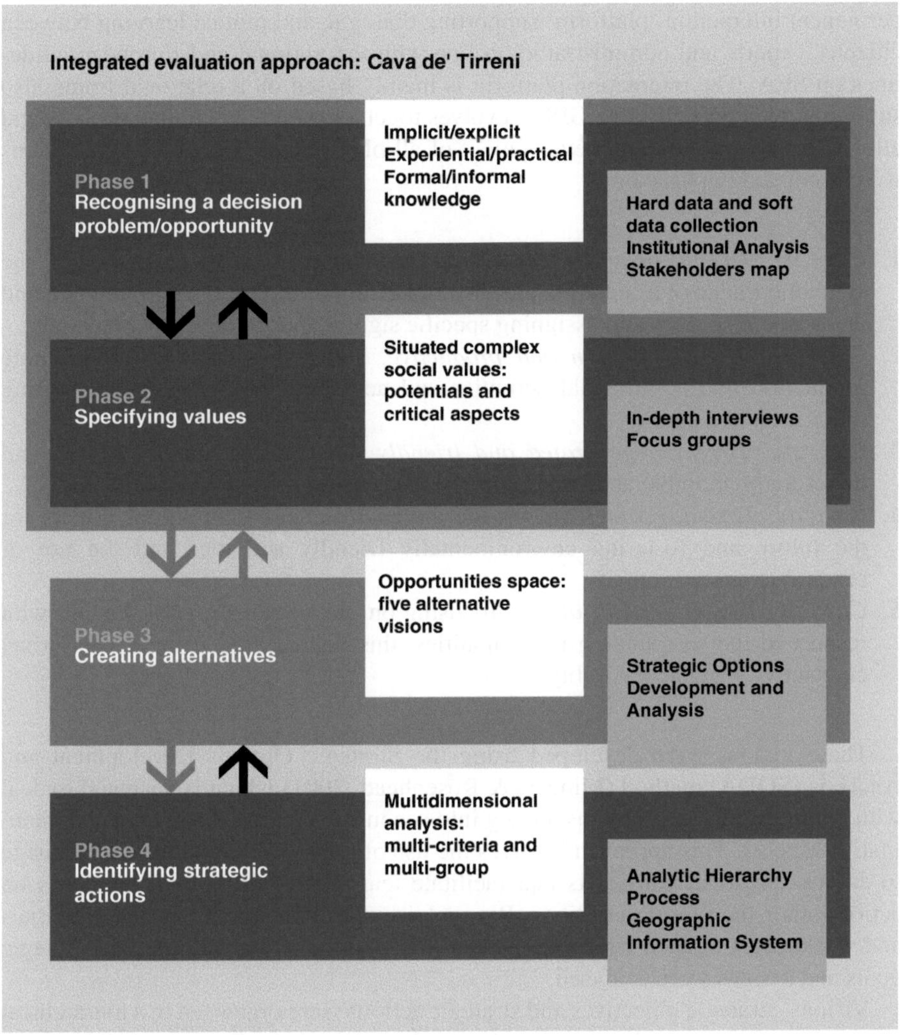

Fig. 21.4 Methodological approach: steps and tools

In ISA the recognition of complex social values is the basis for a collective decision-making process, which includes the steps of problem-setting, problem-posing and problem-solving, and the sharing of different forms of knowledge, and which takes into account issues of justice and equity. Different analyses are combined to manage conflicts and include various levels of uncertainty (see a representation of the evaluation process in Fig. 21.4).

Phase one, implemented similarly to the prior two cases (public meetings, ad hoc interviews and data and information collection), mainly aimed at creating a permanent interaction 'platform' supporting dialogue and mutual learning between citizens, experts and administrators in line with the national and European guidelines on SEA. The interaction platform is mainly based on a relational frame also supported by a participatory GIS; it evolves together with the planning process and allows for the creation and development of all plan-related decisions. Five visions were produced in phase one:

1. *Cava de' Tirreni, beautiful and identity-bearing*, aiming at strengthening the symbolic image of a 'junction city' by enhancing its cultural, environmental and landscape resources and assigning specific significance to their use potentials;
2. *Cava de' Tirreni, modern and productive*, addressing the need to promote production and commercial activities and activating new ones by supporting innovation and quality;
3. *Cava de' Tirreni, regenerated and friendly*, focusing on quality of life and therefore on tangible and intangible actions targeting equity and inclusion;
4. *Cava de' Tirreni, ecological*, considering environmental conditions crucial for the future and fostering environmentally friendly activities and the use of renewable energy sources;
5. *Cava de' Tirreni, territorial hub*, focusing on the attractor role of the city with respect to the surrounding municipalities; this vision aims at improving urban connectivity and accessibility.

These visions were developed using the Strategic Options Development and Analysis (SODA) method (Mingers & Rosenhead, 2001) which is designed to deal with complex, messy problems, taking into account both their qualitative and quantitative aspects. This approach involves the use of cognitive mapping as a language to express personal constructs and facilitate team negotiation, with emphasis on action rather than on descriptions. By applying the SODA method, the potentials and weaknesses of each vision were identified, and the associated specific strategic goals and actions were analysed.

Visions, strategic objectives and strategic actions were organised in a hierarchical structure and for each vision, using the Analytic Hierarchy Process (AHP) (Saaty, 1980, 1992) integrated with the GIS tools (Marinoni & Hoppe, 2006), 'susceptibility to transformation' maps were built, which express the different aptitude of the territory to 'receive' a given strategic action, in the light of its potential environmental impacts, criteria and characteristics. The lesser the territorial and environmental impacts, the greater will be the willingness of that territory to receive the strategic action proposed.

By using the typical approach of the SEA, translating it into a more articulated evaluation process defined ISA, we aimed to integrate social, territorial and environmental aspects in the development of strategies and planning choices, while recognising the important role of stakeholder perceptions and environmental effects within the collective decision-making process for the creation of alternative opportunities.

21.5 Discussion and Final Remarks

The analysis of the three case studies reveals that the four-step process of incremental assessment is strongly influenced by contextual factors.

The territorial dimension, geographical and environmental features, economic, social and human resources are some of the key components in relation to which the dynamics of interaction between knowledge and multi-dimensional values have been explored. They allow possible opportunities and knowledge-multi-dimensional values relationships to be made explicit and be translated into strategies of transformation.

The specific nature of each case study determined the criteria to be followed in the selection of appropriate methods and techniques.

In the Altilia-Saepinum archaeological park the evaluation process made it possible to understand how the multiple values of an archaeological site of great importance but exposed to continuous decay can become drivers of a development process involving the whole region.

Thinking through complex values not only enabled recognition of both use and non-use values peculiar to a specific asset but also mobilised identity values.

Thus, identification of possible transformation opportunities becomes a way for making existing values explicit and, at the same time, creating new, context-sensitive values poised to strengthen the bond with new direct, indirect and potential users.

The techniques selected enabled an exploration which, whilst taking into account the different dimensions of a context, did not compromise the richness of the knowledge framework to be constructed.

In particular the ANP method evidenced the complexity of relationships characterising, on the one hand, the set of decision criteria, factors and objectives and, on the other, the interactions among stakeholders and decision-makers and their influence on the selection of the preferred scenario.

As to the Buccino case, by redefining this municipality as a node in a network of small towns, the evaluation process allowed the identification of local resources and development potentials, thus offering opportunities to towns apparently set on the road to decline.

Relational value, which is the materialisation of synergies and complementarities mobilised and created by the process, acted as a connector between the transformation opportunities and specific strategic goals shared by different stakeholders.

The evaluation of intangible assets takes on a key role and makes it possible to analyse the social concept of value in a multi-dimensional scenario. Soft System Methodology (SSM) methods combined with Multi-Criteria Decision Analysis (MCDA) and Multi-Group Decision Analysis (MGDA) methods enabled the integration of cognitive and evaluation dimensions as well as technical and economic dimensions.

In the Cava de' Tirreni case, the opportunities that emerged from the interactions focused mainly on the preservation of the identity of a context wishing to regenerate itself. This context shows a diversity of values, among which use value long prevailed over the others. Identifying opportunities here means bringing to light intrinsic value in terms of the interdependency of the natural and human landscapes. In this case, the integrated use of SODA, MCDA and GIS shaped the different phases, acting as a powerful combination for providing decision support in strategic decisions. SODA helps decision-makers in devising visions and exploring possible effects, while MCDA and GIS can support an in-depth performance assessment of each strategic action, as well as the design of more robust and better options.

To sum up, these experiences show that, despite the availability of substance-orientated and process-oriented planning tools, achieving the integration of planning and evaluation remains a challenge. However, it can be argued that within the field of integrated approaches, value recognition (use, non-use and intrinsic) assumes a fundamental role and is closely linked to different forms of knowledge, and through their interaction strategic objectives and evaluation criteria are identified, scenarios constructed, decisional rules deduced and sectoral evaluations implemented in order to create and prioritise alternative options.

Notes

1. Strauss (2008) underlines that the concept of frame is used in a variety of different ways. Frames are understood as constructs based on shared cognitive structures that inform perception and identification and condition behaviour (in Foucault's terminology, this is the power/knowledge nexus). Kahneman and Tversky's (1979), Tversky and Kahneman (1986) articulation of prospect theory combines both metaphorical and constructivist elements in the context of the first phase of the decision-making process. The framing and editing phase consists of a preliminary analysis of the problem, which frames the effective acts, contingencies and outcomes (Tversky & Kahneman, 1986, quoted in Strauss, 2008).
2. The first generation includes Benefit-Cost Analysis (BCA), Financial Impact Analysis (FIA) and Cost-Effectiveness Analysis (CEA) and is characterised by reliance on scientific measurement; the second generation combines empirical measurement with some assessment of goals-achievement, applied to Goal Achievement Matrix (GAM) and Multi-Criteria Evaluation (MCE) methods; the third generation looks for objective and value-free ways of assessment and includes methods of impact analysis, as Planning Balance Sheet Analysis (PBSA), Environmental Impact Assessment (EIA), Social Impact Analysis (SIA); the fourth generation is oriented to post-positive intersubjective interaction and Community Impact Analysis (CIA), and its evolution in Community Impact Evaluation (CIE) is the best expression of this category of methods. Other classifications relate evaluation methods to various planning models, which include deliberative planning, interactive-communicative planning, coordinative planning, planning as frame-setting and relational planning, and are associated to different kinds of rationality, as instrumental, substantive and communicative (Alexander, 1998; Khakee, 2003).
3. In the Netherlands there are several of these methods, five of which are (Runhaar et al., 2009): the Milieu Maximalisatie Methode (MMM, or environmental maximisation method), Milieu Op Z'n Plek (the right place or the environment), LOGO (local area typology and environmental quality), MIRUP (environmental tool in spatial plans) and MILO (environmental conditions in the living environment). These five methods have been developed on the basis of the experiences of urban planners with integrating environmental and urban planning.
4. Altilia-Saepinum case study has been carried out within the elaboration of the degree thesis in Architecture of Arch. Maria Carmen Fanelli, on the subject 'The paths of antiquity: heritage to be discovered, knowledge to be utilized', tutor prof. Francesco Bruno, co-tutor Arch. Maria Cerreta, University of Naples Federico II, July 2009.

5. Buccino case study has been carried out within the elaboration of the degree thesis in Architecture of Arch. Vincenzo Cuozzo, on the subject 'Integrated assessments for a valorization sustainable plan: from Volcei to Buccino', tutor prof. Luigi Fusco Girard, co-tutors prof. Pasquale Miano and Arch. Maria Cerreta, University of Naples Federico II, July 2009.
6. The designed interventions are structured into five reference groups: valorisation of the castle, construction of a theatre arena, construction of an archaeological museum area and development of an archaeological-naturalistic path.
7. Cava de' Tirreni case study has been carried out within the elaboration of the new municipality Master Plan. The working group was thus organised: Urban planning and scientific coordination, prof. Carlo Gasparrini; Geomorfology, Dr. Silvana Di Giuseppe; Agronomy, Dr. Maurizio Murolo; Landscape, prof. Vito Cappiello; Economic and financial feasibility, prof. Ettore Cinque; Infrastructures and Mobility, Ing. Vincenzo Cerreta and Ing. Giulio Valfrè, D'Appolonia SpA; SEA, Arch. Maria Cerreta, Arch. Pasquale De Toro, Arch. Saverio Parrella; Graphic design and communication, Arch. Franco Lancio. We thank for support and collaboration the technical staff of Cava de' Tirreni municipality.

Acknowledgements I would like to thank Grazia Concilio and Valeria Monno, warm friends and generous colleagues: this chapter refers to reflections and comments developed in our collaborative work.

References

Ackoff, R. L. (1979). The future of operational research is past. *Journal of the Operational Research Society*, *30*(2), 93–104.

Alexander, E. R. (1998). Conclusion: Where do we go from here? Evaluation in spatial planning in the post-modern future. In N. Lichfield, A. Barbanente, D. Borri, A. Khakee, & A. Prat (Eds.), *Evaluation in planning: Facing the challenge of complexity* (pp. 355–374). Dordrecht: Kluwer Academic Publishers.

Alexander, E. R. (Ed.). (2006). *Evaluation in planning. Evolution and prospects*. Aldershot: Ashgate.

Allmendinger, P., & Haughton, G. (2009). Soft spaces, fuzzy boundaries, and meta-governance: The new spatial planning in the Thames gateway. *Environment and Planning A*, *41*, 617–633.

Allmendinger, P., & Tewdwr-Jones, M. (2006). Territory, identity and space: Planning in a disunited kingdom. London: Routledge.

Amler, B., Etke, D., Eger, H., Ehrich, C., Kohler, A., Kutter, A., et al. (1999). *Land use planning: Methods, strategies and tools*. Eschborn: Deutsche Gesellschaft für Technische Zusammenarbeit. [http://www.iapad.org.publications/ppgis/gtzplup.pdf, access October 2009].

Bana e Costa, C. A., & Vansnick, J. C. (1999). The MACBETH approach: Basic ideas, software and an application. In N. Meskens & M. Roubens (Eds.), *Advances in decision analysis* (pp. 131–157). Dordrecht: Kluwer Academic Publishers.

Bana e Costa, C. A., de Corte, J. M., & Vansnick, J. -C. (2005). On the mathematical foundation of MACBETH. In J. Figueira, S. Greco, & M. Ehrgott (Eds.), *Multiple criteria decision analysis: The state of the art surveys* (pp. 409–442). New York: Springer.

Barbanente, A., & Khakee, A. (2005). Scenarios as an exploratory evaluation approach. Some experiences from southern mediterranean. In D. Miller & D. Patassini (Eds.), *Beyond benefit cost analysis. Accounting for non-market values in planning evaluation* (pp. 225–247). Aldershot: Ashgate.

Bina, O. (2007). A critical review of the dominant lines of argumentation on the need for strategic environmental assessment. *Environmental Impact Assessment Review*, *27*, 585–606.

Bina, O. (2008). Context and systems: Thinking more broadly about effectiveness in strategic environmental assessment in China. *Environmental Management*, *42*, 717–733.

Bond, R., Curran, J., Kirkpatrick, C., & Lee, N. (2001). Impact assessment for sustainable development: A case study approach. *World Development*, *29*(6), 1011–1024.

Campbell, H. (2002). Planning: An idea of value. *Town Planning Review*, *73*(3), 271–288.

Carlsson-Kanyama, A. K., Dreborg, H., Moll, H., & Padovan, D. (2007). Participative backcasting: A tool for involving stakeholders in local sustainability planning. *Futures*, *40*, 34–46.

Cats-Baril, W. L., & Huber, G. P. (1987). Decision support systems for Ill-structured problems: An empirical study. *Decision Sciences*, *18*(3), 350–372.

Checkland, P. (1981). *System thinking, system practice*. Chichester: Wiley.

Checkland, P. (1985). *From optimizing to learning: A development of system thinking for the 1990s. Journal of the Operations Research Society*, *36*(9), 757–767.

Checkland, P. (1999). *Soft system methodology: A 30-year retrospective*. Chichester: Wiley.

Checkland, P. (2001). Soft system methodology. In J. Mingers & J. Rosenhead (Eds.), *Rational analysis for a problematic world revisited: Problem structuring methods for complexity, uncertainty and conflict* (pp. 61–89). Chichester: Wiley.

De Marchi, B., Funtowicz, S. O., Lo Cascio, S., & Munda, G. (2000). Combining participative and institutional approaches with multi-criteria evaluation. An empirical study for water issue in Troina, Sicily. *Ecological Economics*, *34*(2), 267–282.

de Roo, G., & Visser, J. (2004). Slimme Methoden voor Milieu en Ruimte. Een Analyse van Zestien Toonaangevende Milieubeschouwende Methoden ten Behoeve van Planologische Keuzes. [Methods for the Integration of Environment in Spatial Planning]. Groningen: Faculty of Spatial Sciences, Groningen University.

Deakin, M., Mitchell, G., Nijkamp, P., &Vreeker, R. (Eds.). (2007). *Sustainable urban development. The environmental assessment methods* (Vol. 2). London: Routledge.

Fischer, T. B. (2007). *Theory and practice of strategic environmental assessment: Towards a more systematic approach*. London: Earthscan.

Forester, J. (1999). *The deliberative practitioner: Encouraging participatory planning*. Boston: MIT Press.

Friedmann, J. (1987). *Planning in the public domain: From knowledge to action*. Princeton: Princeton University Press.

Funtowicz, S. O., Martinez-Alier, J., Munda, G., & Ravetz, J. (2002). Multi-criteria-based environmental policy. In H. Abaza & A. Baranzini (Eds.), *Implementing sustainable development* (pp. 53–77). Cheltenham: UNEP/Edward Elgar.

Funtowicz, S. O., & Ravetz, J. R. (1991). A new scientific methodology for global environmental issue. In R. Costanza (Ed.), *Ecological economics: The science and management of sustainability* (pp. 137–152). New York: Columbia University Press.

Fusco Girard, L. (1987). *Risorse Architettoniche e Culturali: Valutazioni e Strategie di Conservazione*. Milano: FrancoAngeli.

Fusco Girard, L., Cerreta, M., & De Toro, P. (2005). Integrated planning and integrated evaluation. Theoretical references and methodological approaches. In D. Miller & D. Patassini (Eds.), *Beyond benefit-cost analysis. Accounting for non-market values in planning evaluation* (pp. 175–205). Aldershot: Ashgate.

Fusco Girard, L., Cerreta, M., & De Toro, P. (2008). Valutazione Spaziale Integrata. Il Puc di San Marco dei Cavoti. In F. D. Moccia (Ed.), *Urbanistica Digitale* (pp. 469–487). Napoli: Edizioni Scientifiche Italiane.

Fusco Girard, L., Cerreta, M., De Toro, P., & Forte, F. (2007) The human sustainable city: Values, approaches and evaluative tools. In M. Deakin, G. Mitchell, & P. Nijkamp, & R. Vreeker (Eds.), *Sustainable urban development. The environmental assessment methods* (Vol. 2, pp. 65–93). London: Routledge.

Fusco Girard, L., Forte, B., Cerreta, M., De Toro, P., & Forte, F. (Eds.). (2003). *The human sustainable city. Challengers and perspectives from the Habitat Agenda*. Aldershot: Ashgate.

Fusco Girard, L., & Nijkamp, P. (1997). *Le Valutazioni per lo Sviluppo Sostenibile della Città e del Territorio*. Milano: FrancoAngeli.

Fusco Girard, L., & Nijkamp, P. (2004). *Energia, Bellezza e Partecipazione: la Sfida della Sostenibilità. Valutazioni Integrate tra Conservazione e Sviluppo*. Milano: FrancoAngeli.
Giampietro, M., Allen, T. F. H., & Mayumi, K. (2006). Science for governance: The implications of the complexity revolution. In A. Guimaraes-Pereira, S. Guedes-Vaz, & S. Tognetti (Eds), *Interfaces between science and society* (pp. 82–99). Sheffield: Greenleaf Publishing.
Guba, E. G., & Lincoln, Y. S. (1989). *Fourth generation evaluation*. Newbury Park: Sage.
Healey, P. (2008). Knowledge flows, spatial strategy-making, and the roles of academics. *Environment and Planning C: Government and Policy*, *26*(5), 861–881.
Kahneman, D., & Tversky, A. (1979). Prospect theory: An analysis of decisions under risk. *Econometrica*, *47*(2), 313–327.
Keeney, R. L. (1992). Value-focused thinking: A path to creative decision-making. Cambridge: Harvard University Press.
Keeney, R. L. (1996). Value-focused thinking. Identifying decision opportunities and creating alternatives. *European Journal of Operational Research*, *92*(3), 537–549.
Khakee, A. (2003). The emerging gap between evaluation research and practice. *Evaluation*, *9*(3), 340–352.
Kirdar, Ü. (2003). A better and stronger system of human governance. In L. Fusco Girard, B. Forte, M. Cerreta, P. De Toro, &F. Forte (Eds.), *The human sustainable city. Challengers and perspectives from the habitat agenda* (pp. 225–234). Aldershot: Ashgate.
Krönert, R., Steinhardt, U., & Volk, M. (Eds.). (2001). *Landscape balance and landscape assessment*. Berlin Heidelberg: Springer.
Larner, W., & Le Heron, R. (2002). The spaces and subjects of a globalising economy: A situated exploration of method. *Environment and Planning D: Society and Space*, *20*, 753–774.
Leknes, E. (2001). The role of EIA in the decision-making process. *Environmental Impact Assessment Review*, *21*(4), 309–334.
Lichfield, N. (1996). *Community impact evaluation*. London: UCL Press.
Liew, A., & Sundaram, D. (2009). Flexible modelling and support of interrelated decisions. *Decision Support Systems*, *46*(4), 786–802.
Marinoni, O., & Hoppe, A. (2006). Using the analytic hierarchy process to support the sustainable use of geo-resources in metropolitan areas. *Journal of Systems Science and Systems Engineering*, *15*(2), 154–164.
Matthiesen, U. (2005). *KnowledgeScapes. Pleading for a knowledge turn in socio-spatial research*. Working Paper, Leibniz-Institute for Regional Development and Structural Planning, Erkner.
Mayer, I., van Daalen, S., Els, C., & Bots, P. W. G. (2004). Perspectives on policy analyses: A framework for understanding and design. *International Journal of Technology, Policy and Management*, *4*(2), 169–191.
Medda, F., & Nijkamp, P. (2003). A combinatorial assessment methodology for complex transport policy analysis. *Integrated Assessment*, *4*(3), 214–222.
Miller, D. & Patassini, D. (Eds.). (2005). *Beyond benefit cost analysis. Accounting for non-market values in planning evaluation*. Aldershot: Ashgate.
Mingers, J., & Brocklesby, J. (1997). Multi-methodology: Towards a framework for mixing methodologies. *Omega*, *25*(5), 489–509.
Mingers, J., & Rosenhead, J. (2001). *Rational analysis for a problematic world revisited: Problem structuring methods for complexity, uncertainty and conflict*. Chichester: Wiley.
Munda, G. (2004). *Social multi-criteria evaluation: Methodological foundations and operational consequences*. *European Journal of Operational Research*, *158*(3), 662–677.
Munda, G. (2008). *Social multi-criteria evaluation for a sustainable economy*. Heidelberg: Springer.
Partidario, M. R. (2000). Elements of an SEA framework. Improving the added-value of SEA. *Environmental Impact Assessment Review*, *20*, 647–663.
Pope, J., Annandale, D., & Morrison-Saunders, A. (2004). Conceptualising sustainability assessment. *Environmental Impact Assessment Review*, *24*(6), 595–616.

Richardson, T. (2005). Environmental assessment and planning theory: Four short stories about power, multiple rationality, and ethics. *Environmental Impact Assessment Review*, *25*(4), 341–365.

Rinner, C., & Bird, M. (2009). Evaluating community engagement through argumentation maps: A public participation GIS case study. *Environment and Planning B*, *36*(4), 588–601.

Rittel, H., & Webber, M. (1973). Dilemmas in a general theory of planning. *Policy Sciences*, *4*(2), 155–169.

Rosenhead, J. (Ed.). (1989). *Rational analysis for a problematic world: Problem structuring methods for complexity, uncertainty and conflict*. Chichester: Wiley.

Rosenhead, J. (2005). Controversy on the streets: Stakeholder workshops on a choice a carnival route. In J. Friend & A. Hickling (Eds.), *Planning under pressure: The strategic choice approach* (pp. 298–302). Oxford: Butterworth-Heinemann.

Rotmans, J., van Asselt, M., & Vellinga, P. (2000). An integrated planning tool for sustainable cities. *Environmental Impact Assessment Review*, *20*(3), 265–276.

Runhaar, H., Driessen, P. P. J., & Soer, L. (2009). Sustainable urban development and the challenge of policy integration: An assessment of planning tools for integrating spatial and environmental planning in the Netherlands. *Environment and Planning B*, *36*(3), 417–431.

Saaty, T. L. (1980). *The analytic hierarchy process for decision in a complex world*. Pittsburgh: RWS Publications.

Saaty, T. L. (1992). *Multi-criteria decision-making. The analytic hierarchy process*. Pittsburgh, PA: RWS Publications.

Saaty, T. L. (1996). *The analytic network process: Decision-making with dependence and feedback*. Pittsburgh, PA: RWS Publications.

Saaty, T. L. (2001). *Creative thinking, problem solving and decision-making*. Pittsburgh, PA: RWS Publications.

Saaty, T. L. (2004). The analytic hierarchy and analytic network processes for the measurement of intangibles and for decision-making. In J. Figueira, S. Greco, & M. Ehrgott (Eds.), *Multiple criteria decision analysis: The state of the art surveys* (pp. 346–408). Dordrecht: Kluwer Academic Publishers.

Sandercock, L. (2003). *Cosmopolis II. Mongrel cities of the 21st century*. New York: Continuum.

Schön, D. (1987). *Educating the reflective practitioner*. San Francisco: Jossey-Bass.

Schön, D., & Rein, M. (1994). *Frame reflection: Toward the resolution of intractable controversies*. New York: Basic Books.

Simon, H. A. (1983). *Reason in human affairs*. Stanford: Stanford University Press.

Strauss, K. (2008). Re-Engaging with rationality in economic geography: Behavioural approaches and the importance of context in decision-making. *Journal of Economic Geography*, *8*(2), 137–156.

Susskind, L., McKearnan, S., & Thomas-Larmer, J. (Eds.). (1999). *The consensus building handbook: A comprehensive guide to reaching agreement*. Thousand Oaks, CA: Sage.

Tversky, A., & Kahneman, D. (1986). Rational choice and the framing of decisions. *Journal of Business*, *59*(4), 251–278.

Valentin, A., & Spangenberg, J. H. (2000). A guide to community sustainability indicators. *Environmental Impact Assessment Review*, *20*(3), 381–392.

Wiek, A., & Walter, A. (2009). A transdisciplinary approach for formalized integrated planning and decision-making in complex systems. *European Journal of Operational Research*, *197*(1), 360–370.

Zeleny, M. (1998). Multiple criteria decision-making: Eight concepts of optimality. *Human Systems Management*, *17*(2), 97–107.

Zeleny, M. (2005). *Human systems management: Integrating knowledge, management and systems*. Hackensack: World Scientific Publishers.

Zeleny, M. (2006). Knowledge-information autopoietic cycle: Towards the wisdom systems. *International Journal Management and Decision-Making*, *7*(1), 3–18.

Name Index

M. Cerreta et al. (eds.), *Making Strategies in Spatial Planning*,
Urban and Landscape Perspectives 9, DOI 10.1007/978-90-481-3106-8,

Subject Index

D

T

W

Y

Z

MIX
Papier aus verantwortungsvollen Quellen
Paper from responsible sources
FSC® C105338

If you have any concerns about our products,
you can contact us on
ProductSafety@springernature.com

In case Publisher is established outside the EU,
the EU authorized representative is:
Springer Nature Customer Service Center GmbH
Europaplatz 3, 69115 Heidelberg, Germany

Printed by Libri Plureos GmbH
in Hamburg, Germany